Computational Intelligence for Movement Sciences:
Neural Networks and Other Emerging Techniques

Rezaul Begg, Victoria University, Australia

Marimuthu Palaniswami, The University of Melbourne, Australia

IDEA GROUP PUBLISHING
Hershey • London • Melbourne • Singapore

Acquisitions Editor:	Michelle Potter
Development Editor:	Kristin Roth
Senior Managing Editor:	Amanda Appicello
Managing Editor:	Jennifer Neidig
Copy Editor:	Bernard J. Kieklak, Jr.
Typesetter:	Sharon Berger
Cover Design:	Lisa Tosheff
Printed at:	Integrated Book Technology

Published in the United States of America by
Idea Group Publishing (an imprint of Idea Group Inc.)
701 E. Chocolate Avenue, Suite 200
Hershey PA 17033
Tel: 717-533-8845
Fax: 717-533-8661
E-mail: cust@idea-group.com
Web site: http://www.idea-group.com

and in the United Kingdom by
Idea Group Publishing (an imprint of Idea Group Inc.)
3 Henrietta Street
Covent Garden
London WC2E 8LU
Tel: 44 20 7240 0856
Fax: 44 20 7379 0609
Web site: http://www.eurospanonline.com

Library of Congress Cataloging-in-Publication Data

Computational intelligence for movement sciences : neural networks, support vector machines, and other emerging techniques / Rezaul Begg and Marimuthu Palaniswami, editors.
    p. ; cm.
  Includes bibliographical references and index.
  Summary: "This book provides information regarding state-of-the-art research outcomes and cutting-edge technology on various aspects of the human movement"--Provided by publisher.
  ISBN 1-59140-836-9 (hardcover) -- ISBN 1-59140-837-7 (softcover)
 1. Musculoskeletal system--Physiology. 2. Human locomotion. 3. Artificial intelligence. 4. Neural networks (Computer science)
  [DNLM: 1. Movement--physiology. 2. Artificial Intelligence. 3. Biomechanics. 4. Models, Biological. 5. Neural Networks (Computer) WE 103 C7383 2006] I. Begg, Rezaul. II. Palaniswami, Marimuthu.

QP301.C588 2006
612.7'0285--dc22
                                                    2005032065
British Cataloguing in Publication Data
A Cataloguing in Publication record for this book is available from the British Library.

Computational Intelligence for Movement Sciences:
Neural Networks and Other Emerging Techniques

Table of Contents

Preface

Studies into human movement sciences have been usually undertaken from an interdisciplinary perspective. Individuals and groups who are involved in movement science research come from a number of diverse backgrounds, including: biomechanics, biomedical engineering, health science, exercise science, sports science, computer science, clinical science, physiotherapy, prosthetics and orthotics, to name a few. Research and development in movement sciences are progressing quite rapidly. The main aims of these advances are to gain a better understanding of the normal and abnormal human movement characteristics, and also to develop new and innovative ways of combating the rising health care costs around the globe. Analysis of gait and other human movements has proved very useful in revealing many useful insights into the recognition and assessment of movement abnormalities. In recent times, gait analysis is taken almost as a routine procedure in aiding many diagnostic and rehabilitative procedures. Common application examples include: the design of a rehabilitation program to assist the disabled, the planning and assessment of surgical outcomes, the recognition of gaits due to falls-risk in the elderly and also for the improvement of sports techniques and performance.

Computational intelligence (CI) encompasses approaches primarily based on artificial neural networks, fuzzy logic rules, evolutionary algorithms and support vector machines. These methods have been applied to solve many complex and diverse problems. Recent years have seen many new developments in CI techniques and consequently this has led to an exponential increase in the number of applications in a variety of areas including engineering, finance, social and biomedical. In particular, CI techniques are increasingly being used in biomedical and human movement areas because of the complexity of the biological systems as well as the limitations of the existing quantitative techniques in modelling. The contents of this book cover a wide range of relevant applications in human movement sciences written by leading researchers and academicians in the area. Altogether, the book has 13 chapters organized into the following four sections:

- Section I: Methods and Tools for Movement Analysis
- Section II: Advances in Gait Analysis and Modelling
- Section III: Applications in Rehabilitation and Sport
- Section IV: Computational Modelling for Predicting Movement Forces

Section I has four chapters that are aimed to provide the readers with a comprehensive overview of the various approaches and techniques for analysing human movements. The first chapter provides a comprehensive overview of the traditional movement analysis techniques, potential errors and noise contents in the captured data and some of the major data processing and analysis techniques. Feature extraction is an important process in movement analysis tasks and forms an integral part of a computerized data analysis procedure. An extensive overview of the techniques that could be used to derive features from the processed data is presented. In addition to the laboratory-based measurement techniques, alternative approaches using body-mounted inertial sensing have received considerable attention in recent years, especially for ambulatory monitoring of human motion during various activities. One major advantage of body-mounted inertial sensing in the biomedical domain is its capability to objectively determine a person's level of functional ability in independent living. The next two chapters focus on such techniques and their applications including: sensor configurations and reviews, computational techniques for automatic recognition of activity, quantitative analysis of motor performance, and personal navigation systems. Significant clinical applications (e.g., in orthopedics, Parkinson disease, aging, etc.) as well as potential applications in nanotechnology, materials sciences, and advanced mobile and ubiquitous body movement are discussed. The final chapter in this section presents a brief description of the major computational intelligence techniques for pattern recognition and modelling tasks that often appear in biomedical, health and human movement research. These include techniques such as artificial neural networks, fuzzy logic rules, evolutionary approaches, support vector machines, and also approaches that combine two or more techniques (hybrid).

Section II includes four chapters that focus on applications of neural networks and other CI techniques for analysing, modelling and visualizing gait data. Neural networks have been predominantly used in most gait recognition, classification and modelling tasks. The first chapter in this section examines the use of artificial neural networks to model and probe the control of walking movements. Chapter VI describes self-organising artificial neural networks and their use to reduce the complexity of gait data and to improve visualisation of large amounts of complex data in a two-dimensional map.

The next two chapters focus on recognition of gait changes due to aging and falling behaviour. With significant increase in the number of aging population around the world, and falls in older adults being a major public health issue, researchers are looking for ways to reduce the falls risk incidence in older adults and to improve aging health care. Chapter VII looks into gait pattern changes in the elderly and demonstrates the effectiveness of artificial neural network modelling in mapping gait measurements onto reduced function in the balance control system. Chapter VIII describes an automated gait pattern recognition system using gait classifier based on support vector machines.

Section III explores applications of CI techniques in rehabilitation and sport-related areas. Chapter IX provides a brief description of different methods for the control of man-machine FES systems and talks about a clinical FES system to demonstrate the successful application of these strategies. Specially, application involving FES systems for the restoration of movement to the paralyzed limbs in spinal cord injury patients is discussed. In Chapter X, evolutionary methods are introduced — these are intelligent systems that can further our understanding of human movement and help us devise new treatments. The evolutionary methods resemble the biological processes that led to the development of natural human movement, and they can solve optimization and learning problems that cannot be solved with existing methods. Evolutionary computation is well suited to parallel processing that can reduce the computation time significantly and can be applied extensively to treating human movement disorders. In this chapter, the authors discuss applications of evolutionary methods to explore feasible movement patterns that can be used to prescribe movement therapies to improve the existing functions or design FES control systems to restore the lost movement.

Chapter XI provides applications of self-organising artificial neural networks for analysing sport games, motor activities or rehabilitation. Such processes are often characterized by a complex structure. Measurements considering them may produce a huge amount of data. This chapter presents neural network approaches and examples of application in the field of sport.

Section IV focuses on computational modelling approaches to estimate internal joint forces during movement. Specifically, Chapter XII deals with biomechanical model of the forces about the ankle joint applicable to both unimpaired and neurologically impaired subjects. An EMG-driven hybrid forward and inverse dynamics model of the ankle is employed and optimization procedures are discussed that are used to tune the physiological parameters for the model for each subject.

The final chapter describes computational modelling of the shoulder complex. Following a brief background in anatomy and biomechanics of the shoulder complex the authors present a review of the essential functions of the shoulder and the important features of practical shoulder models. Computational modelling techniques, and also in vivo and in vitro methods for verifying computational models are briefly discussed and a summary of the emerging trends are presented to indicate the clinical impact of computational modeling.

The book contains information regarding state-of-the art research outcomes and cutting-edge technology from the leading scientists and researchers working on various aspects of the human movement. It is hoped that the book will be of enormous help to a broad spectrum of readerships including researchers, professionals, lecturers and graduate students from a wide range of disciplines. It is our belief that the ideas presented in this book will trigger further works and research efforts in this rapidly expanding multidisciplinary area.

Rezaul Begg, Victoria University, Australia
Marimuthu Palaniswami, The University of Melbourne, Australia
Editors

Acknowledgments

The editors would like to thank all the authors for their excellent contributions to this book. Without their support the project could not have been completed. We would like to thank Victoria University for granting outside studies program (OSP) leave for Rezaul Begg to undertake this book project. Also, we acknowledge support from the ARC Research Network on Intelligent Sensors, Sensor Networks and Information Processing (ISSNIP) *for this project.*

The editors would also like to thank the publishing team at the Idea Group Inc., who provided continuous help, encouragement and professional support from the initial proposal stage to the final publication. Special thanks to Mehdi Khosrow-Pour, Jan Travers, Renée Davies, Amanda Phillips and Kristin Roth.

Finally, we thank our families for their love and support in the world of conflicting demands.

Section I

**Methods and Tools
for Movement Analysis**

Chapter I

Overview of Movement Analysis and Gait Features

Russell Best, Victoria University, Australia

Rezaul Begg, Victoria University, Australia

ABSTRACT

This chapter provides an overview of the commonly used motion analysis approaches and techniques and the key features that are extracted from movement patterns for characterizing gait. The ultimate goal of gait analysis should be to provide reliable, objective data on which to base clinical decisions (Kaufman, 1998). Thousands of gait features/parameters have been used over the years. Selection of the correct gait features forms an important part of the research process, and often the success of the research outcomes depends heavily on selecting the most appropriate gait features. Analysis tools based on both statistical and machine-learning techniques use various types of gait features, ranging from the basic and directly measurable parameters to parameters that have undergone significant data processing and treatments. In this chapter, we attempt to introduce the commonly used methods to extract these features for use with the various statistical and computational intelligence analysis tools.

INTRODUCTION

Walking is one of the most common and most important forms of human movement. Gait analysis entails measurement, analysis and assessment of the biomechanical features that are associated with the walking task. Significant technical and intellectual progress has been made in the area of gait analysis over the past few decades, especially because of advances in computing speed which in turn has aided the development of more advanced movement-recording systems that require less data processing time. Improved computing speed has also made feasible and inspired increasingly complex and innovative gait data analysis techniques.

As a consequence of the vastly improved gait analysis techniques over recent decades, there has been an exponential increase in the number of applications of gait analysis, some examples of which include: assessment of treatment outcomes (e.g., following anterior cruciate ligament reconstruction, Knoll et al., 2004, and hip joint surgery, Kyriazis & Rigas, 2002), evaluation of the orthotic and prosthetic alignment on the lower limbs (Johnson et al., 2004) and falls risk assessment (Maki, 1994). Gait analysis is used in disability/injury management programs, for initial assessment of gait function, monitoring of progress (e.g., post-surgery), and discharge testing (e.g., prior to leaving hospital or returning to work). It can also be used to screen healthy individuals and is used in a preventative medicine/health screening context (e.g., falls risk assessment). In general, gait analysis is considered an acceptable tool for kinesiologic analysis of movement disorders, including for evaluating gait and posture disturbances.

In recent times there has been a surge in health care costs around the world associated with gait-related issues. One major problem in this regard, has been identified as falls in the aging population (Fildes, 1994). In Australia, falls have a larger injury-

Figure 1. Direct medical costs of all injuries in Australia (adapted from SRDC, 1999)

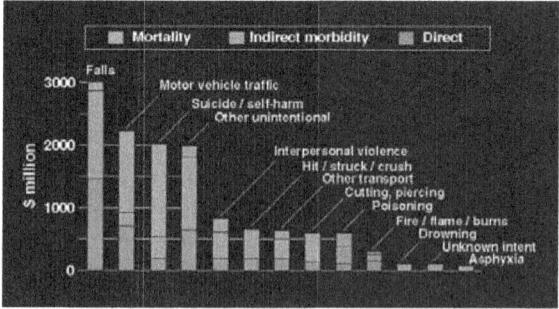

related medical cost (~ $3billion) than any other accident-type, costing even more than motor vehicle traffic accidents (Figure 1).

More than half of all falls are reported to occur during locomotion and occur as a result of tripping, slipping and stair/step negotiation. Gait analysis in the analysis of trips, slips and stair/step negotiation has been proposed by many as a method to help in the reduction of the incidence of falls. Indeed, there is a fast growing interest and a growing number of publications in this area (e.g., Barnett, 2002; Pavol et al., 1999).

The use of gait analysis to evaluate and treat musculoskeletal disorders (e.g., polio, muscular dystrophy, amputation, osteoarthritis, trauma) and neurological disorders (e.g., cerebral palsy, stroke, brain trauma) is arguably the area where the best examples to date of the positive uses of improved gait analysis techniques and technology can be seen. There are now gait analysis laboratories in a number of hospitals in most developed countries. Most of these facilities have now examined the pathological gait of thousands of patients and used this information to direct their surgical, orthotic and therapeutic intervention. Gait analysis has not only altered overall treatment philosophies but has prompted modifications to surgical techniques (Vankoski & Dias, 1999).

In addition to applications of gait analysis, a vast amount of literature focuses on basic or fundamental research into gait. This research is very important because it helps us to understand the processes of gait and to accumulate knowledge about the process. Research such as collecting normative gait cycle data from normal and pathological populations (e.g., Whittle, 2002; Winter, 1991; Sutherland, 2005), analysing the complex stride to stride variability and long-term control of gait (e.g., Hausdorff et al., 1995b; Herman et al., 2005; Taylor et al., 2001; West & Griffin, 1999), and the development of complex computer models of gait (e.g., Anderson & Pandy, 2001) are all crucial to understanding the fundamental characteristics of gait.

Both statistical and artificial intelligence (AI) approaches (e.g., neural networks, fuzzy logic, evolutionary technique/genetic algorithm or support vector machine; see Chapter IV for a description of the AI tools) are used for analysing gait and movement data. The main advantages of using statistical techniques is that they provide insight into the gait models being used and the effects of the various independent variables on the dependent variables can be studied directly. Statistical techniques, however, have significant limitations especially when the problem to be studied is non-linear or complex. Neural networks and other artificial intelligence techniques have the potential to offer a better alternative in some circumstances (Wu et al., 1998; Hahn et al., 2005) and are becoming increasingly popular in gait research. Most of the work in the area has utilized artificial neural networks for gait pattern classification and recognition tasks including categorization of normal and pathological gaits (Holzreiter & Kohle, 1993; Wu & Su, 2000), discriminating the effects on gait of controlled/simulated leg asymmetries (Barton & Lees, 1997), and categorization of normal/healthy and faller gaits (Hahn et al., 2005; Begg et al., 2005). Neural networks have also proved effective in modelling gait variables such as mapping between joint angle and joint moment data (Sepulveda et al., 1993). Other artificial intelligence approaches such as fuzzy clustering (O'Malley et al., 1997), genetic algorithms (Wu et al., 2001) and support vector machines (Begg & Kamruzzaman, 2005) have been used in gait pattern classification tasks to demonstrate their potential as gait diagnostic tools.

Table 1. Gait cycle events for one limb

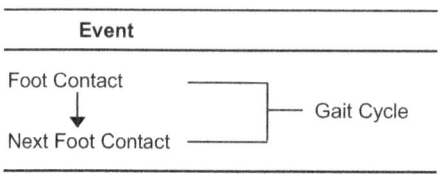

The Gait Cycle

In free speed walking, human gait is a quasi-periodic activity with the left and right limbs out of phase. Control of the whole body, and specifically the lower limbs, is often described in terms of the magnitude of parameters at key events and during key phases that occur during the gait cycle.

One gait cycle for one limb is defined as the duration from one event, usually foot contact, to the next occurrence of the same event on the same limb (e.g., from right limb foot contact to the next right limb foot contact; see Table 1 and Figure 2). Note that the term foot contact is used here and not heel contact. In normal gait, the heel is the first part of the foot to contact the ground but in some pathologies other parts of the foot (e.g., toe) may contact the ground first.

This level of breakdown of a person's gait into gait cycles is the basis of most gait analysis research, whether it be looking at joint angles and moments or the onset of muscle activity via electromyography (EMG). Although the gait cycle has its own duration, and this duration can be an important parameter in itself (e.g., Hausdorff et al.,

Figure 2. Normal gait cycle for the right limb (adapted from Sutherland et al., 1994)

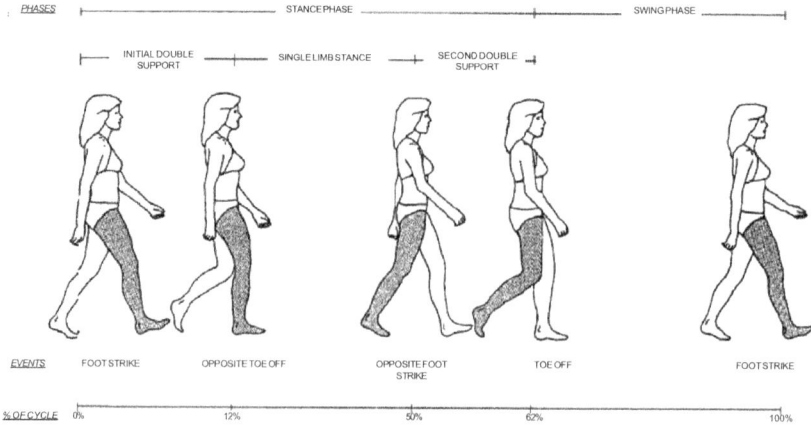

Table 2. Gait cycle events and phases for one limb

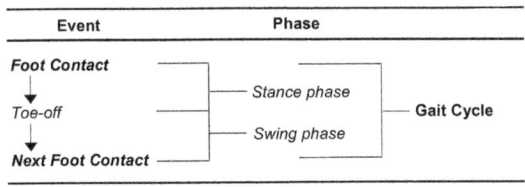

1995b), the gait cycle is usually normalised such that the initial foot contact represents 0% or the start of the gait cycle and the next foot contact of the same limb represents 100% or the end of the gait cycle (Figure 2). This normalisation process makes ensemble averaging of multiple gait cycles, comparisons across the two limbs and comparisons across different people and populations much easier, but it also has its limitations (see discussion of "temporal dependency" issue later).

The gait cycle can be further reduced for each limb into the two major phases of the gait cycle, the stance phase and the swing phase, which alternate for any one limb during the walking task. This involves adding another key event of the gait cycle for the limb, referred to as toe-off (Table 2 and Figure 2).

Looking at the full gait cycle and breaking it down to stance phase and swing phase is the basis for nearly all gait analysis research. Some research examines only the stance phase (e.g., the study of ground reaction forces which can either be presented as a function of time; e.g., Vaughan et al., 1992, normalised to the stance phase duration; e.g., Begg et al., 1998, or normalised within the full gait cycle duration; e.g., Meglan and Todd,

Table 3. Gait cycle events, stance and swing phases and single- and double-support sub-phases for one gait cycle of one limb

Figure 3. Gait cycle events, phases and sub-phases recommended by Sutherland et al. (adapted from Sutherland et al., 1994)

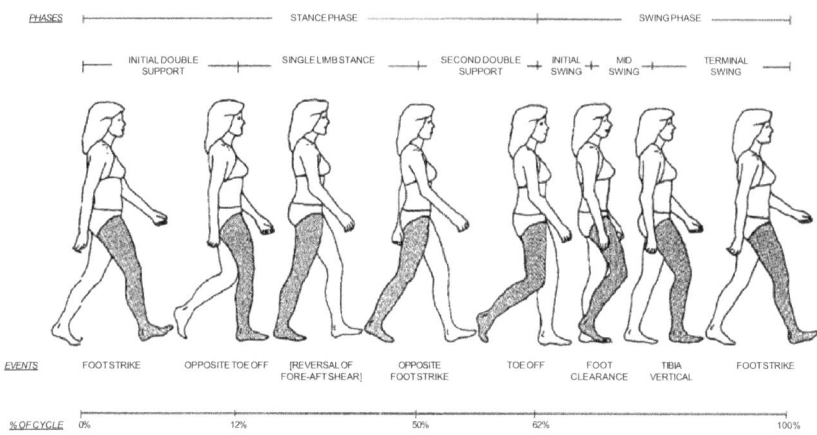

Table 4. Gait cycle events, phases and sub-phases used by Sutherland et al. (1994)

% Gait Cycle	Event	Sub-phase	Phase
0	*Foot Contact*	Early double support phase	
12	Opposite Toe-off		
30*	Reversal of Fore-Aft Shear *	Single support phase	*Stance phase*
50	Opposite Foot Contact		
62	*Toe-off*	Late double support phase	
75	Foot Clearance [#]	Initial swing phase	
85	Tibia Vertical	Mid swing phase	*Swing phase*
100	*Next Foot Contact*	Terminal swing phase	

Gait Cycle

* When the anteroposterior force is zero in mid-stance — not used by Sutherland et al. in further analysis.
[#] When the swinging foot passes the stance foot.

1994) and some research looks at the swing phase only (e.g., foot clearance and tripping research, which can either be presented as a function of time or normalised to swing phase duration; e.g., Johns, 2004, or normalised to the full gait cycle duration; e.g., Winter, 1991).

The stance and swing phases can be further broken down and many more events and phases can be added depending on the requirements of the study. Most clinical gait analysis procedures include the foot contact and toe-off of the contralateral limb as key events in the gait cycle. This allows the stance phase of a gait cycle to be reduced further into sub-phases that includes one single-limb support phase (the phase where the contralateral foot is not in contact with the ground) and two double-limb support phases (where both feet are in contact with the ground) (Figure 2 and Table 3).

Many other events and phases can be added to give further detail to the gait analysis. There is no standard practice at this level of deconstruction of the gait cycle, even in clinical gait analysis, and the choice of events and phases should be based on the needs of the study. Sutherland et al. (1994) recommended using the events and phases described in Figure 3 and Table 4. It is important, however, that all events are reproducible. That is, the events occur in all participants in the study (discussed in more detail later).

GAIT MEASUREMENT TECHNIQUES

A typical gait measurement procedure usually involves recording a number of biomechanical variables, often using a number of different measurement techniques. Biomechanical analysis of human movement requires information about motion (kinematics), human inertia parameters and internal and external forces (kinetics) (Yeadon and Challis, 1994). It is generally agreed that a detailed laboratory-based clinical gait analysis test can be achieved using: (a) a motion measurement system (motion analysis), (b) a force platform (measuring the forces exerted on the ground by the foot) and (c) an electromyography (EMG) data acquisition system. There are, of course, many other measurement techniques that are used in gait analysis in and out of the laboratory setting. This section will cover some of the main equipment and measurement techniques used for gait analysis research.

Kinematics (Motion)

Motion analysis, or kinematics, is concerned with the description of how a body moves in space. Kinematics is the study of the "output" or motion of the activity, independent of the "driving" or "input" forces that change the properties of the motion.

The basic parameters of kinematics, and motion analysis, are (presented as a vector form followed by scalar equivalent in brackets from 3-8):

1.   Time, t in seconds (s)
2.   Position, p in metres (m)
3.   Linear displacement, $\mathbf{s}$ in m (linear distance, s)
4.   Linear velocity, $\mathbf{v}$ in m/s (linear speed, v)
5.   Linear acceleration, $\mathbf{a}$ in m/s^2 (linear acceleration, a)

6.   Angular displacement, $\boldsymbol{\theta}$ in degrees or radians (angle, θ)
7.   Angular velocity, $\boldsymbol{\omega}$ in °/s or rad/s (angular speed, ω)
8.   Angular acceleration, $\boldsymbol{\alpha}$ in °/s² or rad/s² (angular acceleration, α)

All displacement, velocity and acceleration data (linear and angular) can be calculated from position-time information. Nearly all motion analysis methods (e.g., using the category of equipment called motion analysis systems) measure position-time information and then use this information to calculate all other kinematics parameters of interest.

Basic Temporal Spatial Measurement

The simplest form of kinematics/motion measurement involves the recording of average walking speed, which can be achieved using simple equipment such as a stop watch (time) and a tape measure (distance). This is a particularly useful and efficient way of measuring some key gait measures in the field situation (e.g., walking down the street and measuring the time taken to walk a known distance). If the number of steps taken to walk the distance can also be recorded (simply by counting as they walk or taking a video recording) then some other key measures such as average cadence (steps per minute), average stride length and average step length can also be measured easily in the field. This simple analysis allows some key parameters to be measured at the level described in Tables 1 and 2.

The next level of sophistication involves the measurement of stance time, swing time and double-stance time (see Table 3). These times are frequently recorded during movement analysis and various methods of recording these parameters have been reported including "gait mats" and footswitches (Bilney et al., 2003; Ross & Ashman, 1987).

Footswitches

Footswitches are the simplest way to detect/measure the gait events and temporal parameter data described in Table 3. They are adaptable to any situation, including field measurements over large distances and treadmill walking. Footswitches, if placed under both feet, record step-by-step temporal information. This basic, yet essential, gait information is recorded in many gait studies and has been the only data collected in some highly complex research into long-term gait control (e.g., Hausdorff et al., 2004; Herman et al., 2005). Footswitches measure timing of foot placement information and can detect all the events detailed in Table 3 and measure the temporal parameters of gait cycle time, stance time, swing time, single support time and double-stance times.

There are a number of different types of footswitches available (Figure 4) both as commercial products (see Bontrager, 1998) or they can be built in-house (Hausdorff et al., 1995a). The two main types of commercially available footswitches (Bontrager, 1998) are compression closing insoles (e.g., from B & L Engineering) and force sensitive resistor (FSR) switches. FSR systems are available as insoles (e.g., from Noraxon) or as discrete switches that can be placed in user-specified areas beneath the foot (e.g., from Motion Lab Systems; Noraxon). Footswitches typically have contact areas in some or all of the heel, first and fifth metatarsal, and great toe areas. The advantage of discrete switches is that different size insoles are not required to fit a large range of foot sizes.

Figure 4. Footswitches

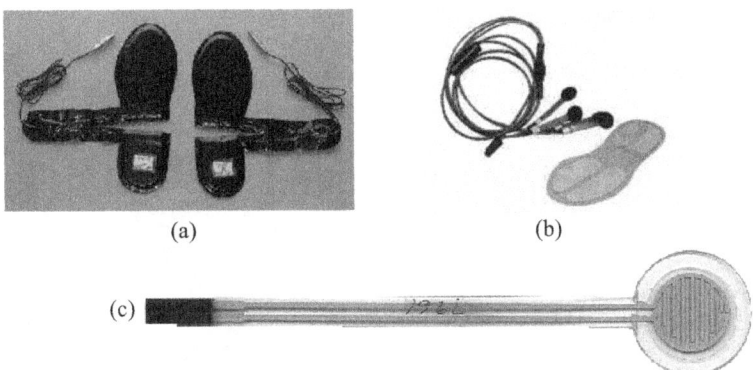

(a)   (b)

(c)

(a) Compression closing insoles (B & L Engineering: www.bleng.com)
(b) FSR insoles and switches (Noraxon: www.noraxon.com/products/sensors.php3)
(c) FSR switch (Motion Lab MA-153: www.emglab.com/fsw.htm)

The disadvantage is in obtaining reliable data because of the difficulty in consistently placing the switches at the required locations under the foot (Bontrager, 1998).

3D motion analysis studies of gait often use force plates which can provide the same temporal event information as footswitches. Force plates (see below) are only useful for one step (if one force plate is available) or one stride (two steps — if two force plates are available) but they are more accurate than footswitches. In comparing data from a force platform and footswitches, Hausdorff et al. (1995a) found the times coincided within ± 10 ms (mean: 0 ± 3 ms) for the start of stance phase and within ± 22 ms (mean: "1 ± 8" ms) for the end of stance phase. For stance duration, the differences ranged from "24 to 28" ms (mean: 1 ± 10 ms). In combination, these measures were used to estimate stance duration to within 3% of force plate determined values for steps with stance durations ranging from 446 to 1594 ms. Estimates of swing and stride duration were within 5% of force plate determined values.

Footswitches provide important gait timing data that are difficult to obtain automatically in any other way, especially over a number of steps (e.g., outside or treadmill walking). They can also be bought or built very cheaply.

Gait Mats

Gait mats measure foot placement position and time information and can detect all the information described above for footswitches plus foot placement spatial parameters (e.g., cadence, stride length, step length, support base and walking speed) on a step-by-step basis. Gait mats consist of a long walking strip (Figure 5) with an embedded array of pressure sensitive switches running across and along the length of the mat (Bontrager, 1998). As a person walks on the mat, the switches close under the feet. Since the position

Figure 5. The GAITRite mat (www.gaitmat.com)

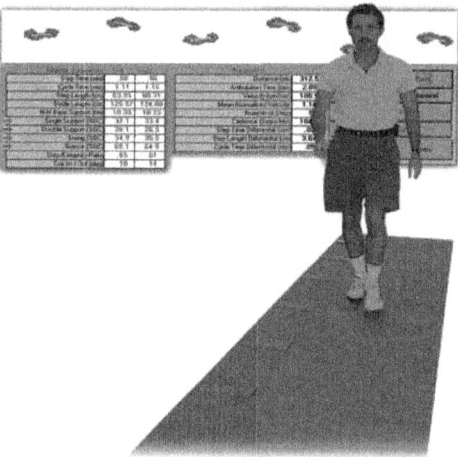

of each switch is known it is possible to calculate the position and timing of each switch closure.

The main advantages of the mat systems are that they measure all the basic temporal and spatial measures of the gait cycle, they are cheap (relative to other motion analysis systems) and portable, they do not interfere with normal gait, there are no attachments to the participant and no preparation of participants required, the mat can be taken to the participants, and the mats can cope with walking aids. The limitations of these mats include measurement of gait only over a set distance, walking speed is determined via foot placement, and the resolution (1.3-1.5cm and 5-12.5ms) of the systems may be insufficient for some data in some circumstances.

The standard GAITRite mat (CIR Systems, Figure 5) and GaitMat II (EQ Inc) have active areas over 3.6m long × 0.6m wide. GAITRite has a better spatial resolution, 1.3cm, with its 13,824 sensors while GaitMat II has 9,728 pressure sensitive switches with a resolution of 1.5cm. The GAITRite mat is 4mm thick whereas the GaitMat II, at 32mm thick, may require a 32mm thick runway at each end if a pre- and post-mat walking area is used (Bontrager, 1998). The temporal resolution of GaitMat II is better at 5ms compared to 12.5ms for GAITRite.

Although basic, temporal spatial measurements can be undertaken using relatively simple and cheap equipment and with relative ease. Nonetheless, these data offer fundamental information regarding the overall movement under investigation. As a result, many studies and projects have shown that basic temporal and spatial data can be useful as indicators for the assessment of movement disorders or improvements (Andriacchi et al., 1977; Kyriazis & Rigas, 2002).

3D Motion Analysis

To accurately measure the motion of the body in 3D space and to obtain a comprehensive overview of the kinematics of human gait as used, for example, for decision making in clinical gait analysis, 3D motion analysis is the most often used procedure. While many parameters and events can be measured using a single video camera and 2D motion analysis (including all the basic temporal spatial measures described in the previous section and, for example, the timing and occurrence of the "tibia vertical" and "foot clearance" events described in Table 4), 3D motion analysis offers all this and much more. In fact, 3D motion analysis is much easier these days than 2D motion analysis, though much more expensive. Modern motion analysis systems generally capture 3D motion data in an automated fashion and in real time. There are four main types of motion analysis equipment, referred to as:

- **Video digitising systems** (e.g., Peak, APAS): Use manual digitising on video pictures, frame by frame, or automated digitising/tracking of reflective markers post-video-capture. Video is the only option for 2D analysis (no commercial 2D non-video-based systems are available) and video is the only option in situations where attaching markers to the study's participants is not possible.
- **Video-based reflective/passive marker systems** (e.g., VICON, Motion Analysis, Elite, Qualysis, Peak, APAS): Use reflective markers (passive markers) attached to the participant. Cameras pick up the reflections from the markers. These systems automatically capture marker positions and most systems present 3D position-time data of markers in real time or near real time.
- **Optoelectronic or active marker systems** (e.g., Optotrak, CODA): The markers themselves are infrared-emitting (active markers) and individually identifiable. These systems automatically capture 3D position-time data of markers in real time.
- **Magnetic tracking systems** (e.g., Ascension, Polhemus): Unlike the other three optical-based systems, use magnetic fields. Instead of markers they have sensors, each of which returns 6DOF data (three coordinates and three angles) in real time.

Table 5. Motion analysis systems

System Type	System Name	Web Address
Passive Marker (PM)	VICON	http://www.vicon.com/
	Motion Analysis	http://www.motionanalysis.com/
	Elite	http://www.bts.it/
	Qualysis	http://www.qualisys.com/
PM and Digitising	Peak	http://www.vicon.com/products/peakmotussoftware.html
	APAS	http://www.arielnet.com/start/apas/default.html
Active Marker	Optotrak	http://www.ndigital.com/certus.php
	CODA	http://www.charndyn.com/
Magnetic	Polhemus	http://www.polhemus.com/
	Ascension	http://www.ascension-tech.com/

Both passive and active marker systems are widely used in gait research with passive marker systems most often used, especially in hospital settings. Magnetic tracking systems are also used, but to a lesser extent. Collection of gait data via manual digitising of video images is rarely used these days. Both passive and active marker systems require markers to be attached to the participants and the markers are either infrared light-emitting diodes (LEDs) (active marker systems) or solid shapes covered with reflective tape (passive marker systems). These systems provide as output the 3D coordinates of each marker. Magnetic systems also require equipment to be attached to the participants, but they are magnetic sensors rather than markers.

Passive Marker Systems

Some optical motion capture systems use reflective markers (passive markers) attached to the participant. These systems are called passive marker systems. Reflections from the markers (Figure 6b) are tracked using multiple video cameras (minimum 2-3 but 6-8 cameras is the recommended minimum). Infra-red flash illuminators surround each camera lens (Figure 6a) sending out pulses of infra-red light that are reflected back into the lens from the markers.

Each camera records a 2D image with the markers appearing as bright dots. Image processing systems isolate the marker dots in the image and record their position (Fischer, 2002). Since they are "passive" markers, each marker trajectory must be identified and tracked and there are a number ways to do this. Markers are often obscured from one or more cameras or their trajectories cross, so the trajectories can be difficult to track. This is why 6 cameras is the recommended minimum for automated passive marker systems. Once the visible markers have been located on the 2D camera images, the coordinates of the centroid of each marker are noted and a series of intersecting "rays" are mathematically projected from each camera position for each marker. Since the camera positions and lens parameters of each camera are known, the rays from the same marker must intersect and the sets of 2D coordinates for each marker can be "reconstructed" and 3D coordinates of each marker can be calculated (Shao et al., 2001). Finally, markers are assigned to existing trajectories, ghost markers are rejected (visual noise, eg. due to shiny reflective surfaces, especially rounded shiny objects like a walking aid or artificial limb), and new and lost trajectories are noted. The next set of camera images can then be processed. Modern systems lose track of markers much less often, and are also able to automatically re-establish them more frequently. While the data processing problem is startlingly large for passive marker systems, steady progress toward its solution means that 3D data capture is now very efficient (Fischer, 2002). The primary sources of problems for passive marker systems include:

- marker slippage
- markers leaving the volume
- bad camera coverage of the volume
- bad volume calibration
- marker occlusion
- ghosting
- bad threshold level

*Figure 6. Passive marker motion analysis systems — cameras, markers and output
(http://www.motionanalysis.com and http://www.vicon.com)*

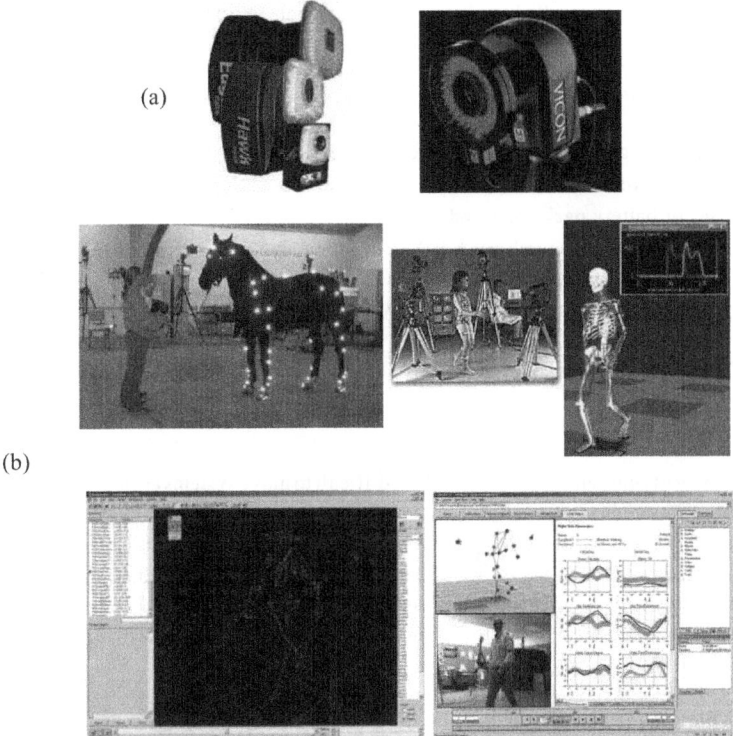

(a) Cameras used for passive marker motion analysis with infra-red flash illumination
*(b) Reflective markers and an example of the animated and graphical output from various
systems*

- bad ring flash or exposure setting
- dirty markers
- reflective surfaces in view
- stray light

With the possible exception of the first five, none of the above problems are
encountered with active marker systems. Unlike active marker and magnetic systems,
setting up and processing of data from passive marker systems require more interaction
with an operator. Despite the complexity of the problem for passive marker systems, they
are still the most commonly used automated 3D motion analysis systems, in part because

the user is not involved in the more complex processes described above. Automated passive marker systems generally have an excellent frame rate (e.g., VICON and Motion Analysis Corp offer a range of camera systems of varying resolution and full frame rates from 166-500Hz).

The main advantages of passive marker systems compared with other systems are:

- the highly configurable marker setup achievable
- large capture volumes
- no cables or battery packs required
- much user-generated research into how best to obtain 3D kinematics data from their passive reflective markers
- most manufacturers provide software for 3D gait kinematics and kinetics

The main disadvantages of passive marker systems compared with other systems are:

- cost
- complexity of setup and calibration make it difficult to move systems around (many labs have an employee specifically employed to manage and run the system)
- data processing complexity (although this is mostly done for the user)
- some markers are mounted on wands to increase their visibility from multiple cameras (this will magnify any errors due to skin movement)
- some extra markers are required to distinguish the right from the left side of the body

Active Marker Systems

Active marker systems also have markers attached to the participant. Markers are light emitting diodes (LED) that are powered and cabled and each LED pulses in a set sequence (active markers). With only one marker flashing at any one time, the system can automatically identify and track each marker. This is a significant advantage over passive marker systems and, it would seem, the ultimate technological solution to the marker tracking issue (Fischer, 2002). However, the down side to this solution is that, after sampling the first marker, it must sample all other markers before it can sample the first marker again. This means the sample rate reduces as the number of markers increases. The sequential pulsing solution of active marker systems means marker occlusion and ghosting are not an issue and, in turn, unlike passive markers, active markers can be placed close together. The two main commercially available active marker systems are OptoTrak (Northern Digital, NDI) and CODA (Charnwood Dynamics). OptoTrak systems are commonly used for gait analysis.

Active marker systems have three cameras mounted in a rigid housing, about 1m long, called a "Scanner" (CODA) or "Position Sensor" (OptoTrak). The OptoTrak is pre-calibrated at the factory which means the calibration of even multiple sensors is extremely simple. Two units are usually required to collect bilateral data and three OptoTrak units seem to be the maximum required to cover all needs (e.g., Corriveau et al., 2004; Sadeghi et al., 2004).

Figure 7. Active marker motion analysis systems — cameras and marker/cable system

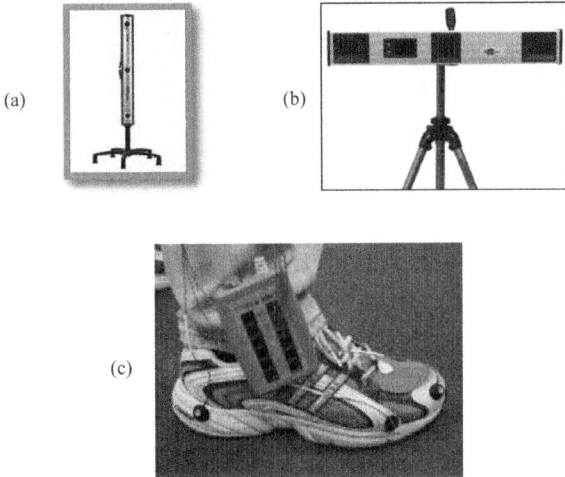

(a)

(b)

(c)

(a) OptoTrak Certus "Position Sensor"
(b) CODA "Scanner"
(c) Optotrak marker/cable system

The CODA system has a constant frame/capture rate of 240Hz which is as good as many passive marker systems and more than adequate for most gait analysis research. The frame/capture rate for the OptoTrak Certus system can be chosen by the user. The maximum frame/capture rate for the Optotrak Certus is calculated by the equation:

Maximum frame/capture rate = Maximum marker flash rate / (number of markers + 2)

so the maximum frame/capture rate reduces as the number of markers increases. For the Optotrak 3020, the maximum frame/capture rate equation is slightly different:

Maximum frame/capture rate = Maximum marker flash rate / (number of markers + 1)

The Optotrak Certus has a maximum marker frequency of 4600Hz which means the maximum frame/capture rate for one marker is 1533Hz, the maximum frame/capture rate for 15 markers (e.g., Newington/Helen Hayes marker set commonly used for clinical gait analysis; Figure 8) is 270Hz, and the maximum frame/capture rate for 40 markers (e.g., "Plug-in-Gait" marker set, Figure 9) is 109Hz. The older (now superseded but still commonly used) Optotrak 3020 has a maximum marker frequency of 3500Hz which means

Figure 8. "Helen Hayes" marker set (www.lifemodeler.com/LM_Manual/A_motion.htm)

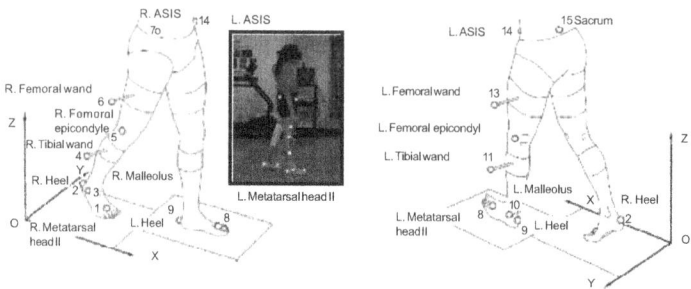

Figure 9. "Plug-in-Gait" marker set (www.lifemodeler.com/LM_Manual/A_motion.htm)

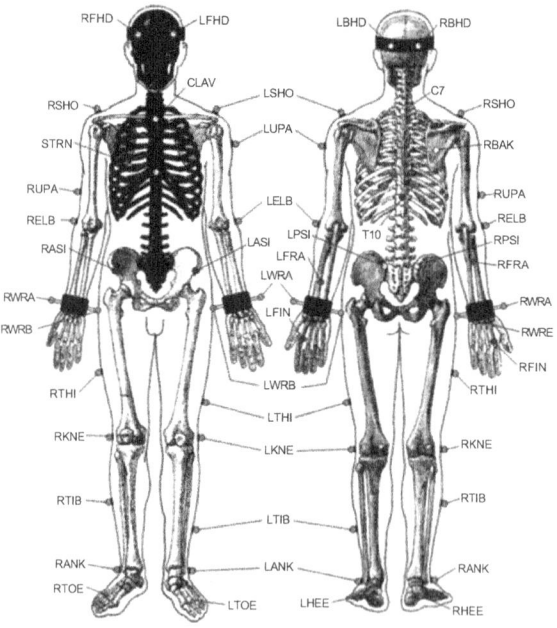

Figure 10. C-Motion proprietary software "Visual3D" (www.c-motion.com)

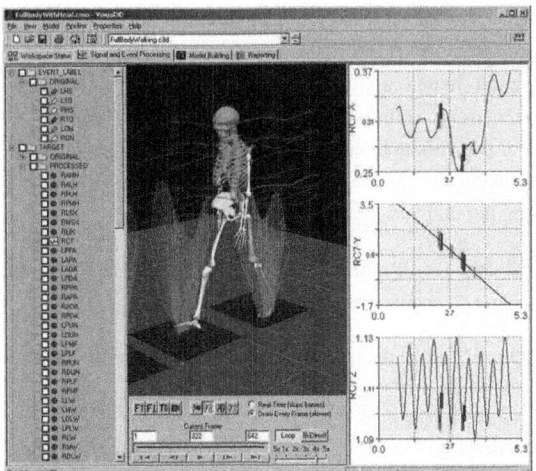

the maximum frame/capture rate for one marker is 1166Hz, the maximum frame/capture rate for 15 markers is 218Hz, and the maximum frame/capture rate for 40 markers is 85Hz.

OptoTrak has a field of view of 3.6m × 2.6m that can be increased with multiple position sensors. It is cabled but there is an optional tetherless strober that eliminates the cable between the wearer and the control unit. The body-worn battery pack that is required with the tetherless strober weighs about 1kg (Bontrager, 1998). Gait kinematics and kinetics software is not provided with OptoTrak but NDI has a partnership with C-Motion who provide the proprietary "Visual3D" software (formerly 'Move3D'), at extra cost, that includes gait analysis (including Helen Hayes marker set compatibility) and inverse dynamics functions (e.g., Wojcik et al., 1999). An example of the output from Visual3D is presented in Figure 10.

Some markers are mounted on wands in passive marker systems to increase their visibility and some marker placements are designed to assist with distinguishing the right from the left side of the body. This is not required for active marker systems. In all other ways, the marker sets used in both systems are essentially the same. Like passive marker systems, active marker systems output 3D marker trajectories. Active marker systems have noise and accuracy issues similar to passive markers. Occlusion of markers and markers passing out of the capture volume (line of sight requirements) are the primary source of data gaps for active marker systems (Fischer, 2002).

The main advantages of active marker systems compared to other systems include:

- Easy setup and calibration of the system
- High sample rates can be achieved when a small number of markers are used
- Excellent spatial resolution (as low as 0.1mm)
- Large and irregular shaped capture volumes can be created with multiple receivers
- No markers on wands or right/left body identifier markers are required
- Cost is considerably less than equivalent passive marker systems
- No marker occlusion and ghosting issues and markers can be placed close together
- Markers are smaller than passive reflective markers

The main disadvantages of active marker systems compared to other systems include:

- More equipment carried on the participant (battery pack, pulsing circuitry, LEDs and cables)
- Field of view of one position sensor system is small
- Large marker sets (e.g., Plug-in-Gait) will reduce the sampling rate achievable, although this is unlikely to be an issue for gait analysis studies
- For long duration tests, heat generated by the LEDs might be a problem (Bontrager, 1998)
- No gait kinematics/kinetics software is provided for OptoTrak (although proprietary software is available from C-Motion at extra cost)

Magnetic Systems

Magnetic tracking devices generate and sense magnetic fields. There is a transmitter that emits magnetic fields and a receiver that detects the magnetic fields emitted by the transmitter. A major attraction, and advantage, of magnetic systems is that each sensor placed on the participant delivers six degrees-of-freedom (X, Y, Z, yaw, pitch, roll). Another significant advantage of these systems is they do not require line of sight between the sensor and receiver. The size of the receivers attached to the participant are quite small:

- Ascension MotionStar Wireless 2; 2.5cm x 2.5cm x 2cm
- Polhemus Fastrak; 2.3cm x 2.8cm x 1.5cm

Use of magnetic systems is restrained both by limitations in system accuracy when a large volume of operation is desired and by their inherent sensitivity to large metallic objects (Périé et al., 2002). While magnetic tracking systems can be calibrated for some magnetic field distortions, data from a sensor when slid across the floor will appear to rise or turn as it approaches a metal object. In an AC varying magnetic field (e.g., Polhemus) certain metals become secondary transmitters because of induced eddy currents. The recently developed DC approach by Ascension also induces eddy

Figure 11. Magnetic tracking systems (left — Ascension MotionStar Wireless 2: from http://www.ascension-tech.com/products/motionstar_10_04.pdf; right — Polhemus system as used by Skill Technologies, from www.skilltechnologies.com/3dgait.htm)

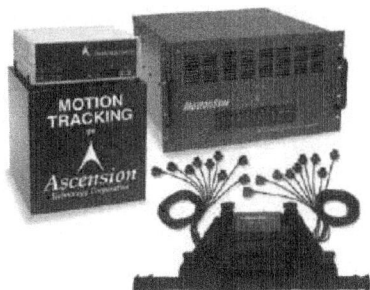 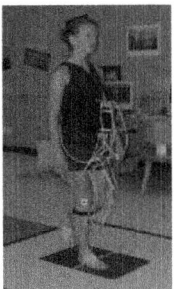

currents, but the eddies settle to zero during the sample time such that their effect is minimized (Fischer, 2002). Though still susceptible to distortion, the DC signal format is significantly less susceptible to distortion errors than the older AC systems (Zhou & Hu, 2004).

Another disadvantage of magnetic tracking systems is said to be their small field of operation. The "Extended Range Transmitter" of MotionStar Wireless 2 has a communication range of about 3m (Zhou and Hu, 2004). This field of operation is actually comparable with, or better than, some active- and passive-marker systems. However, magnetic fields lose strength as the receiver moves away from the transmitter. This means that data can become noisy towards the edge of the field of operation.

Magnetic tracking systems are used most often in animation applications and are not usually the preferred tool for gait analysis research and testing for the reasons stated above (e.g., where metal force plates, treadmills and other metallic objects are regularly used) and the rather cumbersome cables carried by the participant. The accuracy of the magnetic tracking device should be confirmed in each setting because different physical environments will affect the magnetic field differently (McGarry et al., 1999).

3D Kinematics

For passive and active marker systems the basic output is 3D marker coordinates moving in time, called "marker trajectories." The markers form an "exo-skeleton" around the participant which has to be related to an "endo-skeleton" model of the participant (Fischer, 2002). There are many methodological issues involved with converting raw marker coordinate data into useful 3D human body kinematics. In brief, each body segment must be defined by at least three external markers (segments can share markers), joint centres must be defined, and Euler angles computed. Often, anthropometric data must be measured from each participant (e.g., knee width, ankle width). The equations for calculating joint coordinates and segment orientations from external markers are often provided in software with the purchase of commercial 3D motion analysis systems. There

Figure 12. Hip joint angles with maximum, minimum and other features highlighted (From Benedetti et al., 1998)

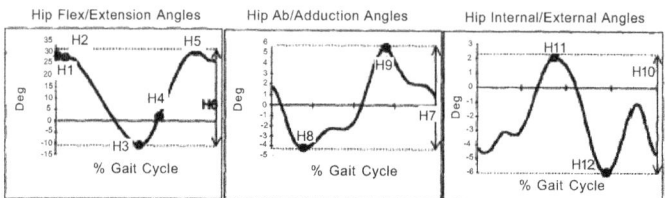

are also many publications and "standards" for joint centre position location and segment orientation definitions in the literature (e.g., Davis et al., 1991; Kadaba et al., 1990; Vaughan et al., 1992; and ISB, 2005).

Two commonly used marker sets for human body kinematics are described in Figures 8 and 9. Once the joint centre coordinates, segment orientations and, if required, segmental centre of mass locations have been derived it is then a relatively straightforward mathematical process to calculate 3D kinematics data (i.e., joint centre displacements, velocities, accelerations, and joint angles, angular velocities and angular accelerations). Joint angles, for example (Figure 12), are calculated at the capture rate of the cameras (usually 50Hz or above) forming a continuous joint angle time-series dataset. Key features of this graph/time-series are often extracted that include maxima and minima (e.g., Benedetti et al., 1998; see Figure 12) or parameter values are measured at the events such as those described in Table 4 and Figure 3. From lower limb joint angles alone it is possible to measure over 3000 parameters (Wolf et al., 2005).

Kinematics variables (e.g., linear displacement, velocity and acceleration) are usually presented with respect to a three-dimensional (3D) coordinate reference system. A coordinate reference system contains three straight lines orthogonal to each other (axes, e.g., X, Y and Z axes) and meeting at one point, the origin (O, see Figure 8). For example, the overall velocity of a body/point is split into three components, one component along each axis, represented as V_x, V_y and V_z. Many different types of axis systems are used, or could be used more, in gait analysis. These include:

- Earth-fixed axis systems
- Body-carried earth axis systems
- Body axis systems
- Principal body axis systems
- Path axis systems

The number of kinematics parameters that can be measured from the break down of gait using the techniques described in this section is in the thousands. In kinematics alone, this is achieved by:

- The break down of the gait cycle into events/phases (e.g., the seven main events and nine phases described in Table 4)
- The break down of the human body into segments and joint centres (e.g., 14 segments, 18 joint centres)
- Measuring the seven key parameters of kinematics (t, $\mathbf{s}$, $\mathbf{v}$, $\mathbf{a}$, $\boldsymbol{\theta}$, $\boldsymbol{\omega}$, $\boldsymbol{\alpha}$), with 3 components per parameter plus resultants
- Analysis using different axis systems (e.g., five described above, three axes per system)

There are many other issues that have not been covered in this chapter regarding each part of the process from marker/sensor placement to data capture to body modeling. This includes how to minimize errors/noise and how and when to filter/smooth data (e.g., accelerations derived from position/time data are notoriously noisy). Most of these issues are dealt with by commercial systems and are often largely hidden from the user. Excellent references enabling the reader to further investigate 3D motion analysis techniques include Allard et al. (1995, 1998), Whittle (2002), Dimnet et al. (1996), Whittle and Blanchi (1997), Small and Whittle (1999), ISB3D (2005), and Sutherland (2005).

Goniometers and Electrogoniometers

Goniometers are devices used to measure joint angles. Goniometers vary from the very simple mechanical linkages (e.g., Lafeyette Co., Figure 13) which are generally used for measuring an angle across one joint in one plane in static positions (Norkin and White, 2003), to the more sophisticated electronic goniometers (electrogoniometers) (e.g., Biometrics Ltd - formerly Penny and Giles, Figure 14). Realistically, only electrogoniometers can be used for clinical gait analysis. They are essentially electric potentiometers or transducers that produce an output voltage proportional to the angular change between the two attachment surfaces. The electrogoniometer is not attached to an external landmark in proximity to the joint centre, but instead has attachments to the two segments spanning the joint, so they literally measure the angle between the sensors as they are

Figure 13. Goniometers (www.lafayetteinstrument.com/rangeofmotion.htm)

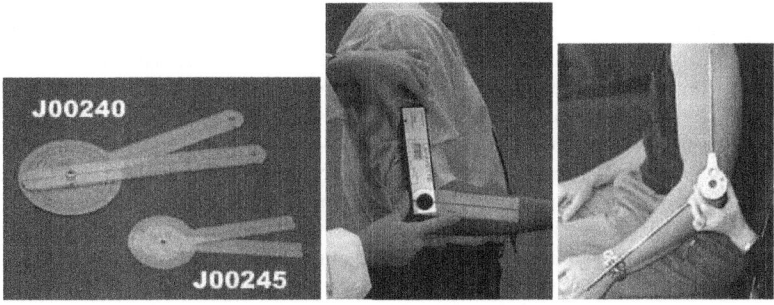

Figure 14. Single axis (left and centre) and biaxial (right) electrogoniometers (www.biometricsltd.com/gonio.htm)

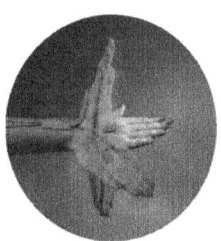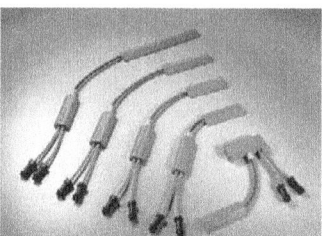

attached to the segments (i.e., the angle between the longitudinal axes of the endblocks, see Figure 14). They operate on the assumption that the attachment surfaces move with (track) the midline of the limb segment on to which they are attached, and thereby, measure the actual angular change at the joint (Bontrager, 1998).

Electrogoniometers can be used to measure angles in a single plane or for bi-planar movements (Nicol, 1989) (Figure 14). For example, Biaxial devices can simultaneously measure sagittal and frontal plane motions. Two single-axis electrogoniometers can be used to record joint angles at two sites, and consequently can be used to construct angle-angle diagrams to illustrate how the two joints are coordinating during a movement task, for example, hip-knee angle-angle diagrams (Whittle, 2002).

Electrogoniometers are relatively inexpensive, easy to use, portable, do not require attachment to joint centres, can be used in the field situation and can provide joint angle information in real time. Their main limitation is that they do not relate to any other reference/axis system (Bartlett, 1994). Some authors have also stated that they can restrict movement, especially when many of them are used on one subject (Bartlett, 1994; Winter, 1990), but if only one or a few are used then they are not much more restrictive than passive marker and active marker motion analysis systems. As with all gait instrumentation, their accuracy should be carefully evaluated to give a general idea of the kinds of errors that might be encountered.

Accelerometers

Accelerometers are transducers that can be attached to various parts of the human subject to measure accelerations in three dimensions. There are three common types of accelerometer (Zhou & Hu, 2004): piezoelectric, which exploit the piezoelectric effect whereby a naturally occurring quartz crystal is used to produce an electric charge between two terminals; piezoresistive, which operates by measuring the resistance of a fine wire when it is mechanically deformed by a proof mass; and variable capacitive, where the change in capacitance is proportional to acceleration. There are many manufacturers of accelerometers. The G-link (Microstrain) and Entran EGA3 triaxial accelerometers are suitable for use in gait analysis (Figure 15).

Figure 15. G-Link wireless (left) and Entran (right) triaxial accelerometers

Velocities can be derived through integration of the acceleration-time data ($v = \int a.dt$), and a second integration will provide the displacement ($s = \int v.dt$), as a function of time. While carrying out these integrations, initial conditions have to be provided (Bartlett, 1994) which can be difficult to determine in some movement tasks. Nonetheless, the acceleration output from accelerometers is instantaneous, and they are useful when basic acceleration information of body segments is of primary interest in an investigation, for real-time biofeedback, and for data collection in the field (Macko et al., 2002; Sekine et al., 2000, 2002; Tamura et al., 1997).

It is perhaps surprising, especially in clinical biomechanics, that more attention has not been given to accelerometry since it is accelerations that are the drivers of motion in the human body, and all the velocities, displacements and changes in velocities and positions of segments are the outputs derived as a result of accelerations (and the forces that produce them). Best results come from accelerometers attached directly to a rigid body (e.g., artificial limbs or bone, Lafortune, 1991) although the former is obviously not possible for routine testing (Bartlett, 1994). As with any item that requires attachment to soft tissue, errors can arise from improper mounting and fixation procedures.

Gyroscopes

Gyroscopes have mostly been used for outdoor gait-recording over a long period (Aminian et al., 2002). They are angular velocity sensors that can be attached anywhere on the body to record angular velocity-time history plots (they are often called rate gyros). Angular velocity-time plots taken from the lower limbs exhibit unique characteristics or patterns which can be processed to derive various spatial-temporal events and parameters of gait such as velocity, step/stride lengths, stance/swing times (Aminian et al., 2002; Pappas et al., 2004). This type of data-recording involving gait is particularly suitable for situations demanding outdoor recording and may also be suitable for the clinic or rehabilitation setting. The MTx (Figure 16; Xsens Motion Technology) is a miniature inertial measurement unit providing output of 3D acceleration, 3D rate of turn (rate gyro) and 3D earth-magnetic field data. Through software it also provides 3D orientation data (in the form of Euler angles and quaternions) in real time.

Figure 16. Xsens MTx gyroscope (www.xsens.com/index.php?mainmenu=products&submenu=human_motion&subsubmenu=MTx)

Although the use of gyroscopes is in its infancy in biomechanics, gyroscopes can provide a relatively inexpensive alternative to standard motion analysis systems in some situations. The body-mounted sensors are inexpensive and portable and allow long-term recordings in clinical, sport and ergonomics settings (Mayagoitia et al., 2002). There appears to be a strong focus on developing better, smaller and cheaper gyroscope systems (Xingang & Jianping, 2003) and it is likely that their use will increase in biomechanics in the next decade (e.g., Catalfamo & Ewins, 2003; Moser et al., 2004). The possibilities for gyroscopes (as well as accelerometers, electrogoniometers and footswitches) in taking gait analysis out of the lab and into the field situation are considerable.

Kinetics (Force)

Kinetics refers to the forces (and moments) that are responsible for changing a body's state of motion. Gait is the result of muscle action exerting forces on the skeletal limb segments in a coordinated way to change motion, and hence locomotion, as we interact with our external environment. It is not possible to measure internal muscular forces but we can measure muscle activity. By measuring external forces in association with kinematics data, calculations of joint forces and joint moments can be made.

Force Platforms

Force platforms, sometimes called force plates (Figure 17), are embedded in a walkway and are commonly used in gait analysis to record foot-ground reaction force and moment time histories.

Foot-ground reaction forces (ground reaction forces, GRF) are, as the name suggests, reaction forces as a result of contact between the foot and the ground, and these form an integral part of human movement analysis (Benedetti et al., 2004). There are two types of force platforms that are widely used: those based on piezo-electric transducers (e.g., Kistler) and those based on strain-gauge transducers (e.g., AMTI, Bertec). From a gait analysis viewpoint there is little, if any, functional advantage of one type of system over the other. While they essentially measure the same information, the

Figure 17. Kistler force platform

Figure 18. Ground reaction force and moment outputs for the AMTI force platform

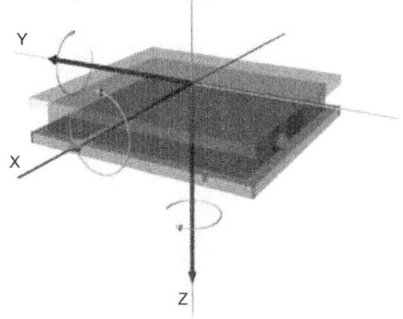

raw outputs of the systems are different. For example, the output from AMTI force plates is three orthogonal force components (F_x, F_y, F_z) and three moments (M_x, M_y, M_z), as depicted in Figure 18. The output from Kistler force plates is four vertical forces and four horizontal forces. In software, all systems convert this raw data to the main information of interest in human movement analysis, that is F_x (horizontal medio-lateral force or medial shear component), F_y (horizontal antero-posterior force or AP shear component), F_z (vertical force component), centre of pressure position (COP) and T_z (vertical torque; moment about a vertical axis passing through the COP position). An example of the normal gait graph for F_x, F_y and F_z is presented in Figure 19.

If mounted correctly, force platforms are very stable devices and the data they produce is critical to gait analysis. The forces measured by force plates, along with gravity, are the only significant external forces acting on the body during gait and, therefore, the only external forces involved in changing the state of motion during gait.

Figure 19. Typical output from a force platform showing vertical (load, F_z), anterior-posterior (AP shear, F_y) and medio-lateral (Medial shear, F_x) force-time histories during normal gait (from CGA, 1997)

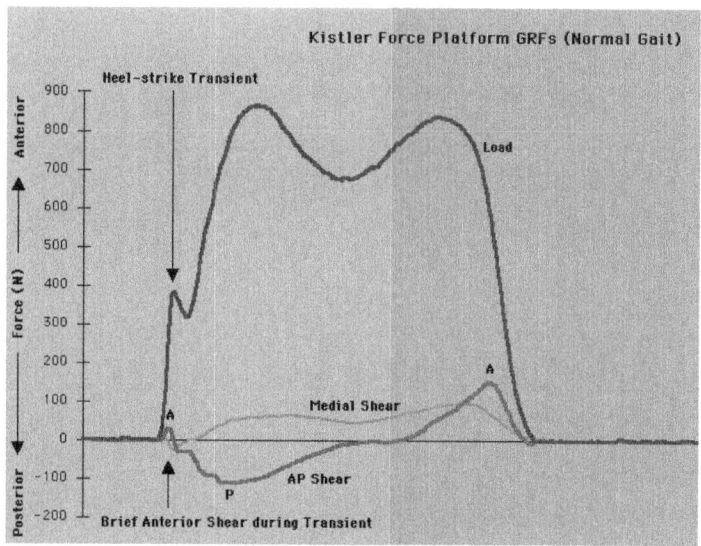

Pressure Mats and Foot Pressure Insoles

Foot pressure systems allow measurement and analysis of the distribution of one-dimensional pressures (vertical for pressure mats and normal to the sensor for pressure insoles) on the plantar surface of the foot. Figure 20 shows a typical output from a foot pressure sensing system. This type of information is useful for treating foot problems and monitoring the effectiveness of orthotic interventions. It has proven particularly useful in the treatment of diabetic neuropathy where feedback from the patient is unreliable or non-existent. Foot pressure sensing can quickly determine the areas beneath the foot that may be subject to tissue breakdown (Cavanagh et al., 1992).

Pressure mats (e.g., Figure 21) are a similar size and shape to a small force platform. Like force plates, pressure mats are placed within a walkway and a participant walks barefoot across the mat. The advantage of the pressure mat is that nothing needs to be attached to the individual but the disadvantage is that only barefoot walking gives reliable pressures felt by the foot (if shoes are worn the pressure between the ground and the underneath of the shoe is measured).

If the pressure on the underside of the foot while wearing a shoe is required or the effects of various orthoses are to be evaluated, then a pressure-sensing insole must be used. Apart from in-shoe measurement, insoles have the added advantage of being able to measure multiple consecutive gait cycles on one or both feet.

Figure 20. Example of the pressure distribution under the foot using the F-Scan system (taken from www.tekscan.com/medical.html)

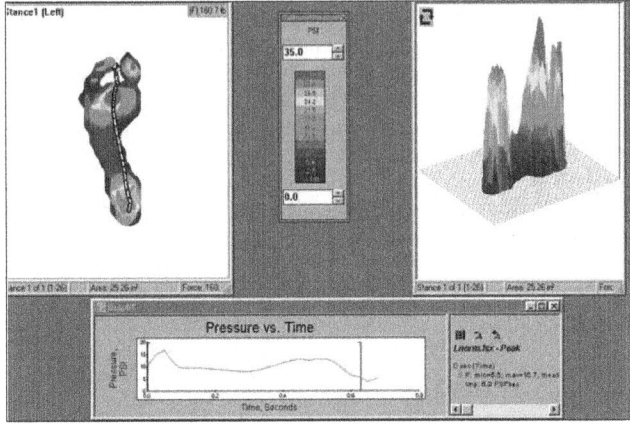

Figure 21. Emed pressure mat (taken from www.novel.de/productinfo/systems-emed.htm) (left) and Fscan www.tekscan.com/medical/specs_mobile.html (centre) and Pedar (right) insoles

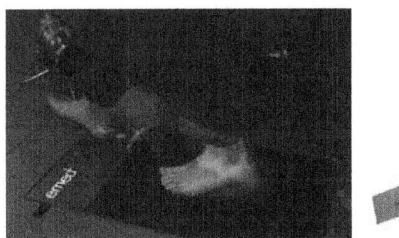  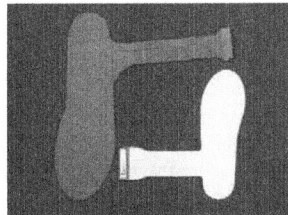

There are two main manufacturers of pressure insoles, Novel (Pedar) and Tekscan (F-Scan), and they are quite different products. The Pedar system uses capacitive transducers and they last many years. There are 12 insole sizes available and, though the initial outlay for the system plus a range of insole sizes is expensive, once the system is bought there are no consumable costs. The F-Scan system uses FSR transducers and each insole lasts for only a few minutes of continuous walking (i.e., insoles are consumables). Each insole is large and can be scissor-trimmed to the size required. Though the initial outlay for F-Scan is less than a comparable Pedar system, once the system is bought there are ongoing consumable costs.

In terms of functionality the two systems again vary considerably. Pedar has 99 sensors per insole whereas F-Scan has 960 (which are reduced as the insole is trimmed). Pedar insoles are reliable over long periods whereas F-Scan insoles have a time-limited lifespan. Pedar insoles are quite thick (no more than a shoe insole you would buy from a shop) but they are comfortable to wear. F-Scan insoles are very thin but they can be a little slippery under foot, which in turn means they can at times move and "crinkle" in the shoe, which can result in unreliable data and individual sensors in the insole breaking. The short lifespan of the F-Scan sensors continually plays on the mind of the user since it is not known precisely when the insole should not be used anymore. Both Pedar and F-Scan insole systems can be calibrated using a pressure bladder calibration system. In comparing the reliability of two systems, Cavanagh et al. (1992) and McPoil et al. (1995) found Novel outperformed F-Scan. However, this was using sensor technology no longer in use. Using more modern versions of the two systems, Rash et al. (1997) found no difference in their ability to measure uniform absolute pressures.

Pressure mats and insoles calculate force and pressure time histories, force and pressure/time integrals and centre of pressure position beneath the foot. Insoles are also particular useful for creating "masks" or "areas" that allow analysis of selected areas under the foot. While pressure mats can measure single foot contact time, if two pressure insoles are worn then all the important temporal event information for the full gait cycle (Table 3) can be measured. While pressure mats can normalise data to foot contact time, if two pressure insoles are worn then data can be collected and normalised to gait cycle time and all the significant temporal events detailed in Table 3 can be detected.

If accurate forces or centre of pressure are required then force platforms are the preferred technology. If centre of pressure is required relative to an earth/lab-based axis system then a force platform (preferred) or pressure mat has this capability. If centre of pressure is required in relation to the orientation and shape of the foot then a pressure mat or insoles are the best solution (unless force platform centre of pressure information is collected and synchronised with 3D kinematics data or a force vector visualisation system is used (e.g., Roberts et al., 1995). If forces are required over multiple, consecutive steps then pressure insoles are the only realistic solution, noting that pressure insoles only measure forces normal to the sensor (not necessarily vertical) and cannot measure shear/horizontal forces.

No matter which system is purchased, there will always be reminders of the limitations of one system compared to the other system. Proponents of the PEDAR system can claim higher repeatability over longer periods and greater insole durability. Proponents of the F-Scan system can claim higher spatial resolution and lower cost.

Electromyography (EMG)

A "complete" clinical gait analysis test usually involves 3D motion analysis, force platform and EMG data. EMG is the analysis of the electrical activity of the contracting muscles. It is useful for detecting those muscles that are working (or are not working) and in what sequence they are working with respect to the needs of the movement. EMG can also give an indication of the intensity of muscle activity. Accordingly, EMG analysis is predominantly focused on the on/off timing of the muscle activity and the intensity or amplitude of the electrical signals.

Surface or in-dwelling/wire electrodes can be used to record the electrical activity of muscle. Surface electrodes are most often used in gait analysis because they are less invasive and easy to use. The main disadvantage of surface electrodes is that deep muscles cannot be monitored because of crosstalk (electrical activity) from other muscles that lie between the deep muscles and the surface electrode. There are two types of surface electrode, active and passive:

- **Passive surface electrodes** — Most of the early EMG research in biomechanics and much of today's EMG research is done using passive surface electrodes. The procedure is very simple and involves an electrode of, for example, the silver/silver chloride type (Figure 22) and a conducting gel placed on a prepared skin surface (e.g., shaved, thoroughly cleaned and abraded to reduce skin resistance) on a muscle site. It is a good idea to measure impedance at the site when using passive electrodes to give an indication of the quality of the site preparation. The EMG signal is sent via a cable to an amplifier and the amplified EMG signal then sent to a computer via an analog to digital converter. The cable and transmission of the small EMG signal to the amplifier is a significant source of noise with passive electrodes. This noise can be reduced by using a pre-amplifier near the electrode site.

- **Active surface electrodes** — Active surface electrodes are so called because they amplify EMG signals at the site and, therefore, require power at the site. Amplification at the site virtually eliminates cable artifacts (Bartlett, 1994). It also means that little or no skin preparation and no gel is required. All commercially available active surface electrodes have high impedance differential amplifier inputs with high common mode rejection ratios. The Delsys DE-3.1 EMG electrodes (pictured in Figure 22) have double differential amplifier inputs which should help minimise crosstalk problems.

Figure 22. Passive EMG electrodes (left: from Konrad, 2005), Delsys DE-3.1 double differential active surface electrode (centre) and Biometrics SX230 active surface electrode (right)

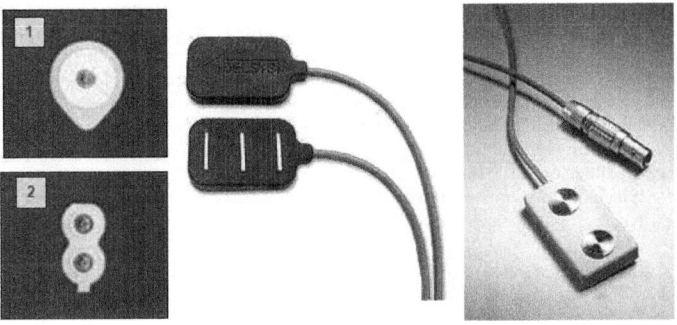

Figure 23. Raw EMG signal (above) and the same signal full-wave rectified (below); Solid line is foot strike and dashed line is toe-off. (From Motion Lab Systems Web site; www.emgsrus.com/emgfull.htm)

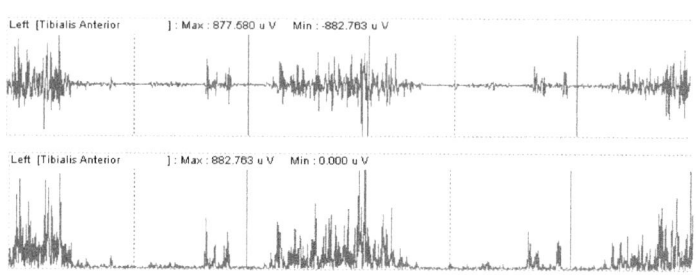

EMG data acquisition is generally achieved in one of two ways: cable or wireless (radio telemetry or data loggers). Cable systems eliminate the need for a battery on the wearer but there is a need for a cable connecting the wearer to the instrumentation. Telemetry systems eliminate the cable but they require a body-worn battery. Wireless systems can also suffer from problems with signal dropout and RF interference. Data loggers eliminate the cable and RF problems. They also require a body-worn battery and are limited in the amount of data that can be acquired before being downloaded to the computer (Bontrager, 1998).

Once the raw amplified EMG signal has been collected the next task is often to work out the timing of muscle activity onset and determining the amount of activity. Of primary interest in EMG gait analysis are the onset, duration, and amplitude of muscle activity of the leg muscles during a gait cycle (Jansen et al., 2003). An example of a raw EMG signal is presented in Figure 23. Often the first stage of analysis of the signal is a process of "full-wave rectification," that is all the negative values are made positive (Figure 23).

There are a number of ways to quantify EMG onset (e.g., Allison, 2003; Nieuwboer et al., 2004; Merlo et al., 2003; Lauer et al., 2003; Hodges and Bui, 1996) that include visual inspection and preset thresholds (the latter often being confirmed by visual inspection). Increasingly, EMG data acquisition systems are being sold with footswitches so that timing of muscle EMG onset can be determined relative to the key events of the gait cycle (e.g., the events displayed in Table 3). Figure 24 shows EMG onset for normal gait as a function of time normalised to the period of the gait cycle. Simple on/off diagrams such as that presented in Figure 24 are commonly used in gait analysis. These diagrams deliberately give no indication of the "amount" of muscle activity. Instead they give a useful indication of binary muscle on/off status. They are also sometimes used because of the difficulty in reliably quantifying the amount of EMG activity and quantifying how the magnitude of an EMG signal relates to muscle tension.

Since the "amount" of EMG signal recorded is a function of factors such as skin resistance, crosstalk, electrode position, walking speed and many other factors, it is very

Figure 24. Muscle EMG onset data during gait (from Rab, 1994)

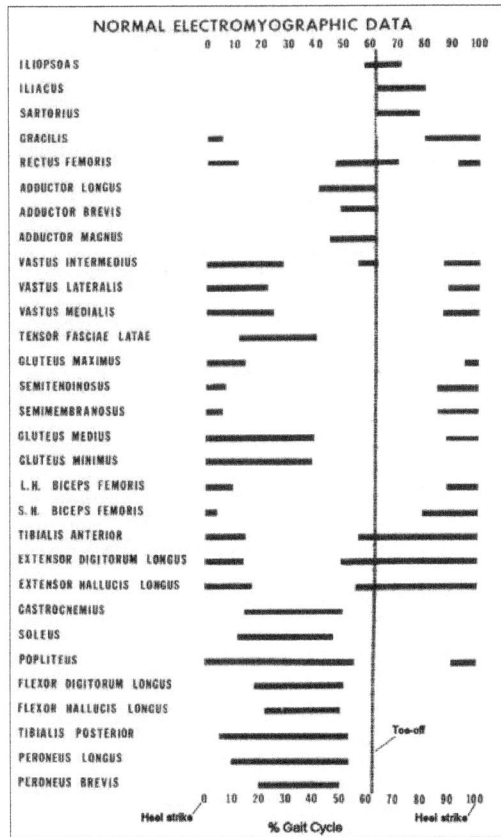

difficult to quantify EMG in a way that allows inferences to be made about the level of muscle activity, muscle tension and comparison across individuals. Standard ways of quantifying the amount of EMG signal after full-wave rectification include average EMG (the average EMG over a specified amount of time — usually a moving average), root mean square EMG (square root of the average signal in a specified amount of time — usually a moving average), integrated EMG (IEMG; the area under the rectified EMG signal — often "time-reset" whereby the integral is computed for a set interval of time and then reset to zero), linear envelope detection (a low pass filter applied to the EMG

Figure 25. From the top — Raw rectified EMG signal, moving average, RMS and linear envelope for the same signal (from Konrad, 2005)

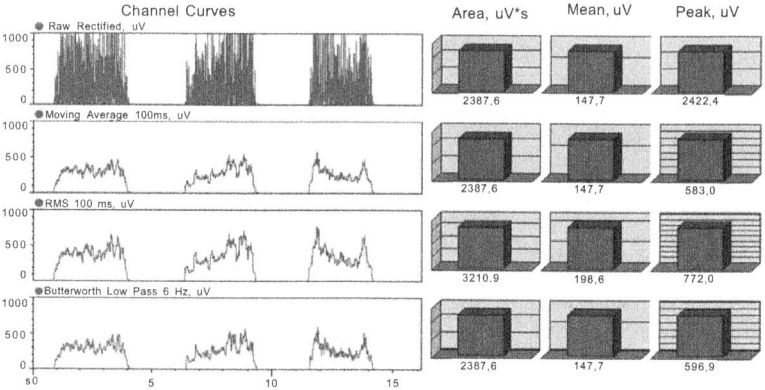

signal — this is often wrongly referred to as IEMG) (Bartlett, 1994). These methods all effectively measure the same thing, namely the amount of EMG activity. Examples of rectified EMG, moving average, RMS and linear envelope EMG displays for the same data are presented in Figure 25.

Where moving averages or "time-reset" methods are used, the time window over which each calculation is made needs to be chosen carefully. EMG data is often collected in association with footswitches (or force platform) so the signals can be time-normalised to the duration of the gait cycle (% gait cycle). Ensemble averaging is then often conducted for multiple gait trials. The time-normalised ensemble averaging process further smoothes the EMG signal and can be subject to the problem of temporal dependency (see discussion of "temporal dependency"). EMG does not provide the quantitative accuracy needed for the assessment of muscle force generation, although the linear envelope is thought to resemble the muscle force curve (Bartlett, 1994).

All of the above relates to time-domain analysis of EMG signals. Another way of analysing the signal is via a *frequency domain analysis*. Converting raw EMG-time data to the frequency domain is a process called harmonic or spectral analysis, and can be done by Fourier analysis. The end result of this is a power spectrum. The square root of the power spectrum forms the amplitude spectrum (Bartlett, 1994).

Frequency domain analysis (often called "spectral analysis") is used for detection of muscle fatigue and muscle fibre health. Median frequency shift of the EMG power spectrum, reflecting a decrease in the muscle fibre conduction velocity, is a commonly used method of assessing muscle fatigue (Figure 26; e.g., Lindström et al., 1970; Mannion & Dolan, 1996; Gefen, 2002; Gefen et al., 2002; Allison & Fujiwara, 2002), although Pullman et al. (2000) suggest the link has not been convincingly established.

Figure 26. Median frequency of normal EMG (line a) reduces when muscle is fatigued (line b) (From Rab, 1994)

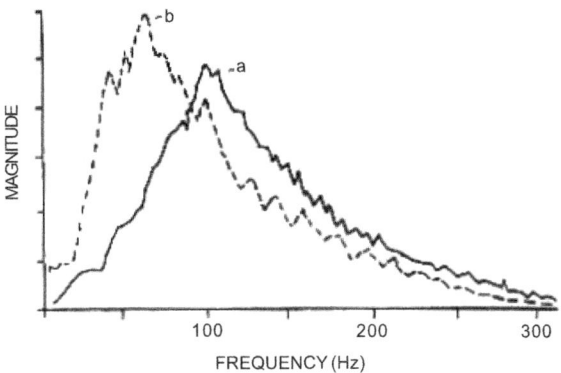

Figure 27. EMG signal and corresponding wavelet visualisation (from Simi Web site www.simi.com/en/products/motion/modules/emg_wave/index.html)

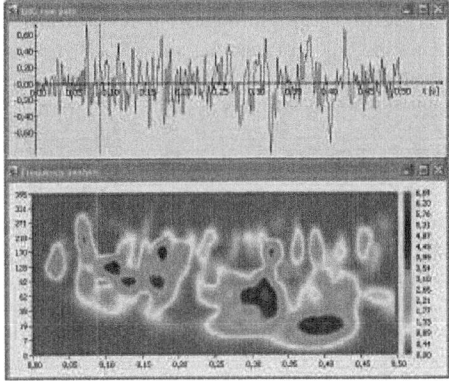

Analysis of EMG using "wavelets" is relatively new to biomechanics research. The wavelet transform may be a powerful tool to analyse EMG signals during gait of asymptomatic and pathological subjects (De Stefano et al., 2003), although this is yet to be fully proven. De Stefano et al. show that some wavelet parameters are capable of classifying asymptomatic and pathological subjects into groups. They concluded that wavelets have the potential to be applied to the monitoring of patients during interventions aimed at improving muscle behaviour, particularly antispasticity treatment such as Botulinum Toxin injections.

The main advantage of wavelets is they incorporate both time and frequency domain analysis, and they allow visualisation of the EMG signal in the time and frequency domains simultaneously. Wavelets are also the best tool for analysing the frequency domain of non-stationary signals (i.e., signals of varying frequency such as fatigue analysis). Wavelet toolboxes are already a part of some EMG analysis software packages (e.g., Noraxon's MyoResearch XP software has a number of wavelet functions). An example of a wavelet transformation from the Simi software package (Simi Reality Motion Systems) is detailed in Figure 27. A more detailed discussion of wavelets and other more advanced analysis tools is included in the following).

Surface EMG (SEMG) has a number of complicating factors that make interpretation and quantification of signals imprecise. It is very easy to apply an electrode on a muscle, display a signal and draw conclusions concerning the pattern, modalities, timing, and intensity of muscle activation. For this reason the use of surface EMG has been the subject of a number of negative reviews in recent years (Merletti & Parker, 2004). These reviews include Haig et al. (1996), Rechtien et al. (1999) and Pullman et al. (2000). The review of Pullman et al. has three main conclusions about SEMG:

- SEMG is considered unacceptable as a clinical tool in the diagnosis of neuromuscular disease at this time
- SEMG is considered unacceptable as a clinical tool in the diagnosis of low back pain at this time
- SEMG is considered an acceptable tool for kinesiological analysis of movement disorders, for differentiating types of tremors, myoclonus, and dystonia, for evaluating gait and posture disturbances, and for evaluating psychophysical measures of reaction and movement time

Joint Forces and Moments (Inverse Dynamics)

By measuring force plate data and kinematics data simultaneously, joint forces and moments can be calculated via a mathematical process known as inverse dynamics. Many motion analysis systems (e.g., VICON) provide inverse dynamics as part of their system, whereas other motion analysis systems (e.g., NDI's Optotrak) have partnerships with software companies (e.g., C-Motion — providers of the Visual3D software, formerly MOVE3D software) that provide inverse dynamics solutions for those hardware providers.

Inverse dynamics is the process of computing the net joint forces, joint moments and joint powers and the calculations require kinematics data (i.e., positions and orientations of joints and segments as well as their linear and angular velocities and accelerations), ground reaction force data and anthropometric data. Joint forces, moments and powers during gait (Figure 28) are critical to the understanding of how gait is produced by an individual, especially for clinical gait analysis, since it includes the internal "driving" forces and moments of gait motion. Joint moments are the result of forces produced by muscles and ligaments acting at a distance from the joint centre. Joint power is the net rate of generating (concentric contraction) or absorbing (eccentric

Figure 28. Net joint forces, moments and powers during gait (from Meglan & Todd, 1994)

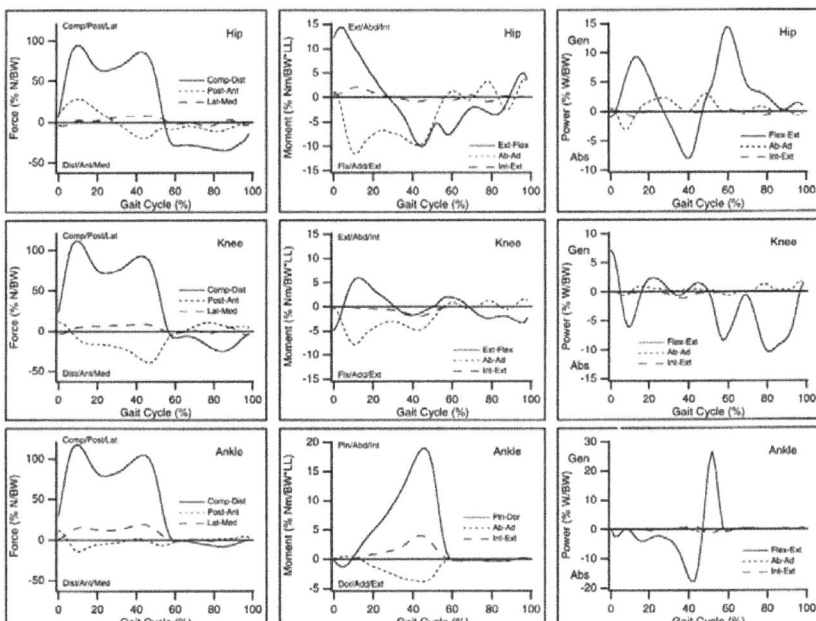

contraction) energy by all the muscles crossing a joint and is calculated as the product of the joint moment and the angular velocity between the two segments crossing the joint (Winter, 1990; Meglan & Todd, 1994).

Significant joint force/moment/power calculation errors are possible if great care is not taken in the placement of joint centre markers, if the calculation of joint centre positions and segment orientations are imprecise, and if the anthropometric data used are not appropriate for the individual being tested (e.g., Delp & Maloney, 1993; Frigo & Rabuffetti, 1998; Leardini et al., 1999; Chiu & Salem, 2005).

In all studies involving inverse dynamics, great care should be taken in the way segment masses and moments of inertia are approximated. Methods of obtaining inertial parameters of body segments can be classified into three groups (Kaufman, 1998):

- **Regression equations** — usually based on measurements derived from populations of cadavers or living subjects (see Durkin & Dowling, 2003; Yeadon & Morlock, 1989)
- **Geometrical approximation** — representing body segment shapes as standard geometric shapes of known mathematical description, usually applied to individu-

als after a range of anthropometric measurements have been taken for that individual that help define the proportions of the geometric shape (see Reid & Jensen, 1990; Cappozzo & Berme, 1990)

- **Direct measurement** — measuring data directly from the individual being studied — usually based on MRI or Dual Energy X-ray Absorptiometer (DEXA) measurements (e.g., Chiu & Salem, 2005)

Direct measurements are by far the preferred method for body segmental parameter (BSP) estimation, probably followed by geometric modelling and finally regression equations. Despite this, regression equations are the most commonly used method because of their convenience, because of the lack of an efficient geometric modelling procedure, and because of the considerable expense, training requirements and general lack of availability of DEXA-like machines.

In isolating the effects of segmental inertia values on inverse dynamics calculations, a comparison of regression equations and direct measurements using a DEXA by Chiu and Salem (2005) found differences of up to 10% in hip joint moment calculations.

Regression equations based on a very large sample size and specific for populations according to age, gender, race and morphology may be required to provide accurate estimations of BSPs for use in inverse dynamics (Durkin & Dowling, 2003). Alternatively, a method for calculating individualised segmental inertial values from a set of anthropometric measurements which requires minimum intervention time would greatly improve this area (Yeadon & Challis, 1993).

Calculation of joint forces and moments is of critical importance to clinical gait analysis, especially in those areas where gait analysis data is used to inform surgical intervention. An excellent overview of inverse dynamics methods and their application to gait analysis is provided by Meglan and Todd (1994).

Energy Consumption/Energetics

There are a number of approaches for measuring energy consumption and energy levels of body segments in movement analysis. Measurement of the whole body energy consumption gives insight into the energy efficiency of gait, and how a particular

Figure 29. The HASDMS-I equipment (Human Activity State Detection and Monitoring System) (home.flash.net/~hpm/pages/HPM.products/ASDMS1.html)

 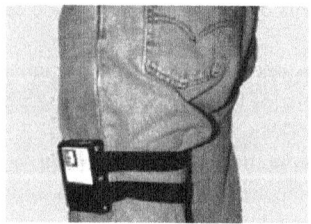

pathology (e.g., cerebral palsy) influences energy consumption and efficiency during an activity. Measurement of the energy consumption can be achieved by: (1) analysing oxygen consumption; (2) recording total body mechanical energy levels (i.e., potential and kinetic energy levels of limbs); or (3) heart rate monitoring using either physiological cost index (PCI) or total heart beat index (THBI), both of which are shown to offer a good estimate of the energy expenditure (Hood et al., 2002). An overview of these techniques is given by Rose et al. (1994) and Sutherland (2005).

A related area is "human activity modelling." In this procedure, a system (e.g., the HASDMS-I system made by Human Performance Measurement; Figure 29) detects and logs selected human "activity states" over prolonged periods (up to 7 days). Activity states usually include lying-sitting (grouped), standing, walking, running and an "unknown" state for patterns that the system is unable to identify (Zhou & Hu, 2004).

GAIT FEATURES

Nowadays, it is usual to generate large quantities of data in a gait analysis study. The data used in gait analysis is most often in the form of time-series with, for example, force platform data, 3D motion analysis data, EMG data and inverse dynamics data (joint forces, moments and power) often calculated 60-1,000 times in every second. It is not conceivable to analyse each measured/calculated data point and nobody does. It is necessary to find ways of reducing the number of parameters, retaining those that are useful and discarding those that are not. This is the topic of this section.

General Principles of Feature Extraction

There are many terms, used almost interchangeably, for the process of extracting the important features from a set of data, including "feature extraction," "feature (subset) selection," "data mining," "information content" and "(feature) dimension reduction." Whatever the term used, the process of feature extraction can be defined as "*a process of identifying valid, useful and understandable patterns in data.*"

The large amount of time-series data that is generated in a gait analysis study is called "high dimensional" data. High dimensional data can be thought of simply as lots of data (there are lots of dimensions to the data). Fast developing gait measuring techniques and methodologies generate more and more data. This in turn results in what is known as the "curse of dimensionality" (more and more data to manage, more dimensions) (Yang & Shahabi, 2004; Chau, 2001a).

The objective of feature extraction is to keep all the useful features of the data and discard all the redundant parts of the data. There are two required outcomes to this process:

1.   to reduce the amount of data to a manageable level (dimension reduction), and
2.   to keep the most important features of the data and eliminate all the redundant features of the data (feature selection).

The idea is to provide a "summary" which can be used to give a meaningful interpretation of the data. The objective of the first step towards feature selection is dimension reduction, to reduce the search space to a lower, more manageable dimensionality and this has to be achieved in a way that retains relevant features of the data and removes irrelevant features. That is, feature selection selects "m" relevant features from the entire set of "n" features such that m ≤ n. Noting the inherent redundancy in data, ideally m <<< n, and ideally m contains all the relevant features and no irrelevant features of the data.

Reducing the amount of data and extracting the key features are most often done in one of two ways in gait analysis, namely *time-series parameterisation* (selecting key features from time-series data/graphs) or *data transformation* (transforming the original data set to a smaller set of numbers while preserving as much information about the original data set as possible, e.g., by Fourier transform or wavelet transform).

Feature Extraction by Time-Series Parameterisation

Time-series parameterisation involves creating a set of "features" or "local events" or "parameters" from time-series data. Features of time-series data in gait analysis are generally of two types:

1.   Amplitude and timing of reproducible "visible" features of the data, including maxima, minima, zero-crossings, etc.
2.   Extracting data at key reproducible events (e.g., extracting joint moments at the events described in Table 4).

The term "reproducible" is important here and it means features must occur, or can be extracted, for all participants in the study. Some time-series "features" that occur in normal gait do not occur in all strides of every individual, and sometimes do not occur at all in other population groups (this is discussed in more detail later in this section).

Figure 30. Toe clearance (mean of 1000 consecutive strides of one individual with standard deviation grey bands) as a function of time normalised to the swing phase of gait (toe-off = 0%, heel strike = 100%)

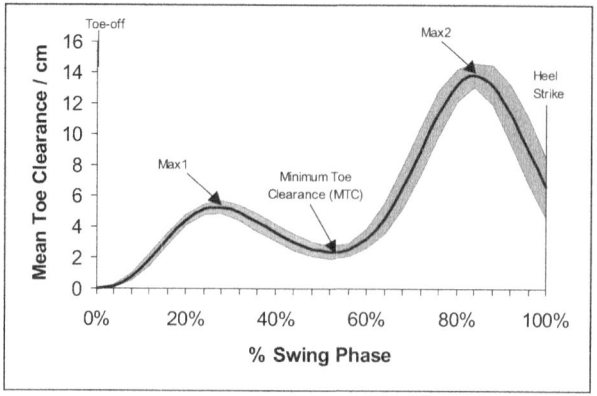

Figure 30 shows an example of time series parameterisation for the vertical toe trajectory during the swing phase of gait from 1,000 gait cycles of one individual. It is feasible from this very simple model that mean, standard deviation (SD), skewness and kurtosis of the amplitude (magnitude) and timing of max1, max2 and MTC (and magnitude at heel strike) could all be chosen as parameters of interest for this data. This gives a total of 28 parameters, and is hardly exhaustive, representing only the 1D position of one anatomical feature of the body.

When choosing gait parameters it is best to choose those that best facilitate comparison across the different conditions and populations being studied (Begg et al., 1998). These parameters are key to the successful outcome of the research and should not be chosen arbitrarily but guided by a thorough knowledge of past research and an even more thorough understanding of the data from the study. It is important to be creative in defining parameters and not to be restricted to the parameters and events that previous researchers have used or be restricted to the parameters prescribed by proprietary software. If the data show a difference/feature that is not adequately described by previously prescribed parameters, then a parameter must be created that best describes that difference/feature of the data. Proprietary gait analysis software is usually based on an excellent and thorough knowledge of gait analysis and is designed to cover 80+% of every researchers needs. However, equipment manufacturers will be the first to tell us that it is not possible for them to write every bit of software that covers everybody's, sometimes exclusive, needs.

Simple *visualisation* (e.g., looking at graphs) is commonly used to gain insights into the structure of data. Experts with domain-specific knowledge and a deep understanding of the data are capable of extracting parameters that are theoretical or observed discriminatory features of the data. Creating a set of "features" of the time-series data is the elusive "art of science" that involves creating parameters that represent what an expert "sees" in the data.

Many commercial gait analysis systems generate a series of "standard features" for the particular gait data they measure. These are sometimes features that theory determines should be important, or features that experts have determined as being important based on visual interpretation of graphical representations of data. Other times they may be just obvious visual features of the data (e.g., maxima/minima). The AMTI force plate software automatically calculates 44 gait parameters based mostly on standard vertical, medio-lateral and anteroposterior ground reaction force and vertical torque time-series graphs. Benedetti et al. (1998) created a list of 124 gait cycle parameters including temporal-spatial parameters and, mostly, amplitude and timing of maxima and minima of time-series data for ground reaction forces (18 parameters compared with AMTI's 44), lower limb joint angles and lower limb joint moments. The Benedetti et al. feature-set is a useful attempt at parameterising gait time-series data but it emphasizes how difficult it is to fulfil the definition of feature extraction, that is, to keep all the relevant features and discard all the redundant features. While the Benedetti et al. feature-set appears comprehensive for the time-series analysed, some of the parameters will inevitably be redundant, non-discriminatory features for some/many studies. Also, inevitably, the list will miss some key discriminatory features. The same applies for any set of "standard features."

Barton and Lees (1995) measured 1,316 foot pressure parameters values, representing the measured pressure values of each cell on a foot pressure platform, for input to a neural network. Wolf et al. (2005) measured 3,670 parameters for a single gait cycle that included parameters derived from lower limb joint angles that were "easy to automatically compute" during 10 phases of the gait cycle (gait phases based on those defined by Perry, 1992). They limited the study to parameters derived from lower extremity joint angles to "reduce the complexity of the problem" so they could focus on the main topic of the research, which was not the parameters themselves. Had they also included other lower limb time-series data (e.g., other kinematics data, inverse dynamics data, force platform data and EMG data) the number of parameters could have easily exceeded 10,000. Wolf et al. (2005) did go the next step and found which of the 3,670 were the best discriminatory (most relevant) features for their application, thus making their study a true discriminatory feature selection process by including both a dimension reduction process and a redundant feature elimination process (although taking only joint angle data meant they possibly did not collect the best data to begin with). Benedetti et al. (1998) and Barton and Lees (1995) did not include a redundant feature elimination process.

It is important to note that "standard" or "easy to automatically compute" features should not be assumed to contain useful discriminatory features for every (or any) study. Also, "standard features" will contain redundant (e.g., non-discriminatory) features for many studies and they should be reduced further to eliminate redundant data. A key issue with most gait analysis research, especially machine learning approaches, is generalization of the results. *Generalization* deals with the robustness of the techniques when dealing with variable data from the same domain. Generalization is improved if all of the key features of the data are included in the gait model and generalization is reduced if irrelevant features are included in the model. The aim, therefore, is to establish which parameters are the most relevant parameters to use and which are redundant. Benedetti et al. (1998) described 124 parameters that are "crucial points of curves (maxima and minima….)." While this process is a form of feature selection, it is by itself a pseudo-random feature selection process because there is no means of knowing which parameters are relevant and which are not. If using the Benedetti et al. feature-set, some method is required to determine which parameters are relevant and which are not. One way to achieve the goal of reducing the number of parameters and keeping only the relevant parameters is to use principal components analysis (discussed later).

Reproducible Features

As mentioned earlier, some time-series "features" that occur in normal gait do not occur in all/some strides of every individual and in other population groups. For example, Figure 31 shows the equivalent toe clearance vs. normalised time as that shown in Figure 30 for both limbs of an amputee. This shows that the key parameter of MTC does not occur in the individual depicted in Figure 31 (note that nearly all amputees do exhibit an MTC minima but the one participant detailed in Figure 31 does not). There are many examples where events or parameters occur in normal gait but do not occur in some or all gait cycles of individuals in other population groups (e.g., heel strike is sometimes toe-strike in some individuals) and if a foot is dragged along the ground for some of what would normally be regarded as the swing phase of gait then the "toe-off" event may be compromised. In this respect, Wolf et al.'s (2005) parameters depend on the phases always occurring

Figure 31. Mean toe clearance as a function of time normalised to the swing phase of gait (toe-off = 0%, heel strike = 100%) for both limbs of one atypical amputee participant compared to a healthy young population; — mean of healthy young population with standard deviation grey bands; ▲ below-knee amputee prosthetic limb; ▲ amputee normal limb

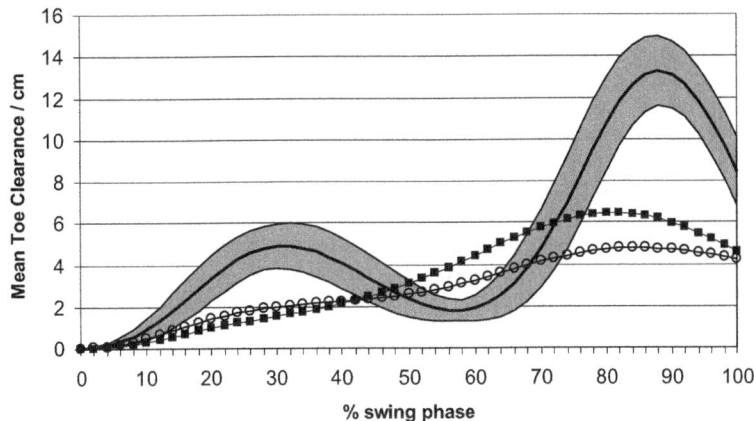

as per a normal gait pattern (or rather the events that determine the phases occurring normally) and they do not rely on maxima and minima turning points to occur. Benedetti et al.'s (1998) parameters do not depend on the phases always occurring as per a normal gait pattern but they do rely on maxima and minima turning points occurring in the population being studied.

Variability of Time-Series Parameters

Figure 30 introduced the issue of variability of parameters. Does an individual's gait data from one gait cycle (or the mean of 2 gait cycles or 10 gait cycles or 1,000 gait cycles) adequately represent the intention of the locomotor system? Parameter estimation deals with finding parameters, such as means and standard deviations, that describe the distribution of a given sample of data points. Means and standard deviations are the most commonly used parameters and they are used to reduce multiple data sets into one set of pseudo-instances. For example, 10-1,000 trials from one individual can be represented as a single mean time-series (a "mean" time-series never actually happens, hence the term pseudo-instances), then parameterisation can occur from this mean time-series and the mean, SD, skewness and kurtosis of each parameter (e.g., MTC in Figure 30) can be calculated, thus significantly reducing the dimensionality of the data collected for each individual or the group. Group (inter-individual) SD is commonly reported in gait studies but the important intra-individual SD is rarely reported. Inter- and intra-individual skewness and kurtosis are also rarely reported (see Best et al., 2001 for an exception to this).

When there are constraints imposed on the free movement of limbs, the distributions of data about the mean value for individuals (and sometimes groups) are unlikely to be symmetrical (i.e., they are unlikely to be normally distributed). This is because the constraint restricts the likelihood of equal variability on either side of the central tendency. Unfortunately, nearly all gait analysis research has neglected to consider variability of parameters and whether the mean of a parameter truly represents the intention of the locomotor control system. For example, one constraint to gait is the ground. This is a physical constraint that disallows any MTC values below zero, and a value of zero may have the constraint penalty of a trip which could in turn lead to a fall. For all individuals in the study by Best et al. (2001), MTC was found to be skewed to the right (i.e., $S > 0$ or skewed away from the ground) and was found to have positive kurtosis. This represents a systematic and intentional deviation from a normal distribution and leads to the conclusion that the median (mode would be better but is not accurately measurable) is more likely to be representative of the central tendency of the data than the mean (the central tendency being representative of the intention of the locomotor control system), and it shows that skewness is a key parameter/strategy that defines how individuals deal with trip avoidance. Similarly, Johns et al. (2004) and Johns (2004) showed the parameter max2, as depicted in Figure 30, is also not normally distributed in healthy individuals. For max2 the opposite, left skew ($S < 0$), was found and this was suggested as a possible consequence of the energy expenditure constraint of gait (i.e., high max2 values are not encouraged because lifting the foot higher results in higher energy expenditure — minimising energy expended per unit distance travelled is often defined as a critically important objective of gait, e.g., Anderson & Pandy, 2001). The implications of these findings are important in two ways. First, a measure of central tendency (mean or median), a measure of variability about the central tendency (e.g., SD or some percentile measure), skewness and kurtosis are all potentially important parameters to consider when representing an individual's gait performance (not just mean and SD). Second, much clinical gait analysis relies on mean $\pm$ SD graphs (e.g., Figure 30) when it has not been established whether the variability in the data is normally distributed. If $S \neq 0$ then mean (and SD bars; e.g., Figure 30) may not appropriately represent the distribution of the data. If $S \neq 0$ then median is a better estimate of central tendency than mean, and when $S \neq 0$ and/or $K \neq 0$ then percentiles are a better measure of the variability about the central tendency than SD. Importantly, if skewness or kurtosis are non-zero, mean and SD have reduced meaning and contain error. No research has been conducted to date looking at the variability of the majority of the most important parameters that are used in gait analysis (e.g., intra- or inter-individual skewness and kurtosis of joint forces/moments/powers at key events of the gait cycle).

Beyond the key descriptive statistics described above for assessing variability, there are other chaos-related measures that are particularly useful in defining the long-term control and variability of an individual's gait, and have been used successfully as key discriminatory features in gait analysis. They include detrended fluctuation analysis (DFA) which characterizes complexity in time series data by quantifying the self-similarity (fractal) features within the patterns of time series plots. The use of DFA is becoming increasingly important in gait analysis but generally requires significant amounts of continuous time-series data (15 minutes of continuous gait or more is the norm) and so is often limited to treadmill-based gait analysis (e.g., Hausdorff et al., 1995b; 2004; Taylor et al., 2001, 2002; Herman et al., 2005) but not always (Terrier et al., 2005).

If variability about the mean is important, as it is in many studies, the 124 gait cycle parameters of Benedetti et al. (1998) should realistically be increased to over 500 by including, for example, intra-individual median, 10th and 90th percentiles, skewness and kurtosis for each parameter. Similarly, Wolf et al. (2005) would see an increase in parameter numbers (if they included joint moments, etc.) to over 40,000 parameters per individual. Benedetti et al. (1998) tested for normality of the group data but, like almost every other study, assumed intra-individual normality.

Temporal Dependency

An important issue with time-series parameterisation, as used in gait analysis, is time normalisation and representing the timing of events as a 0-100 percentage of gait cycle time (e.g., Figure 2), stance time or swing phase time (e.g., Figure 30). The problem with this normalisation process is that the true time dependence in the data is lost (Chau, 2001a). For example, the timing of MTC as presented in Figures 30 (as a percentage of the time between toe-off and foot strike) is in part dependent on the timing of foot strike and the timing of toe-off (i.e., timing of MTC is recorded as "timing of MTC relative to toe-off and foot strike") and is not an independent measure of MTC timing. It is obvious that something is wrong when considering the fact that the SD for the timing of toe-off and heel strike is zero. Furthermore, the SD of the timing of MTC is an overestimate of the true value, because it contains within it the SD of MTC timing plus the SD of the timing of toe-off and foot strike. Similarly, because MTC does not occur at precisely the same time relative to toe-off and heel strike on every gait cycle, there is a degree of rescaling during the percentage conversion (Forner-Cordero et al., 2005). This means, for example, that the minima (MTC) as presented in Figure 30 does not represent the mean of MTC and the SD band does not represent the SD of MTC. Instead it represents the average and SD of different parts of the gait cycle in the region of MTC. It overestimates average MTC and it is an overestimate of SD_{MTC}.

The problem is not necessarily with the time-normalisation itself but the normalisation in association with ensemble averaging, or in association with comparisons made between gait cycles that have different scaling within the 0-100% conversion. To summarise the possible effects of the temporal dependency problem:

- SD values of peaks and troughs from a time-normalised ensemble average graph can be more than twice as large as SD values of actual maxima and minima averages (Forner-Cordero et al., 2005). SD bars may not be appropriate when using 0-100% time normalisation ensemble average graphs.
- Timing information is almost meaningless (especially SD of timing of events) when using 0-100% time normalisation graphs because all timing data becomes dependent on the timing of toe-off and foot strike.
- A 0-100% time normalisation ensemble average graph is essentially a signal smoothing process. Peaks/maxima are reduced and troughs/minima are increased compared to average maxima and average minima values.
- Particular problems will occur when comparing individuals or populations that have a different scaling of events during the percentage conversion.

- The greatest problems will occur when the time-series data has sharp peaks/troughs. In this instance it is possible to contain both peaks and troughs, and everything in between, in an ensemble average calculation.
- If the data is skewed, SD ensemble average graphs are not representative of the central tendency or variability of the data, and median and a percentile range should be considered as an alternative.

Principal Components Analysis (PCA)

It has been shown that it is possible to generate thousands of parameters in an attempt to represent the key features of an individual's or group's gait data. The approach of Benedetti et al. (1998) and Wolf et al. (2005) resulted in many parameters, some of which may be key features and some of which may not. There are few guiding rules for determining which parameters contain useful information within a particular context and which do not (Chau, 2001a). A method is required to further reduce the amount of parameters/data and, specifically, separate the useful parameters from the redundant parameters. This is achieved by principal components analysis.

Principal component analysis (PCA) is a mathematical procedure that transforms correlated parameters into a (hopefully) smaller number of uncorrelated parameter groups called principal components. The objective of principal component analysis is to reduce the dimensionality (number of parameters) of the data set but retain most of the original variability in the data. The first principal component accounts for as much of the variability in the data as possible, and each succeeding component accounts for as much of the remaining variability as possible. The goal of PCA is to find the fewest number of components that explain most of the variance in the data. For example, the PCA studies by Olney et al. (1998), Yamamoto et al. (1983), Sadeghi et al. (1997, 2000), and Laughton et al. (2003) reduce 9-74 original parameters down to 2-4 principal components. This result corresponds to both a significant dimension reduction and a successful extraction of the key features. Different choices of variance threshold will lead to smaller or larger numbers of components and, hence, different interpretations of the data are possible (Chau, 2001a).

If a time-series parameterisation approach similar to Benedetti et al. (1998) or Wolf et al. (2005) is used, then PCA is an obvious next step in reducing the dimensionality of the data, extracting the features that account for the majority of the variance. It should be noted, however, that PCA will eliminate redundancy in the dataset but is limited by quality of the initial parameters presented to it (i.e., if the key features are not presented to the PCA then the end result will obviously have limited meaning). PCA is a very powerful procedure because it reduces the dimensionality of the data, and it retains key features while eliminating redundant features.

PCA is also known as singular value decomposition (SVD), or the empirical orthogonal function expansion (EOF) (Huang et al., 1998). The term SVD (and EOF) is mostly used when every data point in the entire time-series waveform is analysed to find those parts of the waveform containing most of the variance. SVD has been used in gait analysis research to extract fine details that are key discriminatory features in the entire time-series waveform (e.g., Daffertshofer et al., 2004; Stokes et al., 1999; Sadeghi et al.,

2000, 2002; Lamoth et al., 2006). Sadeghi et al. (2000, 2002) applied PCA as a curve structure detection method for walking patterns and provided significant insight into the functional tasks accomplished by the hip and knee during gait.

PCA is more commonly used in gait analysis research to refine a set of time-series parameters. Applying PCA reduces the number of parameters (dimension reduction) to, usually, just a few principal components (key features). The smaller number of new variables (principal components) usually allows easier interpretation of the data's structure (Chau, 2001a). Each of the resulting principal components consist of correlated parameters that, together, often lend themselves to the component being named meaningfully because the correlated parameters often have a common link. Examples of parameter groupings in a principal component analysis that have been found and meaningfully named in gait studies include groupings based on symmetry (Olney et al., 1998; Yamamoto et al., 1983; Sadeghi et al., 1997, 2000), knee function (Laughton et al., 2003), overall gait ability (Yamamoto et al., 1983; Laughton et al., 2003), speed and postural flexion bias (Olney et al., 1998). Each principal component can be used to develop a single equation (or index) that gives a single "score" that represents that component. This score can then be used as a new, dimension-reduced feature parameter. In some cases, PCA is simply used in its dimension reduction capacity by finding which individual parameters provide least additional variance to the data so that they can be discarded (e.g., Schwartz et al., 2000) or most variance so they can be retained (Sadeghi et al., 1997).

Principal components analysis or the principal component score has been used along with other component scores to calculate a "normalcy score" (a measure of deviation from a normal population; e.g., Schutte et al., 2000) and this area of research is gaining considerable momentum (e.g., Schwartz et al., 2000; Romei et al., 2004; Syczewska et al., 2005; and Tingley et al., 2002, using FFT coefficients rather than time-series parameters). "Normalcy" index scores can in turn be used to monitor change in an individual's function, for example, pre- vs. post-surgery (Laughton et al., 2003).

Data entered into a PCA are usually standardized using the zero-mean normalisation procedure (Chau, 2001a — see section on normalisation later in this chapter for more on zero-mean normalisation). Once the principal components in the data are found and if "component scores" from each principal component are calculated, the scores themselves should also be normalised before being used in a "normalcy" score or if entering the data into a neural network. This is to ensure that large values from one parameter do not skew the results (Schutte et al., 2000).

Summary of Time-Series Parameterisation

The significant advantages of time-series parameterisation are that it can encapsulate high level knowledge from experts and the parameters are highly localised (i.e., point specific and not wave/pattern or frequency-based). In this sense it fits well with the definition given above for feature extraction, namely "valid, useful and understandable" features of the data. The particular strength of this approach in some applications is the ease of understanding and applicability of the chosen features, which is particularly important where the end result is clinical intervention (Chau, 2001a). Wolf et al. (2005) commented that they "prefer the extraction of single features instead of measures for complete time series (e.g., Fourier coefficients) to ensure clinical acceptance." Having parameters that have meaning increases the potential for knowledge discovery, interpretability and insight.

The main disadvantage of time-series parameterisation is the subjectivity of the choice of parameters which can lead to key discriminatory features of the gait data not making it on to the parameter list. As such, this approach can leave some, or much, of the potentially useful data (including critical determinants in the data) recorded but unused, especially if a "standard" feature-set is indiscriminately used in place of a domain-specific and knowledge-driven feature selection process.

Time-series parameterisation combined with principal components analysis (PCA) offers a powerful means of key feature extraction from time-series data. PCA has demonstrated a practical ability to provide powerful studies into healthy and pathological human gait (see Chau, 2001a, for a review of some key papers).

The main advantage of PCA is it can reduce the number of parameters down to just a few key features of the data while retaining most of the variance that exists in the data. The main disadvantage of PCA is that it is only capable of searching for linear relationships in the data presented to it. This means that non-linear relationships that exist within the data will be missed. PCA also carries with it the disadvantages inherent with time-series parameterisation (i.e., subjectivity of initial parameter selection). Despite these not inconsiderable disadvantages, PCA is a powerful tool that is significantly under-used in gait analysis research. Because of its efficiency in reducing the dimensionality of datasets and rejecting redundant parameters, PCA is suitable for reducing parameter sets as a precursor to using other powerful tools such as cluster analysis (and fuzzy clustering), neural networks (Chau, 2001b) and normalcy indices (Schutte et al., 2000). In general, PCA is a very efficient method of reducing multidimensional data sets (Lamoth et al., 2004) and it is an excellent tool for situations where interpretability of the results is a key factor.

Feature Extraction by Data Transformation

Time-series parameterisation and PCA have proven to be a powerful and insightful tool in gait analysis research. Instances where an interpretable set of input data is required make time-series parameterisation a popular choice. However, the subjectivity of time-series parameterisation and the inability of PCA to recognize non-linear relationships has led to an increase in the use of methods that transform the entire time-series into a set of coefficients. These coefficients incorporate information about the entire waveform and retain nearly all of the variance in the data. As discussed earlier, PCA (or Singular Value Decomposition; SVD) has been used to model entire time-series waveforms (e.g., Daffertshofer et al., 2004; Yang & Shahabi, 2004; Stokes et al., 1999). The two other main data transformation tools used in gait research for dimension reduction are spectral decomposition (e.g., Fourier analysis, also called Fourier transform or fast Fourier transform, FFT) and wavelet decomposition.

The objective when using FFT or wavelets (and SVD) is to reduce or transform the data to a lower dimensional space in a way that approximately preserves the inherent structure of the data. FFT and wavelets overcome the subjectivity of PCA methods and model some of the non-linearity in waveforms. However, this is at the expense of reduced interpretability of the transformed (dimension-reduced) parameters.

FFT and wavelets are particularly useful for neural network applications (Schöllhorn, 2004) and normalcy index research (e.g., Tingley et al., 2002) and other research areas where a priori knowledge of the input (dimension-reduced) parameters is not required. If the goal of the research is the ability to predict or calculate an output from a series of inputs (e.g., machine learning applications) then the best results will be derived from a tool that can provide a few numbers (inputs) that give a complete overview of the entire time-series waveform.

Spectral Decomposition (Fourier Transform)

The most commonly used method for dimension reduction of an entire waveform, while maintaining the inherent structure of the waveform, is to transform the time-series data into the frequency domain via a Fourier analysis (FFT). The basic idea of spectral decomposition is that any signal, no matter how complex, can be represented by the superposition of a finite number of sine (and/or cosine) waves, where these waves can be represented by Fourier coefficients (Keogh et al., 2000). In this way, an entire waveform can be represented by a series of coefficients. It should be noted that the use of Fourier analysis in the context of this section is its capacity for dimension reduction (and not to interpretively analyse the frequency content of the time-series data).

In a neural network application, Holzreiter and Kohle (1993) used 128 Fourier coefficients in transforming vertical ground reaction forces (from a force platform) of two consecutive left and right footstrikes. In another neural network application, Barton and Lees (1997) transformed hip-knee joint angle-angle diagrams into 128 Fourier coefficients and then subjectively reduced the number of coefficients by suggesting only the lower 8 harmonics, representing the lower frequencies, were necessary to define the essential characteristics of the data. Gait is predominantly a low frequency activity and, as such, many Fourier coefficients representing high frequency signal content have very low amplitude. These low amplitude coefficients can be discarded without much loss of information thereby producing further dimension reduction (Keogh et al., 2000). Tingley et al. (2002) developed a "normalcy index" and went two steps further than Barton and Lees (1997) by transforming sagittal hip, knee ankle angle time-series via Fourier analysis and (1) kept only the lowest 6 harmonics per angle (36 coefficients in total), and (2) further reduced the number of coefficients using PCA. Lakany and Hayes (1997) also used PCA to eliminate irrelevant Fourier coefficients derived from joint angle time-series. Lakany and Hayes also found higher order harmonics to be least relevant. White et al. (2005) conducted a wide-ranging study of the frequency content of force platform gait data in a control group and a group of children with cerebral palsy. Amongst other things, White et al. found that 12-20 harmonics were required to adequately represent the force signal in the control group but 20-47 harmonics were required for the cerebral palsy group.

The biggest advantage of FFT for dimension reduction and feature extraction, and the main reason why FFT has been used in the past, is convenience and ease of use. It is easily implemented, being automated with little or no user intervention required, and because of a lack of alternatives it has attracted a wide following. FFT has been applied with success in a number of applications. Indeed, for strictly linear and stationary data, Fourier analysis should produce excellent and accurate results (Huang et al., 1998). However, there is a now a critical mass of evidence to suggest that Fourier analysis is not the best tool for dimension reduction in gait analysis research, and biomedical research in general (e.g., Huang et al., 1998; Keegan et al., 2003; Arafat, 2003; Verdini et al., 2000) and it may sometimes give misleading results. There is now a wide range of research in gait analysis and other areas that has critically analysed the use of Fourier analysis and discarded it as a tool for feature extraction/dimension reduction.

Perhaps the most objective assessments of spectral decomposition can be found in Huang et al. (1998), Keogh et al. (2000) and Englehart et al. (1999). The main criticisms of Fourier analysis for dimension reduction and feature extraction are as follows:

- Fourier analysis converts data to the frequency domain and, in that process, all time-domain information is lost (e.g., kinematics data subjected to a Fourier analysis loses information about localized time-dependent events), including some of the most important information such as drift, patterns, stride-to-stride variation, and abrupt changes at the beginning and end of specific events (Keegan et al., 2003).

- Small details (e.g., heel strike transient information) are not well represented in Fourier analysis (Verdini et al., 2000; Arafat, 2003). Small details in the data either disappear in the frequency domain (few harmonics in a Fourier transform show low frequency signal characters) or lose their time location information.

- Data collected during walking at a self-selected pace is non-stationary (Chau, 2001a). Indeed, Chau details a range of instances where gait data is non-stationary (data in which the frequency content changes over time). Fourier analysis applied to non-stationary data will make little physical sense (Huang et al., 1998) because FFT is conducted over the whole signal and it assumes the frequency content of the signal is constant over time.

- Both non-stationarity and nonlinearity in the time-series data can induce spurious harmonic components that cause energy spreading in a Fourier analysis (Huang et al., 1998). The consequence is a misleading energy-frequency distribution for nonlinear and non-stationary data.

None of the applied gait research papers listed earlier in this section conducted a critical analysis of the appropriateness of Fourier analysis for the data they were analyzing (e.g., Holzreiter & Kohle, 1993; Tingley et al., 2002; Barton & Lees, 1997; Lakany & Hayes, 1997; White et al., 2005).

Time-series parameterisation attempts to find the important time-based information in the signal. Fourier analysis tells us which frequency components are contained in a signal. Under optimal conditions, FFT has been proven to out-perform time-domain methods of feature extraction and dimension reduction (Englehart et al., 1999). Also, the frequency content of gait data is of interest and should be analysed (e.g., White et al., 2005). However, the uncritical use of Fourier analysis can give misleading results. For gait data that is of finite duration, non-stationary or nonlinear, Fourier analysis is of limited use (Huang et al., 1998).

An alternative to the standard FFT is the spectrogram (also called a Gabor transform of short-time Fourier transform — STFT), which is a limited time window-width Fourier analysis. By sliding a window along the time axis and doing an FFT on each window, a time-frequency distribution can be obtained that gives frequency content as it varies with time (Huang et al., 1998; Graps, 1995). Since STFT still relies on the FFT, the data still has to be piece-wise stationary (Huang et al., 1998). A further problem with this method is that in order to localize a frequency event in time, the window width must be narrow, but, on the other hand, the frequency resolution requires longer time series. These conflicting requirements limit its usability (Huang et al., 1998). The use of STFT in gait analysis is quite limited with the vast majority of STFT work being conducted in the area of EMG analysis (e.g., Hannaford & Lehman, 1986), with a few exceptions (e.g., Lackoviæ et al., 1997), and nearly all applications of STFT being related to the analysis of signals rather than as a dimension reduction tool.

Wavelet Decomposition

The use of wavelets as an analysis tool, as a data smoothing procedure and as a dimension reduction/feature extraction tool in biomechanics and gait analysis has escalated in the past decade. In 2001, Chau (2001b) stated that the limited use of wavelets in gait analysis at that time was an obstacle to objectively gauging their usefulness. While the use of wavelets in gait analysis is still not as prevalent as perhaps it could be, their usefulness in gait analysis, and other areas of biomedical research, is now well proven. Much of the research that has used wavelets in gait analysis research has expressed the need to use wavelets because of the problems associated with FFT, STFT and other methods (e.g., Ismail & Asfour, 1998; Keegan et al., 2003; Verdini et al., 2000) and many researchers simply use wavelets as though they are the new standard, with no reference to FFT, STFT and other methods (e.g., von Tscharner et al., 2003; Sekine et al., 2002; Lauer et al., 2005).

The main advantages of wavelet transforms over Fourier transforms are that they account for both time and frequency domain information (local and global) simultaneously, and they can cope with non-stationary signals. Wavelets also have advantages over Fourier analysis in analyzing situations where the signal contains discontinuities and sharp spikes (Graps, 1995). Wavelets are a time-frequency analysis tool that tell us at what time those frequency components are present in the signal. This information is

Figure 32. Left box — Wavelet shape examples from Keegan et al. (2003); Right box — Top left, Daubechies wavelet; Top right, Coiflet; Bottom right, symlet; Bottom left, Haar wavelet (from Graps, 1995)

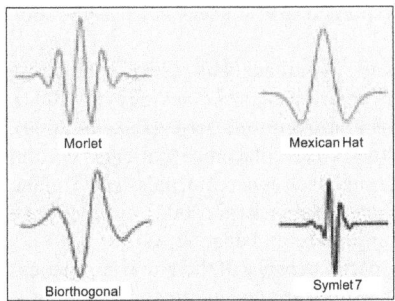
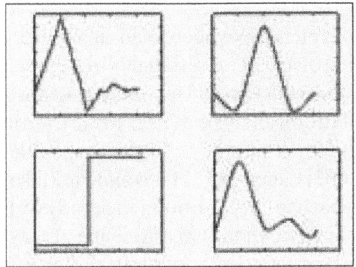

particularly important in analyzing non-stationary signals. Another strength of wavelets is that they can be used to smooth signals locally as well as globally. This means that local high frequency components of a signal can be treated while noise of the same frequency at other parts of the signal can be removed (e.g., Wachowiak et al., 1997, 1998, 2000; Graps, 1995). This particular task is beyond the capabilities of conventional filtering methods (Chau, 2001b). Indeed, wavelets have much to offer in the area of smoothing, especially in applications where specific knowledge is known about the signals (e.g., Zhou & Gotman, 2004).

Wavelets are simply waves of limited duration. A wavelet transformation can be considered as a series of pattern correlations between a selected wavelet of specific size or scale and segments of the original signal (Keegan et al., 2003). Discrete or continuous wavelets can be chosen depending on the application. Wavelets come in many shapes (called mother shapes or prototype shapes or mother wavelets; see Figure 32) and the idea is to fit the shape that best fits a particular segment of the data. Each mother wavelet shape is translated, contracted or dilated as required to improve the fit (Chau, 2001b). The more closely the shape of the scaled wavelet is to that of the segment of the signal, the higher the correlation and the larger the wavelet coefficients for that signal segment.

The first few coefficients of a wavelet contain a global approximation of the data and additional coefficients can be imagined as "zooming-in" to areas of high detail (Keogh et al., 2000). Wavelets of small scale extract information from short-duration events within the signal. Wavelets of large scale extract more global features of the signal. Thus, when wavelets are used for pattern recognition, two important user selections are required: the specific wavelet (i.e., which mother wavelet) and the scale of the selected wavelet (Keegan et al., 2003). The selection of the mother wavelet shape is enhanced by an in-depth knowledge of the signal. For example, Ismail and Asfour (1998) used a fourth order Daubechies wavelet (Db4; fourth order leads to 8 coefficients) for EMG processing because that shape resembles the pattern of the motor unit action potential (MUAP). One of the disadvantages of wavelets is that they do require user inputs in terms of the shape and the scale functions.

Compared to frequency-based techniques, a wavelet transformation, because of its ability to capture localized time-dependent events as well as global signal characteristics, is a better choice for processing kinematics data (Keegan et al., 2003). Englehart et al. (1999) showed that wavelets perform statistically significantly better than STFT, which in turn performed significantly better than time-domain methods. Keegan et al. (2003) classified lameness in horse gait using wavelets and a neural network and recorded an 85% correct classification rate. This was compared to a previous study (Schobesberger and Peham, 2002) using Fourier transforms that recorded a 79% successful classification rate.

Wavelets have been used in a variety of ways in gait analysis research including modeling of acceleration signals in free walking (Tamura et al., 1997; Sekine et al., 2002), dimension reduction and feature extraction of lower limb angle time-series (Lakany, 2000), force platform analysis of heel strike transient in total knee replacement patients (Verdini et al., 2000), in the study of different population groups such as cerebral palsy (De Stefano et al., 2003; Lauer et al., 2005) and Parkinson's disease (Sekine et al., 2002), and they have proven particularly useful in the study of EMG signals (e.g., Ismail & Asfour, 1998).

It is difficult to find criticisms of wavelets, partly because of their appropriateness over a wide range of applications, partly because there are few applications that have used wavelets where they are not the best available method, and partly because all published papers using wavelets (or any chosen method for that matter) will claim their results show potential clinical relevance. The problems associated with wavelets are most often smaller versions of the problems associated with Fourier analysis. First, wavelets are not absolutely local (i.e., they do not give instantaneous frequency values). For example, the commonly used Morlet wavelet shape has a length of 5.5 waves. Consequently it suffers, though on a much smaller scale, from the same problems as Fourier analysis (e.g., it does not deal with intra-wave non-stationarity). Second, once the mother wavelet is selected, that shape has to be used for all the data (Huang et al., 1998). Third, another criticism of wavelets put forward by Huang et al. is that sometimes the interpretation of the wavelet can be counterintuitive. A change that occurs locally will be found in the high frequency range because the higher the frequency the more localized the basic wavelet will be. If a local event occurs in the low-frequency range, its effects will still be found in the high-frequency range (Huang et al., 1998). There is also little or no guide to which mother wavelet shape should be chosen (Chau, 2001b). Despite these problems, wavelet analysis is still the best available non-stationary data analysis method (Huang et al., 1998).

There is no doubt that using wavelets means adopting a new mind-set in processing data compared with Fourier analysis and time-series parameterisation. Most of the basic wavelet theory has now been done and the future of wavelets lies in extending the range of their applications (Graps, 1995).

Wavelets do have their own limitations but it seems they are minimal in comparison to Fourier analysis and other methods. Nevertheless, an improvement on wavelets has been developed called the Hilbert spectrum (or Huang-Hilbert spectrum; Huang et al., 1998). The Hilbert spectrum, a two-step process involving empirical mode composition (i.e., PCA) and a Hilbert transform (a $90°$ phase shift transformation on a signal), provides instantaneous frequency information at high resolution and, therefore, would seem to be a significant advance to the general area of data analysis and data visualisation. The Hilbert spectrum method has been patented (Huang & NASA, 2001) which may have slowed down its uptake. There are a few papers emerging using the Hilbert spectrum with biomedical signal analysis (e.g., Sun et al., 2004) and EMG (e.g., Andrade et al., 2004), but none yet in gait analysis. Recently a wavelet-based alternative to the Hilbert spectrum has been proposed by Olhede and Waldon (2004). Neither Olhede and Waldon's technique or the Hilbert spectrum are currently available for general use but the ability to obtain one set of coefficients capable of representing instantaneous time and frequency information will have useful application in gait analysis and machine learning applications.

Comparison of Feature Extraction Methods

The main methods discussed in this section are:

- time-series parameterisation
- time-series parameterisation with principal components analysis (PCA)
- singular value decomposition (SVD)
- Fourier transform (including short-time Fourier transform, STFT)
- Wavelet transform

The first three methods involve modeling the time domain information, Fourier transform is a frequency domain analysis and wavelets represent a combined time-frequency analysis.

In a study comparing time domain, frequency domain and time-frequency domain methods, Englehart et al., (1999) demonstrated for an EMG pattern recognition problem that PCA should be used with each method (time-series parameterisation, Fourier analysis and wavelets) if effective dimension reduction and feature extraction is the goal. They demonstrated that when using PCA and linear discriminant analysis (LDA), there is a significant improvement in performance in the progression time-domain →STFT→ Wavelet, the best performance being exhibited by wavelets.

Each of the methods has its advantages and disadvantages and each method can be used in some circumstances, for example:

- **Time-series parameterisation** — used when an understanding of the parameters being used as inputs and outputs is a priority (best used in association with PCA).
- **Fourier analysis** — used when a pure frequency domain analysis is the objective (as long as data is stationary, etc.). It may help to use PCA and LDA to assist in retaining only the relevant coefficients. Fourier analysis has been superseded as a dimension reduction / feature extraction tool by wavelets.
- **Wavelet analysis** — best single method for feature extraction / dimension reduction when the most comprehensive time-frequency domain information is required. It may help to use PCA and LDA to assist in retaining only the relevant coefficients.

It should be noted that any published paper using one of the above methods will inevitably claim that their method will have scientific merit or clinical relevance. Chau (2001b) recommended using a combination of time domain methods (with PCA) along with time-frequency analysis (wavelet transform). This makes sense for a number of reasons:

1. If knowledge/understanding of parameters is required (e.g., in clinical/surgery applications) then time-series parameterisation is the best tool. Time-series parameterisation in association with PCA and LDA is recommended in this situation. In some situations, wavelet analysis may be used in combination with time-series parameters to ensure all important information is accounted for (noting that time-series parameterisation is subjective and may miss important information).
2. If knowledge/understanding of parameters is not required (e.g., neural network applications) then wavelet transform parameters can be used. The number of parameters could be reduced further by using PCA and LDA. To account for the limitations of wavelets (especially longer wavelets like the Morlet wavelets), the use of time-series parameters in combination with wavelet parameters can also be considered, at least until it has been shown that time-series information adds nothing to the process. Wavelets do not have the local time resolution of time-series parameters.

NORMALISATION OF GAIT DATA

There are many sources of systematic variation in measured data. Normalisation is the term used to describe the process of removing some unwanted aspects of this variation allowing valid comparisons to be made across subjects or experiments. Some common examples of normalisation that occur in gait analysis include:

- Normalisation/scaling of ground reaction forces to body weight
- Normalisation/scaling of stride length to body height (or leg length)
- Normalisation/transformation of time to a percentage of gait cycle time

Three methods of normalisation will be discussed called data transformation, anthropometric scaling and normalisation to a reference value.

Normalisation by Data Transformation

This is conceptually the simplest form of normalisation. The goal of normalisation by data transformation is to represent all data sets in the same scale (e.g., between 0 and 1). This is a particularly important procedure that is carried out prior to a PCA and prior to submitting data to a neural network. Neural networks generally train better when the different datasets are transformed to the same range, for example $[-1,1]$. Some of the more common methods of normalisation by data transformation will be presented here:

1. **Min-Max normalisation (also called normalisation to range or normalisation to an interval):** This type of normalisation transforms the data into a desired range, usually $[0,1]$ or $[0,100]$. It ensures that all data are transformed proportionally into an output range with predefined upper and lower limits. The transformation equation is:

$$x'_i = \frac{(x_i - x_{min}) \times (L_{max} - L_{min})}{(x_{max} - x_{min})} + L_{min}$$

where x_i is the raw data, x'_i is the transformed data, x_{min} is the smallest value, x_{max} is the highest value, L_{max} is the upper and L_{min} the lower limit of the new output range. Transforming data to $[0,1]$ using this method is regularly used prior to submitting data to a neural network (e.g., Barton & Lees, 1997; Schöllhorn, 2004.). In many gait studies, time from one heel strike to the next heel strike of the same limb (i.e., one gait cycle) is normalised using this method by transforming data to $[0,100]$ so time is presented as a percentage of the gait cycle (e.g., Sutherland et al., 1994; Meglan & Todd, 1994; Benedetti et al., 1998; Tingley et al., 2002).

2. **Decimal scaling:** This type of scaling transforms the data into a range between $[-1,1]$, maintains the relative magnitude of data values and maintains the sign (negative or positive). Put simply, decimal scaling only moves the decimal point but it does preserve most of the original character of the value. The transformation equation is:

$$x'_i = \frac{x_i}{10^k}$$

where k is the smallest k such that $\max|x'_i| \le 1$; e.g., for the initial range [-823, 26], k is 3, and the transformed range becomes [-0.823, 0.026].

3.   **Zero-mean normalisation:** This is perhaps the most commonly used form of normalisation in gait analysis research. It is better known as converting data to z-scores. The transformation equation is:

$$x'_i = \frac{x_i - \mu_x}{\sigma_x}$$

where μ_x is the mean and σ_x is the standard deviation of the data set. The transformed dataset has the properties $\mu_{x'} = 0$ and $\sigma_{x'} = 1$. This procedure is sometimes, more correctly, named "zero-mean and unit standard deviation normalisation." This is because the term "zero-mean normalisation" is sometimes used for the simpler transformation of $x'_i = (x_i - \mu_x)$. Zero-mean normalisation works well but transforms the data into a form that is unrecognisable from the original data. It is best suited for transforming scores when one individual needs to be compared with another or with the population. An example of its use might be to obtain a mean and SD for a healthy population then calculate a z-score for a new patient. A z-score of 0 to |2| might be considered as being within the "normal" range and a score of >|2| might be considered outside of the "normal" range (e.g., Simon, 2004). The greater the |z-score|, the greater the deviation from "normal" or average. This method is used to normalise data before a PCA (e.g., Olney et al., 1998; Lamoth et al., 2006). A limitation of this method is that if the data are not normally distributed,

Figure 33. The mean toe clearance data presented in Figure 30 has been min-max normalised between 0 and 1 and plotted against time normalised to the swing phase of gait (toe-off = 0%, heel strike = 100%)

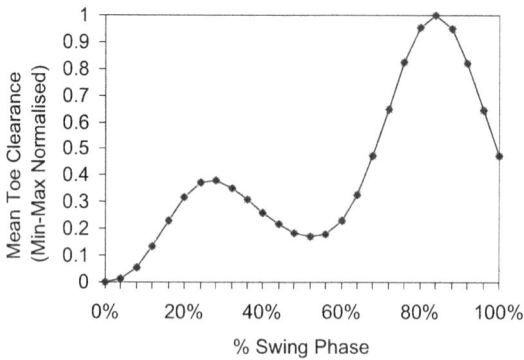

z-score normalisation does not retain the input distribution at the output. This is because mean and standard deviation are parameters that are best used with normally distributed data. If data is not normally distributed then the equivalent transformation is the "median and median absolute deviation (MAD)" normalisation. The transformation equation for the *median and MAD normalisation* is:

$$x'_i = \frac{x_i - median_x}{median\left(\left|x_i - median_x\right|\right)}$$

A search of the literature did not find any gait analysis research using decimal scaling or "median and MAD" normalisation, nor did it find studies that tested for the normality of the data subjected to zero-mean normalisation. There are also no studies that have looked at different data transformation strategies on the same gait analysis data to assess the performance of the different methods. Jain et al. (2005) assessed zero-mean, median and MAD and min-max normalisation procedures, along with other normalisation methods, and found zero-mean and min-max normalisation procedures to perform well. They did, however, also demonstrate the sensitivity of these normalisation procedures to the presence of outliers. The median and MAD normalisation generally performed relatively poorly but they did conclude that a normalisation scheme should be chosen after analyzing the score distributions.

Normalisation by Anthropometric Scaling

There are some anthropometric features of humans that obscure other information about a person's gait. For example, heavier people generally generate larger ground reaction forces and larger joint forces, moments and powers than lighter people, whereas taller people tend to have a larger gait stride length, a faster natural gait velocity and a lower cadence (e.g., Pierrynowski & Galea, 2001). This is true both within an individual as they grow (Sutherland et al., 1994) and within a population. To counter this it is commonplace, in order to compare people against others in a population group, to "factor out" this unwanted variance between individuals by scaling gait data to anthropometric features. This process is commonly called normalisation. Two common examples of anthropometric scaling of gait data are: a) presenting ground reaction forces as a proportion of body weight and, b) to present stride length as a proportion of stature (either body height or limb length). In order to compare people with a "normal" group it is usual practice to correct for unequal weight and stature (Hof, 1996).

The goal of anthropometric scaling is to reduce inter-subject variability, specifically to reduce inter-subject gait value variation due to subject-specific anthropomorphic measures (e.g., height and weight). The gait literature provides several, sometimes conflicting, strategies to perform this task (Pierrynowski & Galea, 2001). One method is described by Pierrynowski and Galea as the "ad hoc" method. That is, scaling data to an anthropometric measure that simply seems logical and/or meaningful. Examples of this include dividing ground reaction forces and joint moments by body mass (e.g., Winter, 1990; Ferber et al., 2002), and dividing walking speed by body height (creating a measure in statures/s; e.g., Siegel & Metman, 2000; Stansfield et al., 1999). Alternatives to these approaches include dividing joint moments by weight and height (e.g., Andriacchi &

Table 6. Dimensionless numbers related to gait mechanics (from Hof, 1996)

Quantity	Symbol	Dimension	Dimensionless number
Mass	m	M	$\hat{m} = \dfrac{m}{m_0}$
Length, distance	l, x	L	$\hat{l} = \dfrac{l}{l_0}$
Time	t	T	$\hat{t} = \dfrac{t}{\sqrt{l_0/g}}$
Frequency	f	T^{-1}	$\hat{f} = \dfrac{f}{\sqrt{g/l_0}}$
Speed, velocity	$v = \dot{x}$	$L\,T^{-1}$	$\hat{v} = \dfrac{v}{\sqrt{g.l_0}}$
Acceleration	$a = \ddot{x}$	$L\,T^{-2}$	$\hat{a} = \dfrac{a}{g}$
Force	F	$M\,L\,T^{-2}$	$\hat{F} = \dfrac{F}{m_0.g}$
Moment	M	$M\,L^2\,T^{-2}$	$\hat{M} = \dfrac{M}{m_0.g.l_0}$
Work, energy	W, E	$M\,L^2\,T^{-2}$	$\hat{W} = \dfrac{W}{m_0.g.l_0}$
Power	P	$M\,L^2\,T^{-3}$	$\hat{P} = \dfrac{P}{m_0.g^{1/2}\,l_0^{3/2}}$
Angle	ϕ	(is already dimensionless)	
Angular velocity	$\omega = \dot{\phi}$	T^{-1}	$\hat{\omega} = \dfrac{\omega}{\sqrt{g/l_0}}$
Angular acceleration	$\alpha = \ddot{\phi}$	T^{-2}	$\hat{\alpha} = \dfrac{\alpha}{g/l_0}$
Moment of inertia	I	$M\,L^2$	$\hat{I} = \dfrac{I}{m_0\,l_0^2}$

m_0 = *body mass*, l_0 = *leg length*, g = *gravitational acceleration*

Strickland, 1985; Andriacchi et al., 1997), dividing joint moments by weight and leg length (e.g., Kadaba et al., 1990), and dividing walking speed by leg length (e.g., Begg et al., 1998).

An alternative approach presented by Hof (1996) suggested that all measured parameters could be normalised into dimensionless numbers by dividing each parameter by a factor of the same dimensions, each factor being some combination of body mass, leg length and gravitational acceleration (Table 6). The general concept here was to present data in the form of non-dimensional numbers. Hof's method was not limited to leg length and body mass (these are the example measures used in Table 6) but the recommendation was that any relevant measure could be used. A number of researchers have since incorporated Hof's method, and developed it further to incorporate other relevant anthropometric features (e.g., Syczewska et al., 2005, used Hof's speed and step length normalisation strategy and also analysed step width normalised by the anterior superior iliac spine width).

Hof's (1996) normalisation to dimensionless quantities method makes sense on a number of grounds, especially, for example, normalising gait timing parameters as a function of leg length (this makes sense because we know that cadence and step time is related to stature). The disadvantage of Hof's method is that in some cases it changes meaningful dimensional numbers into less meaningful numbers. It is also important to note that the original dimensional numbers contain important information and some of this information is lost by normalising. For example, the actual force exerted by a person on the ground is important, as is the actual speed at which s/he walks.

A general limitation of normalising data by anthropometric scaling is that, while reducing unwanted variance in the data, it will inevitably introduce a small amount of unwanted error too. This is because anthropometric proportions are themselves variable. There are people with long legs in proportion to body height, people with short legs in proportion to body height, and every person has legs of different length (some significantly different). It is also important to make the normalising anthropometric quantity as relevant as possible. Overcoming and moving body weight is a major goal of gait so body weight is obviously a relevant normalising quantity for ground reaction forces and joint moments of the lower limb. However, is body weight still the most relevant anthropometric measure for normalising lower limb joint moments produced during the swing phase of gait where the objective is to swing the lower limb, not the whole body? Similarly, what about the two phases of double limb support in the gait cycle where both limbs are sharing the load and contributing to the ground reaction forces that are accelerating the body? Anthropometric proportionality also varies across some populations. Such issues are rarely considered.

Moisio et al. (2003) compared the two most commonly used methods of normalising lower limb joint moments by: (a) body mass (Winter, 1991) and, (b) body mass x body height (Andriacchi & Strickland, 1985). Moisio et al. (2003) found that height and mass/weight were correlated in their population of 172 normal adults. They suggested this might have been a factor behind their finding that both methods were successful in reducing the effects of height and weight in peak lower limb joint moment variance. Indeed, because of the relationship between height and weight, the body mass method may partially account for the variance due to height but probably under-corrects for height, whereas the body weight x height method may over-correct for height by accounting for the variation in height twice. Moisio et al. also found that the body weight x height normalisation method was more effective in reducing differences due to gender than the body mass method.

In a study of limited sample size (n=10), Pierrynowski and Galea (2001) compared the ad-hoc method and the dimensionless quantity method of normalising using everything from stride length and cadence to joint moments and powers. They found both methods performed very well at reducing unwanted variance in the data due to height and mass, and suggest that either method could be used with confidence to normalise gait data.

The practical application of using anthropometric scaling to reduce inter-subject variation is clear. Reduced variation increases the ability of a statistical tool to detect a difference in a gait data that is not related to height and weight (Pierrynowski & Galea, 2001). Data should not always be normalised and the method of normalisation should not always be the same. Normalisation by anthropometric scaling of data is common in gait analysis but there is much debate on the topic of what is the best way to normalise the data. The fact that this debate is continuing suggests there may be no single correct way to normalise. One of the most important practices when normalising data is that everyone should record how they normalised and transformed their data and present that information in their published paper.

Normalisation to Reference Values

EMG amplitude data is particularly difficult to compare across different people, different populations and even for the same person on different days. To overcome this, EMG is often normalised as a percentage of the maximum EMG amplitude recorded from that muscle site during a maximum voluntary (isometric) contraction, %MVC (Konrad, 2005). In this case the largest EMG amplitude during MVC is the "reference value" for all EMG amplitudes recorded subsequently. Problems with this method include: (a) different level of 'MVC' are achievable by different individuals, largely because of differing degrees of neurally-mediated inhibition and, (b) isometric MVC is somewhat detached from the activity being analysed.

An alternative reference value for EMG signal normalisation is to present EMG amplitude as a percentage of the maximum amplitude recorded during the gait cycle (e.g.,

Figure 34. On the vertical axis, raw EMG is normalised as a percentage of the maximum absolute recorded voltage. On the horizontal axis, time is normalised as % gait cycle (from Motion Lab Systems Web site: http://www.emgsrus.com/emgfull.htm)

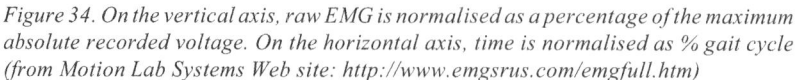
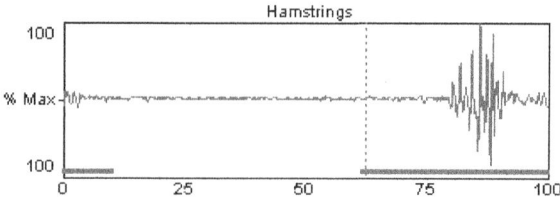

Ferber et al., 2002). An example of latter procedure is detailed in Figure 34. In this case the largest absolute EMG amplitude during the gait cycle is the reference value for all other EMG amplitudes recorded during the cycle. Figure 34 looks similar to a min-max normalisation but this is not the case because the minimum (negative) value is not used in defining the bounds of the normalised data. If the data were rectified and then subjected to the "largest EMG amplitude" reference procedure as detailed above, this would be the same as a min-max normalisation (because the minimum in the rectified EMG signal is zero). Another alternative is to use the mean signal amplitude during a burst of activity as a reference value from which to normalise the signal (Shiavi, 1990). In general, the methods detailed in this paragraph are convenient but they do not allow comparison of relative intensities among muscles and across subjects. This normalisation procedure is most often used in situations where poor patient cooperation makes muscle testing difficult (Perry, 1998).

Normalising to reference values is not as common in gait analysis as other normalisation procedures but other examples where it is used include:

- Representing centre of pressure position data during the stance phase of gait as a proportion of the distance between the rearmost and frontmost position of the foot when the foot is planted on the ground (Trew & Everett, 2001).
- Measurement of energy expenditure in terms of oxygen uptake is becoming increasingly common in clinical gait analysis. Oxygen uptake is often presented as a percentage of maximum oxygen uptake ($\%VO_{2max}$).

ACKNOWLEDGMENTS

Data supplied by Amanda Johns and Tuire Karaharju-Huisman for Figures 30 and 31 respectively are acknowledged. We also greatly acknowledge the help of Dr. Gill Best in revising and proof reading the final version of the manuscript.

REFERENCES

Allard, P., Capozzo, A., Lundberg, A., & Vaughan, C. (Eds.). (1998). *Three-dimensional analysis of human locomotion*. UK: John Wiley & Sons.

Allard, P., Stokes, I. A. F., & Blanchi, J-P. (Eds.). (1995). *Three-dimensional analysis of human movement*. Champaign, IL: Human Kinetics.

Allison, G. T. (2003). Muscle onset detection technique for EMG signals with ECG artefact. *Journal of Electromyography and Kinesiology, 13*, 209-213.

Allison, G. T., & Fujiwara, T. (2002). The relationship between EMG median frequency and low frequency band amplitude changes at different levels of muscle capacity. *Clinical Biomechanics, 17*, 464-469.

Aminian, K., Najafi, B., Bula, C., Leyvraz, P. F. & Robert, Ph. (2002). Spatio-temporal parameters of gait measured by an ambulatory system using miniature gyroscopes. *Journal of Biomechanics, 35*(5), 689-699.

Anderson, F. C., & Pandy, M. G. (2001). Dynamic optimization of human walking. *Journal of Biomechanical Engineering, 123*, 381-390.

Andrade, A. D. O., Kyberd, P. J., & Nasuto, S. J. (2004, June 18-21). Time-frequency analysis of surface electromyographic signals via Hilbert spectrum. In *Proceedings of the XVth ISEK Congress,* Boston (p. 68).

Andriacchi, T. P., Natarajan, R. N., & Hurwitz, D. E. (1997). Musculoskeletal dynamics, locomotion, and clinical applications. In V. C. Mow & W. C. Hayes (Eds.), *Basic orthopedic biomechanics* (pp. 37-68). Philadelphia: Lippincott-Raven.

Andriacchi, T. P., Ogle, J. A., & Galante, J. O. (1977). Walking speed as a basis for normal and abnormal gait measurements. *Journal of Biomechanics, 10*(4), 261-268.

Andriacchi, T. P., & Strickland, A. B. (1985). Gait analysis as a tool to assess joint kinetics. In N. Berme, A. E. Engin, & K. M. Correia da Silva (Eds.), *Biomechanics of Normal and Pathological Human Articulating Joints* (pp. 83-102). Dordrecht, The Netherlands: Martinus Nijhoff.

Arafat, S. (2003). *Uncertainty modeling for classification and analysis of medical signals.* Unpublished Ph.D. Dissertation, University of Missouri-Columbia.

Barnett, R. L. (2002). Slip and fall theory - Extreme order statistics. *International Journal of Occupational Safety and Ergonomics, 8,* 135-159.

Bartlett, R. M. (1994). *Biomechanical analysis of performance in sport.* Leeds, UK: British Association of Sport and Exercise Sciences.

Barton, J. G., & Lees, A. (1995). Development of a connectionist expert system to identify foot problems based on under-foot pressure patterns. *Clinical Biomechanics, 10,* 385-391.

Barton, J. G., & Lees, A. (1997). An application of neural networks for distinguishing gait patterns on the basis of hip-knee joint angle diagrams. *Gait and Posture, 5,* 28-33.

Begg, R. K., Hassan, R., Taylor, S., & Palaniswami, M. (2005, January 4-7). Artificial neural network models in the assessment of balance impairments. In *Proceedings of the International Conference on Intelligent Sensing and Information Processing,* Chennai, India. IEEE Press.

Begg, R. K., & Kamruzzaman, J. (2005). A machine learning approach for automated recognition of movement patterns using basic, kinetic and kinematic gait data. *Journal of Biomechanics, 8,* 401-408.

Begg, R. K., Sparrow, W. A., & Lythgo, N. D. (1998). Time-domain analysis of foot-ground reaction forces in negotiating obstacles. *Gait and Posture, 7,* 99-109.

Benedetti, M. G., Catani, F., Leardini, A., Pignotti, E., & Giannini, S. (1998). Data management in gait analysis for clinical applications. *Clinical Biomechanics, 13*(3), 204-215.

Best, R., Begg, R., & Dell'Oro, L. (2001, July 8-13). Minimum foot clearance during walking: mean, SD, skewness and kurtosis. In *Proceedings of the International Society of Biomechanics Congress,* Zurich, Switzerland.

Bilney, B., Morris, M., & Webster, K. (2003). Concurrent related validity of the GAITRite® walkway system for quantification of the spatial and temporal parameters of gait. *Gait and Posture, 17*(1), 68-74.

Bontrager, E. (1998). Instrumented gait analysis (review of instrumented gait analysis systems). In J. A. De Lisa (Ed.), *Gait analysis in the science of rehabilitation.* Retrieved January 9, 2005, from http://www.vard.org/mono/gait/contents.pdf

Cappozzo, A., & Berme, N. (1990). Subject-specific segmental inertia parameter determination: A survey of current methods. In N. Berme & A. Cappozzo (Eds.), *Biome-

chanics of Human Movement: Applications in Rehabilitation, Sports and Ergonomics (pp. 179-85). Worthington, OH: Bertec.

Catalfamo, P., & Ewins, D. J. (2003, September 15-17). Evaluation of the gyroscope as a sensor in paediatric In *Proceedings of the FES IPEM 9th Annual Scientific Meeting.* Bath, UK.

Cavanagh, P. R., Bewitt, F. G., & Perry, J. E. (1992). In-shoe plantar pressure measurement: A review. *Foot, 2,* 185-194.

CGA. (1997). *Clinical gait analysis Web site.* Retrieved January 6, 2005, from http://guardian.curtin.edu.au:16080/cga/teach-in/transient/

Chau, T. (2001a). A review of analytical techniques for gait data. Part 1: Fuzzy, statistical and fractal methods. *Gait and Posture, 13*(1), 49-66.

Chau, T. (2001b). A review of analytical techniques for gait data. Part 2: neural network and wavelet methods. *Gait and Posture, 13*(2), 102-120.

Chiu, L. Z. F., & Salem, G. J. (2005). Net joint moment calculation errors during weightlifting: Dempster versus DEXA. In *Proceedings of the 4th Meeting of the Southern California Conference on Biomechanics.* California State University, Fullerton.

Corriveau, H., Hébert, R., Raýc, M., & Prince, F. (2004). Evaluation of postural stability in the elderly with stroke. *Archives of Physical Medicine and Rehabilitation, 85,* 1095-1101.

Daffertshofer, A., Lamoth, C. J. C., Meijer, O. G., & Beek, P. J. (2004). PCA in studying coordination and variability: a tutorial. *Clinical Biomechanics, 19,* 415-428.

Davis, R. B., Ounpuu, S., Tyburski, D., & Gage, J. R. (1991). A gait analysis data collection and reduction technique. *Human Movement Science, 10,* 575-587.

Delp, S. L., & Maloney, W. (1993). Effects of hip center location on the moment-generating capacity of the muscles. *Journal of Biomechanics, 26,* 485-499.

De Stefano, A., Allen, R., Burridge, J. H., & Yule, V. T. (2003). Application of complex wavelets for EMG analysis during gait of asymptomatic and pathological subjects. *International Journal of Wavelets, Multiresolution and Information Processing, 1*(4), 425-448.

Dimnet, J., Junqua, A., & Allard, P. (Eds.). (1996). Special issue on 3-D analysis of human movement. *Human Movement Science, 15*(3).

Durkin, J. L., & Dowling, J. J. (2003). Analysis of body segment parameter differences between four human populations and the estimation errors of four popular mathematical models. *Journal of Biomechanical Engineering, 125*(4), 515-522.

Englehart, K., Hudgins, B., Parker, P. A., & Stevenson, M. (1999). Classification of the myoelectric signal using time-frequency based representations. *Medical Engineering and Physics, 21,* 431-438.

Ferber, R., Osternig, L. R., Woollacott, M. H., Wasielewski, N. J., & Lee, J-H. (2002). Gait mechanics in chronic ACL deficiency and subsequent repair. *Clinical Biomechanics, 17,* 274-285.

Fildes, B. (1994). *Injuries among older people: Falls at home and pedestrian accidents.* Melbourne, Australia: Dove.

Fischer, R. (2002). Motion capture process and systems. In M. Jung, R. Fischer & M. Gleicher (Eds.), *Motion capture and editing: Bridging principle and practice.* Natick, MA: A.K. Peters.

Forner-Cordero, A., Koopman, H. J. F. M., & van der Helm, F. C. T. (2005). Describing gait as a sequence of states, *Journal of Biomechanics* (In Press).

Frigo, C., & Rabuffetti, M. (1998). Multifactorial estimation of hip and knee joint centres for clinical application of gait analysis. *Gait and Posture, 8*, 91-102.

Gefen, A. (2002). Biomechanical analysis of fatigue-related foot injury mechanisms in athletes and recruits during intensive marching. *Medical and Biological Engineering and Computing, 40,* 302-310.

Gefen, A., Megido-Ravid, M., Itzchak, Y., & Arcan, M. (2002). Analysis of muscular fatigue and foot stability during high-heeled gait. *Gait and Posture, 15*, 56-63.

Graps, A. (1995). An introduction to wavelets. *IEEE Computational Science and Engineering, 2*(2), 50-61.

Hahn, M. E., Farley, A. M., Lin, V., & Chou, L. S. (2005). Neural network estimation of balance control during locomotion. *Journal of Biomechanics, 38,* 717-724.

Haig, A. J., Geblum, J. B., Rechtien, J. J., & Gitter, A. J. (1996). Technology assessment: the use of surface EMG in the diagnosis and treatment of nerve and muscle disorders. *Muscle Nerve, 19*, 392-395.

Hannaford, B., & Lehman, S. (1986). Short time Fourier analysis of the electromyogram: fast movements and constant contraction. *IEEE Transactions on Biomedical Engineering, 33,* 1173-1181.

Hausdorff, J. M., Ladin, Z., & Wei, J. Y. (1995a). Footswitch system for measurement of the temporal parameters of gait. *Journal of Biomechanics, 28*(3), 347-351.

Hausdorff, J. M., Peng, C. K., Goldberger, A. L., & Stoll, A. L. (2004). Gait unsteadiness and fall risk in two affective disorders: a preliminary study. *BMC Psychiatry, 4*(1), 39.

Hausdorff, J. M., Peng, C. K., Ladin, Z., Wei, J. Y., & Goldberger, A. L. (1995b). Is walking a random walk? Evidence for long range correlations in stride interval of human gait. *Journal of Applied Physiology, 78,* 349-358.

Herman, T., Giladi, N., Gurevich, T., & Hausdorff, J. M. (2005). Gait instability and fractal dynamics of older adults with a cautious gait: why do certain older adults walk fearfully? *Gait and Posture, 21*(2), 178-385.

Hodges, P. W., & Bui, B. H. (1996). A comparison of computer based methods for the determination of onset of muscle contraction using EMG. *Electroencephalography and Clinical Neurophysiology, 101,* 511-519.

Hof, A. L. (1996). Scaling gait data to body size. *Gait and Posture, 4*(3), 222-223.

Holzreiter, S. H., & Köhle, M. E. (1993). Assessment of gait patterns using neural networks. *Journal of Biomechanics, 26*, 645-651.

Hood, V. L., Granat, M. H., Maxwell, D. J., & Hasler, J. P. (2002). A new method of using heart rate to represent energy expenditure: The Total Heart Beat Index. *Archives of Physical Medicine and Rehabilitation, 83*(9), 1266-1273.

Huang, N. E., & NASA. (2001). Computer Implemented Empirical Mode Decomposition Method, Apparatus, and Article of Manufacture for Two-Dimensional Signals. *U.S. Patent App. No. 09/150,671,* filed September 10, 1998, allowed July 2001.

Huang et al. (1998). The empirical mode decomposition and the Hilbert spectrum for nonlinear and non-stationary time series analysis. In *Proceedings of Royal Society,* London (Vol. 454, pp. 903-995).

ISB. (2005). *Standards for coordinate systems, joint centres and segments.* Retrieved January 9, 2005, from http://isbweb.org/o/content/view/39/58/

ISB3D. (2005). *3D technical group of ISB.* Retrieved from www.utpb.edu/3D-HumanMovement

Ismail, A. R., & Asfour, S. S. (1998, November 1-4). Continuous wavelet transform application to EMG signals during human gait. In *Proceedings of the 32nd Asilomar Conference on Signals, Systems and Computers* (Vol. 1, pp. 325-329). Pacific Grove, CA.

Jain, A., Nandakumar, K., & Ross, A. (2005). Score normalization in multimodal biometric systems. *Pattern Recognition,* (In Press).

Jansen, B. H., Miller, V. H., Mavrofrides, D. C., & Stegink Jansen, C. W. (2003). Multidimensional EMG-based assessment of walking dynamics. *IEEE Transactions on Neural Systems and Rehabilitation Engineering, 11*(3), 294-300.

Johns, A. (2004). *The probability of tripping in the swing phase of walking gait.* Unpublished Honours Thesis, Victoria University, Australia.

Johns, A., Best, R., & Begg, R. K. (2004, December 9-10). Trip and energy constraints interplay during gait. In *Proceedings of the 5th Australasian Biomechanics Conference,* Sydney, Australia (pp. 48-49).

Johnson, G. R., Ferrarin, M., Harrington, M., Hermens, H., Jonkers, I., Mak, P., & Stallard, J. (2004). Performance specification for lower limb orthotic devices. *Clinical Biomechanics, 19*(7), 711-718.

Kadaba, M. P., Ramakrishnan, H. K., & Wootten, M. E. (1990). Measurement of lower extremity kinematics during level walking. *Journal of Orthopedic Research 8,* 383-392.

Kaufman, K. (1998). Future directions in gait analysis. In J. A. De Lisa (Ed.), *Gait analysis in the science of rehabilitation.* Retrieved January 9, 2005, from http://www.vard.org/mono/gait/contents.pdf

Keegan, K., Arafat, S., Skubic, M., Wilson, D., & Kramer, J. (2003). Detection of lameness and determination of the affected forelimb in horses by use of continuous wavelet transformation and neural network classification of kinematic data. *American Journal of Veterinary Research, 64,* 1376-1381.

Keogh, E., Chakrabarti, K., Pazzani, M., & Mehrotra, S. (2000). Dimensionality reduction for fast similarity search in large time series databases. *Knowledge and Information Systems, 3,* 263-286. Retrieved January 9, 2005, from http://www.cs.ucr.edu/~eamonn/kais_2000.pdf

Knoll, Z., Kiss, R. M., & Kocsis, L. (2004). Gait adaptation in ACL deficient patients before and after anterior cruciate ligament reconstruction surgery. *Journal of Electromyography and Kinesiology, 14*(3), 287-294.

Konrad, P. (2005). *The ABC of EMG.* Scottsdale, AZ: Noraxon.

Kyriazis, V., & Rigas, C. (2002). Temporal gait analysis of hip osteoarthritic patients operated with cementless hip replacement. *Clinical Biomechanics, 17*(4), 318-321.

Lackoviæ, I., Bilas, V., & Šantiæ, A. (1997). A computer based vertical force monitoring and analysis system for gait evaluation. In *Proceedings of the 10th IEEE Symposium on Computer-Based Medical Systems* (pp. 238-243).

Lafortune, M. A. (1991). Three-dimensional acceleration of the tibia during walking and running. *Journal of Biomechanics, 24*(10), 877-886.

Lakany, H. M. (2000). A generic kinematic pattern for human walking. *Neurocomputing, 35*, 27-54.

Lakany, H. M., & Hayes, G. (1997). An algorithm for recognizing walkers. In *Proceedings of the 1st International Conference on Audio- and Video-based Biometric Person Authentication*, Switzerland (pp. 111-118).

Lamoth, C. J. C., Daffertshofer, A., Meijer, O. G., & Beek, P. J. (2006). How do persons with chronic low back pain speed up and slow down? Trunk-pelvis coordination and lumbar erector spinae activity during gait. *Gait and Posture, 23,* 230-239.

Lamoth, C. J. C., Daffertshofer, A., Meijer, O. G., Moseley, G. L., Paul, I. J. M. W., & Beek., P. J. (2004). Effects of experimentally induced pain and fear of pain on trunk coordination and back muscle activity during walking. *Clinical Biomechanics, 19,* 551-563.

Lauer, R. T., Laughton, C. A., Orlin, M., & Smith, B. T. (2003). Wavelet decomposition for the identification of EMG activity in the gait cycle. In *Proceedings of the 29th Annual IEEE Bioengineering Conference* (pp. 142-143).

Lauer, R. T., Stackhouse, C., Shewokis, P. A., Smith, B. T., Orlin, M., & McCarthy, J. J. (2005). Assessment of wavelet analysis of gait in children with typical development and cerebral palsy. *Journal of Biomechanics, 38*(6), 1351-1357.

Laughton, C. A., Fertig, K. P., Liggins, A. B., Lange, G. W., McCarthy, J. J., & Smith, B. T. (2003). The use of principal components analysis to assess ambulatory function before and after surgical lengthening of the hamstrings. In *Proceedings of the 8th GCMAS Meeting.* Alfred I duPont Hospital for Children. Retrieved January 9, 2005, from http://gait.aidi.udel.edu/gaitlab/gcma/info/abstracts/P17.abs20093.pdf

Leardini, A., Cappozzo, A., Catani, F., Toksvig-Larsen, S., Petitto, A., & Sforza, V. (1999). Validation of a functional method for the estimation of hip joint centre location. *Journal of Biomechanics, 32,* 99-103.

Lindström, L., Magnusson, R., & Petersen, I. (1970). Muscular fatigue and action potential conduction velocity changes studies with frequency analysis of EMG signals. *Electromyography, 10,* 341-56.

Macko, R. F., Haeuber, E., Shaughnessy, M., Coleman, K. L., Boone, D. A., Smith, G. V., & Silver, K. H. (2002). Microprocessor-based ambulatory activity monitoring in stroke patients. *Medicine & Science in Sports & Exercise, 34,* 394-399.

Maki, B. E., Holliday, P. J., & Topper, A. K. (1994). A prospective study of postural balance and risk of falling in an ambulatory and independent elderly population. *Journal of Gerontology, 49*(2), M72-84

Mannion, A. F., & Dolan, P. (1996). Relationship between myoelectric and mechanical manifestations of fatigue in the quadriceps femoris muscle group. *European Journal of Applied Physiology, 74,* 411-419.

Mayagoitia, R. E., Nene, A. V., & Veltink, P. H. (2002). Accelerometer and rate gyroscope measurement of kinematics: an inexpensive alternative to optical motion analysis systems. *Journal of Biomechanics, 35,* 537-542.

McGarry, W. B., Zuelzer, W. A., & Wayne J. S. (1999). Evaluation of knee kinematics using a magnetic tracking device. In *Proceedings of the 23rd Annual Meeting of the American Society of Biomechanics,* University of Pittsburgh. Retrieved from http://asb-biomech.org/onlineabs/abstracts99/151/

McPoil, T. G., Cornwall, M. W., & Yamada, W. (1995). A comparison of two in-shoe plantar pressure measurement systems. *The Lower Extremity, 2*(2), 95-103.

Meglan, D., & Todd, F. (1994). Kinetics of human locomotion. In J. Rose & J. G. Gamble (Eds.), *Human walking*. Baltimore: Williams and Wilkins.

Merletti, R., & Parker, P. (2004). *Electromyography: Physiology, engineering, and non-invasive applications*. UK: John Wiley & Sons.

Merlo, A., Farina, D., & Merletti, R. (2003). A fast and reliable technique for muscle activity detection from surface EMG signals. *IEEE Transactions on Biomedical Engineering, 50*(3), 344-353.

Moisio, K. C., Sumner, D. R., Shott, S., & Hurwitz, D. E. (2003). Normalization of joint moments during gait: a comparison of two techniques. *Journal of Biomechanics, 36*, 599-603.

Moser, D., Catalfamo, P., Ghoussayni, S. N., & Ewins, D. J. (2004). Real time description of lower limb motion for nonanalytical neuroprosthetic control applications. In *Proceedings of the 9th Annual Conference of the International FES Society*, Bournemouth, UK. Retrieved January 9, 2005, from http://www.ifess.org/cdrom_target/ifess04/control%20techniques/oral/moserD.pdf

Nicol, A. C. (1989). Measurement of joint motion. *Clinical Rehabilitation, 3*, 1-9.

Nieuwboer, A., Dom, R., Weerdt, W. D., Desloovere, K., Janssens, L.. & Stijn, V. (2004). Electromyographic profiles of gait prior to onset of freezing episodes in patients with Parkinson's disease. *Brain, 127*(7), 1650-1660.

Norkin, C. C., & White, D. J. (2003). *Measurement of joint motion: A guide to goniometry*. Philadelphia: F.A. Davis.

Olhede, S., & Walden, A. T. (2004). The Hilbert spectrum via wavelet projections. In *Proceedings of the Royal Society London, A460* (pp. 955-975).

Olney, S. J., Griffin, M. P., & McBride, I. D. (1998). Multivariate examination of data from gait analysis of persons with stroke. *Physical Therapy, 78*(8), 814-828.

O'Malley, M. J., Abel, M. F., Damiano, D. L., & Vaughan, C. L. (1997). Fuzzy clustering of children with cerebral palsy based on temporal-distance gait parameters. *IEEE Transactions on Rehabilitation Engineering, 5*, 300-309.

Pappas, I. P. I., Keller, T., Mangold, S., Popovic, M. R., Dietz, V., & Morari, M. (2004). A reliable gyroscope-based gait-phase detection sensor embedded in a shoe insole. *IEEE Sensors Journal, 4*(2), 268-274.

Pavol, M. J., Owings, T. M., Foley, K. T., & Grabiner, M. D. (1999). Gait characteristics as risk factors for falling from trips induced in older adults. *Journals of Gerontology Series A: Biological Sciences and Medical Sciences , 54*(11), M583-590.

Périé, D., Tate, A. J., Cheng, P. L., & Dumas, G. A. (2002). Evaluation and calibration of an electromagnetic tracking device for biomechanical analysis. *Journal of Biomechanics, 35*(2), 293-297.

Perry, J. (1992). *Gait analysis. Normal and pathological function*. NJ: Slack.

Perry, J. (1998). EMG dynamics. In J. A. De Lisa (Ed.), *Gait analysis in the science of rehabilitation*. Retrieved January 9, 2005, from http://www.vard.org/mono/gait/contents.pdf

Pierrynowski, M. R., & Galea, V. (2001). Enhancing the ability of gait analyses to differentiate between groups: scaling gait data to body size. *Gait and Posture, 13*(3), 193-201.

Pullman, S. L., Goodin, D. S., Marquinez, A. I., Tabbal, S., & Rubin, M. (2000). Clinical utility of surface EMG: report of the Therapeutics and Technology Assessment Subcommittee of the American Academy of Neurology. *Neurology, 55*, 171-177.

Rab, G. T. (1994). Muscle. In J. Rose & J. G. Gamble (Eds.), *Human walking*. Baltimore: Williams and Wilkins.

Rash, G. S., Quesada, P. M., & Jarboe, N. (1997). Static assessment of Pedar and F-Scan inshoe pressure sensors; revised. In *Proceedings of the 21ˢᵗ Annual Meeting of the American Society of Biomechanics*. Retrieved January 9, 2005, from http://asb-biomech.org/onlineabs/abstracts97/72/

Rechtien, J., Gelblum, B., Haig, A., & Gitter, A. (1999). Technology review: Dynamic EMG in gait and motion analysis. *Muscle and Nerve, 22(Suppl 8)*, S223-S238.

Reid, J. G., & Jensen, R. K. (1990). Human body segment parameters: a survey and status report. *Exercise and Sport Sciences Reviews, 18*, 225-241.

Roberts, G., Best, R., Begg, R., & Wrigley, T. (1995). A real-time ground reaction force vector visualisation system. In *Proceedings of the Australian Conference of Science & Medicine in Sport*, Hobart, Australia (pp. 230-231).

Romei, M., Galli, M., Motta, F., Schwartz, M., & Crivellini, M. (2004). Use of the normalcy index for the evaluation of gait pathology. *Gait and Posture, 19*, 85-90.

Rose, J., Ralston, H. J., & Gamble, J. G. (1994). Energetics of walking. In J. Rose & J. G. Gamble (Eds.), *Human walking*. Baltimore: Williams and Wilkins.

Ross, J. D., & Ashman, R. B. (1987). A thin foot switch. *Journal of Biomechanics, 20*(7), 733-734.

Sadeghi, H., Allard, P., Barbier, F., Sadeghi, S., Hinse, S., Perrault, R., & Labelle, H. (2002). Main functional roles of knee flexors/extensors in able-bodied gait using principal component analysis (I). *Knee, 9*, 47-53.

Sadeghi, H., Allard, P., & Duhaime, M. (1997). Functional gait asymmetry in able-bodied subjects. *Human Movement Science, 16*, 243-258.

Sadeghi, H., Prince, F., Sadeghi, S., & Labelle, H. (2000). Principal component analysis of the power developed in the flexion/extension muscles of the hip in able-bodied gait. *Medical Engineering and Physics, 22*, 703-710.

Sadeghi, H., Prince, F., Zabjek, K. F., & Labelle, H. (2004). Simultaneous, bilateral, and three-dimensional gait analysis of elderly people without impairments. *American Journal of Physical Medicine and Rehabilitation, 83*(2), 112-123.

Schobesberger, H., & Peham, C. (2002). Computerized detection of supporting forelimb lameness in the horse using an artificial neural network. *Veterinary Journal, 163*, 77-84.

Schöllhorn, W. I. (2004). Applications of artificial neural nets in clinical biomechanics. *Clinical Biomechanics, 19*, 876-898.

Schutte, L. M., Narayanan, U., Stout, J. L., Selber, P., Gage, J. R., & Schwartz, M. H. (2000). An index for quantifying deviations from normal gait. *Gait and Posture, 11*, 25-31.

Schwartz, M. H., Novacheck, T. F., & Trost, J. (2000). A tool for quantifying hip flexor function during gait. *Gait and Posture, 12*(2), 122-127.

Sekine, M., Tamura, T., Akay, M., Fujimoto, T., Togawa, T., & Fukui, Y. (2002). Discrimination of walking patterns using wavelet-based fractal analysis. *IEEE Transactions on Neural Systems and Rehabilitation Engineering, 10*(3), 188-196.

Sekine, M., Tamura, T., Togawa, T., & Fukui, Y. (2000). Classification of waist-acceleration signals in a continuous walking record, *Medical Engineering and Physics, 22*(4), 285-291.

Sepulveda, F., Wells, D.M., & Vaughan, C. L. (1993). A neural network representation of electromyography and joint dynamics in human gait. *Journal of Biomechanics, 26*(2), 101-109.

Shao, J., Fraser, C. S., & Wrigley, T. V. (2001). Object point tracking in photogrammetric measurement of human movement. *Photogrammetric Recording, 17*(97), 103-117.

Shiavi, R. (1990). Quantitative representation of electromyographic patterns generated during human locomotion. *IEEE Engineering in Medicine and Biology Magazine, 9*(1), 58-60.

Siegel, K. L., & Metman, L. V. (2000). Effects of bilateral posteroventral pallidotomy on gait of subjects with Parkinson disease. *Archives of Neurology, 57*, 198-204.

Simon, S. R. (2004). Quantification of human motion: gait analysis - benefits and limitations to its application to clinical problems. *Journal of Biomechanics, 37*, 1869-1880.

Small, C., & Whittle, M. (Eds.) (1999). Special Issue on 3-D Analysis of Human Movement – III. *Human Movement Science, 18*(5).

SRDC. (1999). *Paradigm shift – Injury: from problem to solution – New research directions.* Strategic Research Development Committee, NH&MRC. Canberra, Australia: AGPS.

Stansfield, B. W., Hazlewood, M. E., Hillman, S. J., Lawson, A. M., Loudon, I. R., Mann, A. M., & Robb, J. E. (1999). Gait is predominantly characterized by speed of progression, not age, in 5 to 12 year old normal children. *Gait and Posture, 10*(1), 57.

Stokes, V., Lanshammar, H., & Thorstensson, A. (1999). Dominant pattern extraction from 3-D kinematic data. *IEEE Transactions on Biomedical Engineering, 46*(1), 100-106.

Sun, L., Shen, M., Chan, F. H. Y., & Beadle, P. J. (2004). A novel method of time-frequency representation and its application to biomedical signal processing. In *Proceedings of the 7th International Conference on Signal Processing* (Vol. 1, pp. 236-239).

Sutherland, D. H. (2005). The evolution of clinical gait analysis part III - kinetics and energy assessment. *Gait and Posture, 21*(4), 447-461.

Sutherland, D. H., Kaufman, K. R., & Moitoza, J. R. (1994). Kinematics of normal human walking. In J. Rose & J. G. Gamble (Eds.), *Human walking.* Baltimore: Williams and Wilkins.

Syczewska, M., Dembowska-Baginska, B., Perek-Polnik, M., & Perek, D. (2005) Functional status of children after treatment for a malignant tumour of the CNS: A preliminary report. *Gait and Posture,* (In Press).

Tamura, T., Sekine, M., Ogawa, M., Togawa, T., & Fukui Y. (1997). Classification of acceleration waveforms during walking by wavelet transform. *Methods of Information in Medicine, 36*, 356-359.

Taylor, S. B., Begg, R. K., & Best, R. J. (2001). Inter and intra-limb processes of gait control. In B. Lithgow & I. Cosic (Eds.), *Proceedings of the 2nd Conference of the IEEE Engineering in Medicine and Biology Society (Victorian Chapter)*, Melbourne, Australia, (pp. 222-226).

Taylor, S. B., Begg, R. K., & Best, R. J. (2002). An investigation of two methods for quantifying inter- and intra-limb control/coordination processes of gait. In A.

Zayegh (Ed.), *Proceedings of the 4th International Conference on Modelling and Simulation*, Victoria University, Australia (pp. 434-439).

Terrier, P., Turner, V., & Schutz, Y. (2005). GPS analysis of human locomotion: further evidence for long-range correlations in stride-to-stride fluctuations of gait parameters. *Human Movement Science, 24*(1), 97-115.

Tingley, M., Wilson, C., Biden, E., & Knight, W. R. (2002). An index to quantify normality of gait in young children. *Gait and Posture, 16*(2), 149-158.

Trew, M., & Everett, T. (Eds.) (2001). *Human movement*. Edinburgh: Churchill Livingstone.

Tscharner, V. V., Goepfert, B., & Nigg, B. (2003). Changes in EMG signals for the muscle tibialis anterior while running barefoot or with shoes resolved by non-linearly scaled wavelets. *Journal of Biomechanics, 36*, 1169-1176.

Vankoski, S., & Dias, L. (1999). Clinical motion analysis. *Journal of Children's Memorial Hospital Chicago*. Retrieved from http://www.childsdoc.org/99Spring/clinicalmotionanalysis.asp

Vaughan, C. L., Davis, B. L., & O'Connor, J. C. (1992). *Dynamics of human gait*. Chapaign, IL: Human Kinetics.

Verdini, F., Leo, T., Fioretti, S., Benedetti, M. G., Catani, F., & Giannini, S. (2000). Analysis of ground reaction forces by means of wavelet transform. *Clinical Biomechanics, 15*(8), 607-610.

Wachowiak, M., Rash, G., Desoky, A., & Quesada, P. (1997). Wavelet transforms for smoothing kinesiological data. In *Proceedings of the 21st Annual Meeting of the American Society of Biomechanics*. Clemson University, South Carolina. Retrieved January 9, 2005, from http://asb-biomech.org/onlineabs/abstracts97/73/

Wachowiak, M., Rash, G., Quesada, P., & Desoky, A. (1998). Comparison of wavelet-based and traditional noise removal techniques. In *Proceedings of the 3rd North American Congress on Biomechanics*, University of Waterloo, Canada. Retrieved January 9, 2005, from http://www.asb-biomech.org/onlineabs/NACOB98/60/index.html

Wachowiak, M., Rash, G., Quesada, P., & Desoky, A. (2000). Wavelet-based noise removal for biomechanical signals: A comparative study. *IEEE Transactions on Biomedical Engineering, 47*(3), 360-368.

West, B. J., & Griffin, L. (1999) Allometric control, inverse power laws and human gait. *Chaos, Solitons and Fractals, 10*, 1519-1527.

White, R., Agouris, I., & Fletcher, E. (2005). Harmonic analysis of force platform data in normal and cerebral palsy gait. *Clinical Biomechanics, 20*(5), 508-516.

Whittle, M. W. (2002). *Gait Analysis: An Introduction*. Oxford: Butterworth-Heinemann.

Whittle M. W., & Blanchi, J-P. (Eds.) (1997). Special Issue on 3-D Analysis of Human Movement – II. *Human Movement Science, 16*(2-3).

Winter, D. A. (Ed.). (1990). *Biomechanics and motor control of human movement*. New York: John Wiley & Sons.

Winter, D. A. (1991). *The biomechanics and motor control of human gait: Normal, elderly, and pathological*. Canada: Waterloo Press.

Wojcik, L. A., Etebari, A., & Ferrante, J. A. (1999). Development of a virtual interface for

Bioengineering Conference, Montana. Retrieved from http://166.82.6.12/bio1999/data/pdfs/a0021388.pdf

Wolf et al. (2005). Automated feature assessment in instrumented gait analysis. *Gait and Posture,* (In Press).

Wu, W-L., & Su, F-C. (2000). Potential of the back propagation neural network in the assessment of gait patterns in ankle arthrodesis. *Clinical Biomechanics, 15*(2), 143-145.

Wu, W-L., Su, F-C., Cheng, Y. M., & Chou, Y. L. (2001). Potential of the genetic algorithm neural network in the assessment of gait patterns in ankle arthrodesis. *Annals of Biomedical Engineering, 29*(1), 83-91.

Wu, W-L., Su, F-C., & Chou, C. K. (1998). Potential of the back propagation neural networks in the assessment of gait patterns in ankle arthrodesis. In E. C. Ifeachor, A. Sperduti & A. Starita (Eds.), *Neural networks and expert systems in medicine and health care,* 92-100. World Scientific Publishing.

Xingang, L., & Jianping, Y. (2003). The state of the art of micromachined gyroscopes. *Advances in Mechanics, 33*(3), 301-311.

Yamamoto, S., Suto, Y., Kawamura, H., Hashizume, T., Kakurai, S., & Sugahara, S. (1983). Quantitative gait evaluation of hip diseases using principal component analysis. *Journal of Biomechanics, 16*(9), 717-726.

Yang, K., & Shahabi, C. (2004). A PCA-based similarity measure for multivariate time series. In *Proceedings of 2nd ACM International Workshop on Multimedia Databases,* Washington DC (pp. 65-74).

Yeadon, R. M., & Challis, J. H. (1994). The future of performance-related sports biomechanics research. *Journal of Sports Sciences, 12,* 3-32.

Yeadon, R. M., & Morlock, M. (1989). The appropriate use of regression equations for the estimation of segmental inertial parameters. *Journal of Biomechanics, 22,* 683-689.

Zhou, H., & Hu, H. (2004). A survey: Human movement tracking and stroke rehabilitation. *Technical Report CSM 420: ISSN 1744-8050.* University of Essex, UK. Retrieved from http://cswww.essex.ac.uk/technical-reports/2004/CSM-420.pdf

Zhou, W., & Gotman, J. (2004). Removal of EMG and ECG artifacts from EEG based on wavelet transform and ICA. In *Proceedings of the 26th Annual International Conference of the IEEE EMBS.* San Francisco.

Chapter II

Inertial Sensing in Biomechanics: A Survey of Computational Techniques Bridging Motion Analysis and Personal Navigation

Angelo M. Sabatini, ARTS Lab, Scuola Superiore Sant' Anna, Italy

ABSTRACT

Sensing approaches for ambulatory monitoring of human motion are necessary in order to objectively determine a person's level of functional ability in independent living. Because this capability is beyond the grasp of the specialized equipment available in most motion analysis laboratories, body-mounted inertial sensing has been receiving increasing interest in the biomedical domain. Crucial to the success of this certainly not new sensing approach will be the capability of wearable inertial sensor networks to accurately recognize the type of activity performed (context awareness) and to determine the person's current location (personal navigation), eventually in combination with other biomechanical or physiological sensors — key requirements in applications of wearable and mobile computing as well. This chapter reviews sensor configurations and computational techniques that have been implemented or are considered to meet the converging requirements of a wealth of application products, including ambulatory monitors for automatic recognition of activity, quantitative analysis of motor performance, and personal navigation systems.

INTRODUCTION

At present, human motor performance can be accurately assessed with several measuring instruments, the use of which is spread in many motion analysis laboratories throughout the world. The most important technology used to detect and track human body motion is video-motion sensing. In common with other motion tracking technologies, such as infrared, electromagnetic, ultrasound, video-motion sensing is externally referenced, in the sense an external source — optical, magnetic, acoustic — is needed to determine position and orientation information concerning the moving object of interest (Meyer, Applewhite, & Biocca, 1992). Usually, this source is effective over a relatively small working space. In addition to the range restriction, interference, distortions and occlusions can easily result in erroneous location and orientation information, thereby leading, in critical situations, to a complete loss of track (You, Neumann, & Azuma, 1999). The availability of dedicated laboratory setups is, therefore, a prerequisite for the application of externally referenced sensing techniques. However, from a clinical viewpoint, motor performance measured in laboratory settings may not accurately reflect functional ability in daily-life environments, since behavior of patients in laboratory is not necessarily representative of their daily-life behavior. There is a need for ambulatory monitoring systems that are able to provide objective assessment of human functional ability in the absence of the behavioral modifications induced by performing within constraining laboratory settings.

The capability of inertial sensors of sensing their own motion is the sourceless feature that makes them so attractive for the development of ambulatory monitoring systems (Verplaetse, 1996). Body-mounted sensors of this kind make it possible to determine position and orientation information based on the measurement of physical quantities (acceleration, angular velocity), which are directly related to the motion of the body part where they are positioned. Being internally referenced, inertial sensors can then be proposed to detect and track body motion over a virtually unrestricted working space. Until recent years, inertial sensors have only found use to monitor the motion of man-made vehicles, including spaceships, planes, ships, submarines, cars, and, more recently, wheeled and legged robots. Recent advances in microelectromechanical systems (MEMS) technologies have led to the development of a new generation of inertial sensors (Bachmann, Yun, McKinney, McGhee, & Zyda, 2003), the specifications of which — in terms of encumbrance, robustness, power consumption, measuring performance and cost — seem to be appropriate for applications in the biomedical field.

In inertial systems, the main problem is that position and orientation are found by time-integrating the signals from accelerometers and gyroscopes, as well as any sensor drift and noise superimposed to them. As a result, position and orientation errors tend to grow unbounded. This problem is especially acute when low-cost MEMS inertial sensors are used. Their sensitivity and bias stability are, in fact, orders of magnitude less than the sensitivity and bias stability of the high-grade inertial sensors that are embedded in military and aviation navigation systems (Foxlin, 2002). Another drawback of inertial sensors is that they are not well-suited for determination of absolute position and orientation. In order to be accurate, the integration process needs to be started from accurately known initial conditions, which inertial sensors are unable to provide at all (position and velocity), or can provide to just a limited extent (orientation). Hence, the use of inertial systems is most effective in those applications which involve relative motion.

In this chapter, we are not interested in the viewpoint of those who aim at designing new and better sensors. Rather, we intend to survey the main computational techniques that have been investigated so far, in the effort to take measurements from available sensors and construct the best possible characterization of human body motion. The traditional approach to data processing for navigational purposes has involved the development of filtering algorithms to fuse measurements from inertial sensors and other sensors, such as global positioning system (GPS) receivers and Earth's magnetic field sensors. It has also involved the exploitation of suitable environment maps so as to deal with the person's location uncertainty during indoor and outdoor navigation. Other computational techniques have been designed with a stronger biomechanical inspiration. They specifically aim at obtaining valuable information about either absolute or relative motion of body parts from simple configurations of single or multiple inertial sensors, in the effort to gain a deeper understanding of how we humans perform functional activities, including gait, balance and postural transitions, in normal or pathological conditions.

The borderline between the use of inertial sensors as functional and navigational sensors in biomechanics is becoming increasingly fuzzy. As it will be seen, a better prediction of the errors incurred in the mathematical reconstruction of the walked path can be achieved by exploiting signatures which characterize how certain functional activities such as walking are performed. Conversely, the knowledge of temporo-spatial features of the walked path, reconstructed from navigational sensors, has relevant functional implications, when, for instance, an objective assessment of the person's physical activity in daily-life conditions is desired. Ideally, we would like to determine accurately the current person's location and recognize what he or she is doing and how, without incurring in any significant limitation in the extension of the measurement space and in the duration of the observation records.

At present, ambulatory monitoring systems based on sole inertial sensing cannot accomplish this ambitious goal, although they can already be used with remarkable success. In this chapter we will survey how inertial sensing has been applied in biomechanically-oriented works, with particular emphasis on the sensor configurations and adopted signal processing techniques. In doing that, we hope to give convincing support to the claim that another valuable analytical weapon is ready for inclusion in the repertoire of tools available to measure and study human motor performance.

BACKGROUND

Introductory Remarks

An inertial sensor is a motion sensor the references of which are internal, with the exception of initialization (Curey, Ash, Thielman, & Barker, 2004). It can be used for the determination of position and orientation by measuring physical quantities, such as linear accelerations and angular velocities. An inertial sensor assembly (ISA) is a structure that contains multiple inertial sensors (accelerometers and/or gyroscopes) in fixed orientations relative to one another. An inertial measurement unit (IMU) measures linear and angular motion in the three-dimensional space without external references. Using the outputs from an IMU, an inertial navigation system (INS) estimates a body's position and orientation, in connection with a gravitational field model and the operation of a reference clock.

Earth's magnetic field sensing is formally externally referenced, although it can be used in practice as though it is internally referenced. This is because the Earth's magnetic field is available anywhere, and no environmental modifications are required to operate an Earth's magnetic field sensor, or magnetic sensor in short. The importance of magnetic sensors in human-machine interface applications and human body motion tracking systems has been widely acknowledged (Foxlin, 2002), in part because of the recent market availability of miniaturized tri-axis inertial/magnetic sensors packages. Using a geomagnetic field model, magnetic sensors can help specify the location of an INS during initialization, and mitigate the INS error growth during flight. Magnetic sensors will be included in the sensor configurations of interest to our review in this chapter.

Inertial and Magnetic Sensors

Accelerometers are sensors that measure the applied acceleration acting along a sensitive axis. A range of different transducers are available to measure the acceleration. In virtually all applications in human movement science piezoresistive accelerometers are used (Verplaetse, 1996). Piezoresistive materials have the property of modifying their electrical resistance when they are strained or deflected. In terms of functionality, body-mounted piezoresistive accelerometers respond to either static accelerations, such as gravity, or dynamic accelerations, such as linear/rotational accelerations due to body movements (Figure 1). When exposed just to gravity, a body-mounted tri-axis accelerometer can be used to estimate the inclination relative to the vertical of the body segment where the device is positioned.

Gyroscopes, gyros in short, are sensors that measure the rate of rotation about a sensitive axis. The most popular design of a gyro for applications in the field of human movement science is of the vibrating type (Verplaetse, 1996). Vibrating gyros use the Coriolis acceleration effect to sense when they rotate. In the absence of rotation, the vibrating element embedded in the sensor continuously vibrate within a plane. In the presence of a rotation about an axis perpendicular to the plane of vibration, the vibrating element deviates from its plane of vibration under the Coriolis acceleration effect and the amplitude of the out-of-plane vibration is proportional to the rate of rotation (Abbott & Powell, 1999). Gyros with a vibrating element are more sensitive to temperature and shock

Figure 1. A piezoresistive single-axis accelerometer measures the projection (in the direction of the sensitive axis $\hat{\mathbf{v}}$) of an equivalent acceleration resulting from the sum of the acceleration $\vec{\mathbf{a}}$ due to the accelerometer's own motion and the equivalent gravity acceleration acting on its mechanical structure.

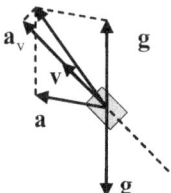

than accelerometers, due to mechanical fastening of the vibrating beam inside the sensor case. Compared with accelerometers, gyros have also relatively higher cost and larger size.

Magnetic sensors resolve the components of the local magnetic field along a sensitive axis. A magnetic compass is an electronic device that measures its heading relative to magnetic north by measuring the direction of Earth's local magnetic field (Abbott & Powell, 1999). The Earth's magnetic field has a component parallel to the Earth's surface that always points toward magnetic north, hence its projection on the horizontal plane can be used to determine compass direction (Caruso, 1997). Common designs of magnetic sensors integrate three magnetic sensors with sensitive axes mutually perpen-

Figure 2. Inclination of the compass device relative to the Earth's local horizontal plane as defined by gravity direction (a). Earth's magnetic field resolved in (x, y, z) coordinates (b).

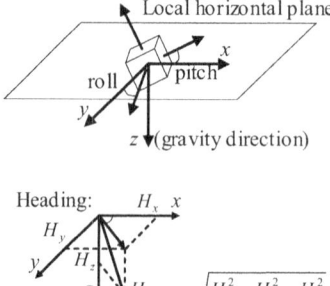

dicular in the same package, so as to reconstruct the horizontal component, provided that the inclination of the sensor case is estimated from gravity sensors (tilt compensation) (Figure 2).

In applications in the field of human movement science, the capability to sense the Earth's magnetic field is the horizontally-sensing principle needed to complement the vertically-sensing principle of accelerometers, so as to enable three-dimensional monitoring of orientation (Kemp, Janssen, & van der Kamp, 1998).

The Strapdown Approach to INS Design

Consider a rigid body in space and define a coordinate system attached to the body — the body frame, $\mathcal{B}$. In a strapdown INS, the sensitive axes of the ISA sensors are permanently aligned along the axes of $\mathcal{B}$ (Curey et al., 2004). Orientation determination means to specify the orientation of the axes of $\mathcal{B}$ with respect to the Earth's reference frame $\mathcal{E}$; position determination means to specify the origin of $\mathcal{B}$ relative to $\mathcal{E}$.

Suppose that a vector is resolved along the axes of $\mathcal{B}$ and $\mathcal{E}$, to yield the 3×1 column-vectors $\vec{\mathbf{x}}^B$ and $\vec{\mathbf{x}}^E$, respectively. The transformation between the 3×1 column-vector $\vec{\mathbf{x}}$ as expressed in the body frame and in the reference frame is:

$$\vec{\mathbf{x}}^B = \overset{E \to B}{C} \vec{\mathbf{x}}^E \tag{1}$$

where $\overset{E \to B}{C}$ is the direction cosine matrix for the transformation from $\mathcal{E}$ and $\mathcal{B}$. Furthermore, suppose a single-axis accelerometer is attached to the body at point $\vec{\mathbf{p}}(t)$, and the orientation of its sensitive axis is denoted by the unit vector $\vec{\mathbf{v}}(t)$. In principle, the accelerometer measures the projection of the acceleration of point $\vec{\mathbf{p}}(t)$, along $\vec{\mathbf{v}}(t)$:

$$\vec{\mathbf{a}}^E(t) = \left[\vec{\omega}(t) \times \vec{\mathbf{p}}(t) + \vec{\omega}(t) \times \vec{\omega}(t) \times \vec{\mathbf{p}}(t) + \vec{\mathbf{a}}_o(t) + \vec{\mathbf{g}} \right] \cdot \vec{\mathbf{v}}(t) \tag{2}$$

where $\vec{\mathbf{a}}_o(t)$ is the acceleration of the origin of $\mathcal{B} \vec{\mathbf{o}}(t)$, $\vec{\mathbf{g}}$ is the gravitational acceleration, and $\vec{\omega}$ is the angular velocity of $\mathcal{B}$ relative to $\mathcal{E}$, resolved in $\mathcal{B}$. In (2), $\times$ and $\cdot$ denote the standard vector cross-product and dot-product, respectively.

The determination of the direction cosine matrix, also called attitude matrix, is usually performed by numerically integrating non-linear systems of first-order differential equations from initial conditions that are assumed to be known (initialization, or alignment) (Bortz, 1971). In this approach, the differential equations involve the components of $\vec{\omega}$, which are usually measured by a tri-axis gyro. Alternatively, the attitude can be determined by vector matching in $\mathcal{E}$ and $\mathcal{B}$ (Wertz, 1978).

Vector matching, also known as Wahba's problem, calculates the attitude solution by matching two non-zero, non-colinear vectors that are known in $\mathcal{E}$ and measured in $\mathcal{B}$. The Earth's gravitational field and the Earth's magnetic field are two these vectors. Vector matching solves the alignment process and forms, in principle, an alternative to the use of gyros in determining the attitude matrix (Gebre-Egziabher, Elkain, Powell, & Parkinson, 2000). While suitable to track slow movements, this gyro-free method is not suited for fast movements, yielding quite large orientation errors during the motion.

To understand why, let $\vec{\mathbf{o}}(t)$ be the point whose position in $\mathcal{E}$ is to be determined. It is apparent from (2) that the output of a body-mounted accelerometer depends on where it is placed, its orientation relative to the subject, the subject's posture and activity. In the absence of motion, the accelerometer output is determined by its orientation relative to the vertical. Knowing the accelerometer orientation relative to the subject, the accelerometer can be used to determine the subject's orientation relative to the vertical. In the presence of motion, the resulting signal depends on the subject's posture and activity.

The orientation estimate is used twice in the strapdown approach to position determination. First, the accelerometer senses the body's acceleration $\vec{\mathbf{a}}_o^B(t)$ as the superposition of the sensed acceleration in $\mathcal{B}$, $\vec{\mathbf{a}}^B(t)$ and the projection of the gravitational acceleration $\vec{\mathbf{g}}$ on $\mathcal{B}$:

$$\vec{\mathbf{a}}_o^B(t) = \vec{\mathbf{a}}^B(t) - \overset{E \rightarrow B}{C}(t)\,\vec{\mathbf{g}} \tag{3}$$

The next step requires integration of $\vec{\mathbf{a}}_o^E(t) = \overset{B \rightarrow E}{C}(t)\vec{\mathbf{a}}_o^B(t)$ to derive the position estimate:

$$\vec{\mathbf{o}}^E(t) = \int_0^t dt' \int_0^{t'} \overset{B \rightarrow E}{C}(t')\,\vec{\mathbf{a}}_o^B(t'')dt'' \tag{4}$$

Any numerical integration routine can be used to solve (4). The flow of information underlying the strapdown approach is illustrated in Figure 3.

Drift in the orientation determination comes from gyro biases, defined as the output produced by gyros at rest. Uncompensated gyro biases lead to a constant rate of drift after integration. More than the biases in itself, their time-stability is a matter of concern. Time-stable biases can be measured when the INS is stationary and compensated for. There is also drift in the linear position determination which arises from several sources. First, there are the effects of accelerometer biases, the compensation of which is not so straightforward as the compensation of the gyro biases (Lötters, Schipper, Veltink, Olthius, & Bergveld, 1998). Second, since position is obtained by double integration, uncompensated accelerometer biases result in a position drift error which grows quadratically in time. Third, the orientation errors incurred by gyros lead to imperfect gravity cancellation during the double integration.

Additionally, the capability of geomagnetic compassing to correct drift is poor in many environments, because the Earth's magnetic field is weak and easily masked by magnetic disturbances within or near the measurement space.

In conclusion, it is critical to achieve high accuracy in position/orientation determination by strapdown INSs incorporating low-cost inertial/magnetic sensors, the stand-alone accuracy and run-to-run stability of which are poor. Different applications may involve different accuracy requirements relative to the duration of each observation run. In the absence of special precautions, the requirements of human motion tracking applications are shown to be violated when the duration of the observation run exceeds just several seconds (Foxlin, 2002).

Figure 3. Strapdown approach to the design of an inertial navigation algorithm

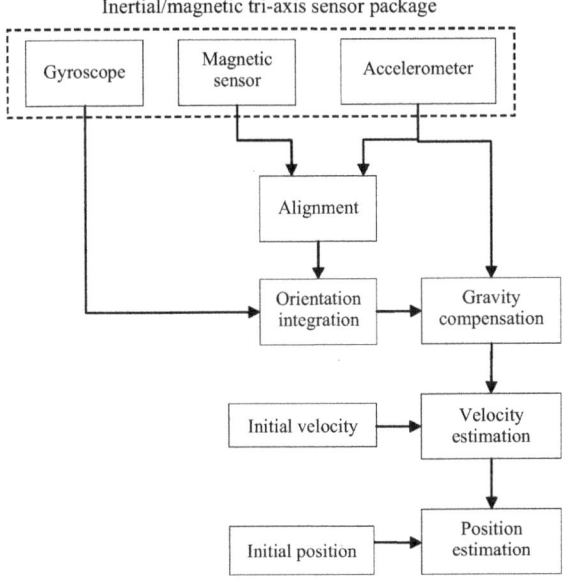

Inertial/magnetic tri-axis sensor package

There appear to be different means to deal with these problems, as stated below. An approach is to use externally referenced aids, such as GPS, and carry out the integration process underlying the combined use of GPS and INS technologies with Kalman filters (KFs). Another approach is to exploit idiosyncrasies of the human motion dynamics by designing algorithms that can keep the drift rate low (Foxlin, 2002; Sabatini, 2005). Alternatively, in some applications, the difficulty inherent in the strapdown approach can be circumvented by using signal features the construction of which do not require time integration from uncertain initial conditions.

TAXONOMY OF APPLICATIONS

Ambulatory Monitoring of Human Motion

The use of accelerometry to perform ambulatory measurements of human motion is well established (Gage, 1964; Smidt, Arora, & Johnston, 1971; Morris, 1973). Morris (1973) suggests to calculate the general displacement of a rigid body with one point fixed — a rotation about an axis through the fixed point — from three acceleration measurements, followed by three additional acceleration measurements to resolve the motion of the fixed point with respect to the Earth's reference frame. Valuable considerations are made by this author concerning the problem of IMU alignment and drift correction. In the case of shank motion during level walking, the redundancy of the information on shank motion during the foot-flat phase of the walking cycle is exploited to calculate the initial orientation (alignment). Because of the cyclical nature of leg motion, the initial angular velocity and rotation, translation acceleration and velocity can be made equal to the values of these functions at the beginning of the next cycle (drift correction).

IMUs composed of several single-axis accelerometers with their sensitive axes properly positioned in space relative to one another have been proposed and lead, for the estimation of the components of angular velocity, to systems of nonlinear coupled differential equations (IMU with six accelerometers) and linear uncoupled differential equations (IMU with nine accelerometers) (Padgaonkar, Krieger, & King, 1975). The angular velocity components are then used to estimate the attitude matrix (Bortz, 1971). Double-integration of gravity-compensated accelerometer signals yields positioning information. Giansanti, Macellari, Maccioni, and Cappozzo (2003) analyze these IMU configurations in depth, and conclude that severe restrictions exist in the time duration over which motion tracking is feasible by accelerometry methods in routine biomechanical applications.

The determination of angular acceleration is probably the commonest computational task solved by accelerometry for the ambulatory monitoring of human motion. The main application is the clinical assessment of gait. Under the simplifying assumption of a gait motion planar model, a minimal configuration set composed of two leg-mounted single-axis accelerometers with parallel sensitive axes would be enough. In the attempt to circumvent the problem of integration drift, pairs of accelerometers on each segment are used to resolve the relative angle between two segments, namely the joint angle, without integration (Willemsen, van Alsté, & Boom, 1990). The potential of such an accelerometric angle sensor is highlighted in (Willemsen, Frigo, & Boom, 1991), where the most important error sources are analyzed in detail, namely not fulfilling the gait motion planar model and the rigid-body condition by external fixation of body-mounted sensors.

Luinge and Veltink (2004) propose to use one tri-axis accelerometer to measure inclination during dynamic tasks without requiring additional sensors. A Kalman-based filtering algorithm is designed to separate the different acceleration components: gravity, linear acceleration, bias, measurement noise, in accordance with a model of the linear acceleration components which relies on reasonable assumptions about the dynamics of human motion. The resulting inclination error is shown to be significantly smaller than the inclination error obtained by the method of low-pass filtering the accelerometer signals, especially as the speed of motion increases. Usually, low-pass filtering is used to separate the gravitational component from the linear acceleration components (Veltink, Bussmann, de Vries, Martens, & van Lummel, 1996; Foerster & Fahrenberg, 2000; Williamson & Andrews, 2001). Unfortunately, since the procedure of bias compensation works only in the direction of gravity, the bias estimate in all measurement directions is accurate only when the sensor is rotated over large angles.

As for the computation of joint angles, an alternative to the use of accelerometers is represented by the use of gyros (Miyazaki, 1997; Williamson & Andrews, 2001). Two approaches are possible to compensate the gyro bias during integration: a) high-pass filtering of the signals (Miyazaki, 1997) — this technique is not recommended for non-cyclical motions, for instance postural transitions, for whom a zero frequency component of angular velocity cannot be expected; (b) automatic nulling and resetting (Williamson & Andrews, 2001). The underlying basis for the implementation of automatic nulling is that, when the gyro is at rest, the angular velocity is null and the bias can then be estimated by averaging. The underlying basis for the implementation of automatic resetting is that, for cyclical motions, the values of angular rotation would be identical at each instance of a specific event occurring in the gait cycle, for instance foot-flat (Sabatini, Martelloni, Scapellato, & Cavallo, 2005). Automatic nulling and resetting imply the capability of discriminating rest and dynamic conditions by algorithmic means — a precursor to more sophisticated methods for automatic recognition of activity, as discussed later in this section. Another problem of gyroscopic sensing is that gyros measure changes of orientation rather than absolute orientation. Gravimetric tilt sensing and geomagnetic compassing are then required (Kemp et al., 1998), however alignment would be performed only in the time intervals between successive flights of the involved body segment (Figure 4) (Sabatini, 2005).

The need for alignment exists unless the change of orientation is the quantity of interest. This is the case when, for instance, the change of orientation is introduced in simple symmetric gait models of extended limbs to determine stride length (Miyazaki, 1997).

Assessment of Gait

A full kinematic analysis of the body motion based on body-mounted sensors would be impossible without incurring in the "Christmas tree" effect (Meijer, Westerkerp, Verhoeven, Koper, & ten Hoor, 1991), even in the limit of a simplified two-dimensional analysis (Mayagoitia, Nene, & Veltink, 2002). Fortunately, in clinical practice, the values of simple temporo-spatial parameters of gait, including stride time/length, cadence and walking speed, turn out to be good indicators of how well a person walks, and of how effective are different rehabilitation techniques or exercise programs in facilitating ambulatory recovery (Weir & Childress, 1997; Maki, 1997). Valuable information concerning a person's gait can also be obtained by measuring temporo-spatial features related to the displacement of particular anatomical points, such as the vertical displacement of the sacrum during walking (Weir & Childress, 1997).

Simple parameters of gait can be determined from waist, thigh, heel accelerations (Evans, Duncan, & Gilchrist, 1991; Auvinet, Chaleil, & Barrey, 1999; Aminian et al., 1999a; Zijlstra & Hof, 2003); from thigh, shank, heel, foot instep angular velocities (Miyazaki, 1997; Tong & Granat, 1999; Pappas, Popovic, Keller, Dietz, & Morari, 2001; Aminian, Najafi, Büla, Leyvraz, & Robert, 2002; Sabatini et al., 2005). The parameterization of gait requires the detection of subsequent foot contacts. The commonest approach to the detection of foot contacts with inertial sensors is based on a search for distinctive and stable signal features occurring at the time of heel-strike (when the foot first touches the floor) and toe-off (when it last takes off) during a gait cycle. Usually, the detection and time-localization of these features require (a combination of) peak or zero-crossing detectors applied to signals contaminated by relatively high-frequency noise components. In a number of papers (Aminian et al., 1999; 2002; Paraschiv-Ionescu, Buchser, Rutschmann, Najafi, & Aminian, 2004) the use of wavelet packages is proposed to

Figure 4. Strapdown approach to the design of an inertial navigation algorithm, specifically intended for gait analysis applications (Sabatini, 2005)

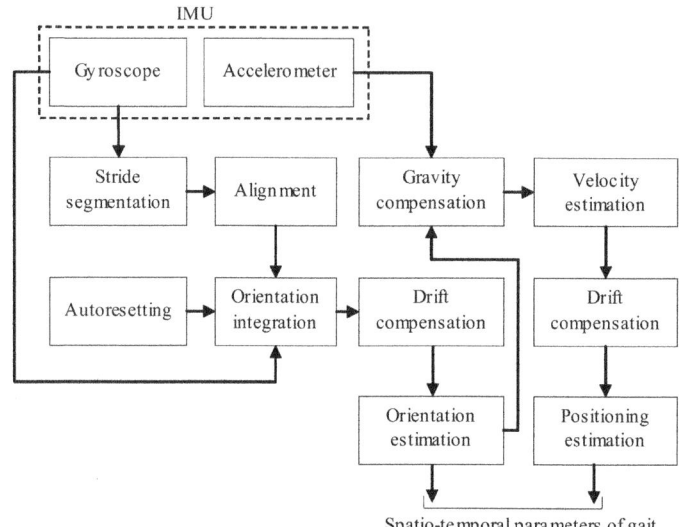

Spatio-temporal parameters of gait

perform the process of gradually focusing on the exact location of heel-strikes and toe-offs. In particular, the claim of these authors is that wavelet packages are well adapted for gait events identification, such as heel-strike and toe-offs, because they allow detection of a specified frequency at a specified time. In general, for biomechanical signal processing applications, the superiority over standard filtering techniques of wavelet denoising and smoothing of kinematic data related to complex motions is documented (Ismail & Asfour, 1999). Another good point in favour of wavelets is that the existence of fast algorithms for carrying the multiresolution analysis would allow to implement gait parameterization in real-time conditions, although this point has not been yet investigated.

The spatial parameters of gait can be determined by so-called indirect methods or by integrating the strapdown equations (direct method). Indirect methods are based on the recognition of few important features of human's walking patterns. These include the good correlation existing between the stride length and the cadence, or the influence of walking speed on the amplitudes of the acceleration signals at different point on the legs. Recent studies demonstrate that during walking trunk accelerations exhibit patterns with fixed relationships to spatio-temporal gait parameters. Zijlstra and Hof (2003) analyze the 3D-acceleration pattern of the lower trunk at different walking speeds, and develop algorithms to determine the instant of foot contacts during each step. Based on a simple inverted pendulum model of the body's center of mass (BCOM) trajectory, the step length is predicted by the peak-to-peak trunk vertical displacement, which is obtained from double integration of the high-pass filtered vertical component of trunk acceleration. An advantage of computing peak-to-peak displacements is that sensor calibration and initialization issues are insignificant, although inaccurate placement of the sensor device in the frontal plane is a matter of concern for achieving good test-retest reliability, as discussed by Moe-Nillsen (1998), who prefer to work in the domain of root mean square (RMS) values from 3D-waist accelerometer data.

In the attempt to include the incline of the walked surface in the number of variables that are known to correlate with walking speed, the parameterization of body accelerations at the waist and heel is performed in (Aminian, Robert, Jéquier, & Schutz, 1995) on the basis of simple features, such as means, standard deviations, correlation coefficients between various acceleration components, step time, and so forth. Two artificial neural networks (ANNs) are designed, one for predicting the walking speed (averaged over a single stride), one for predicting the incline of the walked surface, according to standard design procedures for training and testing multilayer perceptrons. Subject-specific training is performed by presenting examples of treadmill walking at different combinations of walking speed/incline, so as to account for individual walking styles. Testing is performed during overground walking. Results show that only marginal improvements are due to using ANN technology in the walking speed prediction, as compared with the results of a stepwise linear regression analysis. Conversely, owing to the possible presence of strong nonlinearities in the complex relationships between the acceleration parameters selected for the study and the incline, the ANN use is shown to remarkably improve the incline estimate. It is worthy noting that ANNs exhibit significant generalization abilities in this application, although the training phase must be necessarily subject-specific.

Common to all indirect methods, intra-individual physiological variability and environmental conditions heavily influence the accuracy of the relationships exploited to infer the quantities of interest. Hence, frequent calibration procedures may be needed, which requires additional sensors (Perrin, Terrier, Ladetto, Merminod, & Schutz, 2000; Terrier, Ladetto, Merminod, & Schutz, 2000). Another limitation of indirect methods is that they are approaches which suffer from changes in walking style. It is expected that, in particular situations, such as crowded streets and many indoor environments, the human displacements occur with frequent walking strategy adaptations. Finally, the generalization implied by indirect methods is critical when walking is performed over irregular terrain or in the face of frequent changes of inclines.

At the expense of greater computational costs, direct methods would be more appropriate than indirect methods. Anecdotal evidence highlights the remarkable accuracy of direct methods for the estimation of gait spatio-temporal parameters (Sagawa, Sato, & Inooka, 2000; Veltink et al., 2003). In support of this claim, a recent study analyzes the performance of a direct method during level/uphill/downhill treadmill walking (Sabatini et al., 2005).

Assessment of Physical Activity and Energy Expenditure of Daily-Life Activities

Physical activity is an important part of human behavior and it may be related to various aspects of health and disease; intense research is therefore undergoing to find means for its objective assessment (Mejier et al.,1991). The complex nature of physical activity makes this goal very difficult to pursue. At present, the energy expenditure due to physical activity is widely accepted as the standard reference for physical activity. Together with heart rate monitoring, motion sensors such as accelerometers are considered suitable to obtain a quantitative estimation of the energy expenditure incurred in daily-life activities.

The use of accelerometers is based on relationships between accelerometer output and energy expenditure or oxygen uptake that are documented in several studies on gait analysis and ergonomics (Bouten, Koekkoek, Verduin, Kodde, & Janssen, 1997; Bussmann, Hartgerink, van der Woude, & Stam, 2000). Regardless where the right location of body-mounted accelerometers is for the best prediction of the energy expenditure, or which the effects of placement and orientation are (Bouten, Sauren, Verduin & Janssen, 1997), we cannot expect extremely accurate estimates of energy expenditure with the use of accelerometers (Bussmann et al.,2000). Albeit dynamic activities such as walking greatly contribute to daily-life physical activity, and are a prerequisite to many other activities, the contribution to energy expenditure due to static activities is also relevant, for instance weight bearing in the upright posture, work of isometric contractions, and so forth. Of course, motion sensors are totally blind to this part of the energy budget.

Since double integration of acceleration to retrieve position and compute external work is not believed accurate enough for long-term monitoring, the general consensus is that the summation of the time integrals of acceleration components in various measurement directions, and the time integral of the acceleration vector norm result in the most accurate prediction of energy expenditure during dynamic activities (Bouten et al.,1997). More precisely, the various acceleration components are high-pass filtered to subtract gravity influence; the numeric integration is carried out, then the integral(s) are normalized to the duration of the integration interval. Possibly, energy expenditure estimation can be improved by adopting nonlinear regression models to capture the relationship between the acceleration statistical feature and the energy expenditure (Chen & Sun, 1997). The improvement is especially important in estimating light-intensity activities, whose energy expenditure is generally underestimated by linear models (Bouten et al.,1997a).

Besides walking speed, the assessment of the energy expenditure during walking is influenced by many other parameters, including the incline. Without an external measurement of the slope, the standard method of analysis of body accelerations cannot accurately predict the energy cost of uphill/downhill walking (Terrier, Aminian, & Schutz, 2001). Use of additional processing techniques, such as ANNs (Aminian et al.,1995), have shown success in estimating the incline along which the subject is walking. Use of a direct method based on strapdown integration from a foot IMU is discussed in Sabatini, Martelloni, Scapellato and Cavallo (2004), in connection with the use of empirical models for determining the energy cost of walking.

Automatic Recognition of Activity

Another problem existing with the assessment of physical activity by motions sensors as previously described is that these methods do not provide information on the type of activity. In order to assist a clinician in diagnosis, choosing the most suitable

Table 1. Summary of a collection of relevant past and ongoing work on automatic recognition of activity using inertial sensors

Reference	Activity	Sensor number	Sensor type and placement	Subjects	Accuracy [%]
Aminian et al., 1999b	Dynamic/sitting/standing/lying	2	1D-acc: thigh, sternum	5	89.3
Bao and Intille, 2004	Activities of daily living	10	2D-acc: shank, thigh, hip upper leg, wrist	20	41-97
Bussmann et al., 1995	Sitting/standing/lying; Walking level/upstairs/downstairs, cycling	6	2D-acc: shank, thigh; 1D-acc: sternum, shoulder	5	N/A
Bussmann et al., 2004	Inactive; Active: walking upstairs/downstairs, cycling, other movements; Locomotion: walking level	3	2D-acc: shank; 1D-acc: shank	12	98.5
Foerster et al., 1999	Sitting/standing/lying; Walking level/upstairs/downstairs, cycling	4	1D-acc: shank, thigh sternum, wrist	24	95.8
Kiani et al., 1997	Sitting/standing/lying; Walking level (at different speeds)	4	1D-acc: thigh(s); 2D-acc: sternum	11	98
Lee and Mase, 2002	Sitting/standing; Walking level (at different speeds), walking upstairs/downstairs	4	(2D-acc, 1D-gyro): hip compass: waist	8	91.8
Mäntyjarvi et al., 2001	Walking level/upstairs/downstairs	6	3D-acc: hip(s)	6	83-90
Mathie et al., 2004	Sitting/standing/lying; Walking level	3	3D-acc: lower back	26	97
Paraschiv-Ionescu et al., 2004	Sitting/standing/lying; Walking level (with estimation of gait parameters)	5	(2D-acc, 1D-gyro): sternum 1D-gyro: thigh, shank	21	98
Najafi et al., 2003	Sitting/standing/lying; Walking level	3	(2D-acc, 1D-gyro): sternum	15	> 90
Randell and Muller, 2000	Sitting/standing; Walking level/upstairs/downstairs, running	2	2D-acc: hip	10	85-90
Uiterwal et al., 1998	Sitting/standing/lying; Walking	3	3D-acc: waist	1	86-93
Van Laerhoven and Cakmacki, 2000	Sitting/standing; Walking level/upstairs/downstairs, running, cycling	4	2D-acc: knee(s)	1	45-96
Veltink et al., 1996	Sitting/standing; Walking level (at different speeds); Walking upstairs/downstairs, cycling (at different speeds)	3	2D-acc: sternum; 1D-acc: thigh	10	83

The column "Sensor number" specifies the overall number of input data channels.; the column "Sensor type and placement" specifies how many single-axis, dual-axis, tri-axis accelerometers, gyros, magnetic sensors are used and where they are positioned.

therapy for a patient, monitoring patient progress and assessing the effects of treatment, the "quantity" of motion is certainly an important issue, but aspects of mobility and locomotion related to the specific activity where energy is spent ought to be elucidated as well. Ambulatory monitors have been designed with the goal to identify and classify sets of postures and activities (Table 1).

Most systems have used only accelerometers (Bussmann, Veltink, Koelma, van Lummel, & Stam, 1995; Foerster & Fahrenberg, 2000; Mathie, Celler, Lovell, & Coster, 2003), while other systems have used accelerometers together with another type of sensor to improve discrimination. Najafi et al. (2003) and Parischev-Ionescu et al. (2004) are interesting works where inertial sensors of different types are integrated into ambulatory monitors where activity recognition and some (limited) form of gait analysis are jointly performed.

There are a number of difficulties in developing ambulatory monitors (Kiani, Snijders, & Gelsema, 1997; Foerster & Fahrenberg, 2000). The problems concern the algorithms that have to be designed to provide a reliable detection and discrimination of motions patterns and posture and the optimal sensor configuration set (type, number, placement). Algorithms for the detection of postures and motion patterns are a crucial aspect of this approach. The research is still intense on methods to achieve adequate data reduction and discrimination power, in the face of the generally ample variety of strategies existing in performing a particular daily-life motor activity and the large inter- and intra-subject variability. Optimality means that the highest discrimination ability would be achieved for a spectrum of motor activities as broad as possible, while keeping sensor number at minimum so as to increase subject's compliance and usability, and to reduce overall system complexity and cost.

An example of a decision tree for discriminating a set of postures and mobility-related activities by an accelerometry-based system is sketched in Figure 5.

Each node of the tree has multiple branches leading to all of the movements of interest at the next level of the hierarchy. At each node, all the possible classifications are considered, including the fallback case, namely the case that is accepted after that all other classifications have been discarded, and the most likely candidate is retained. Decision trees of this kind are presented in several works (Bussmann et al.,1995; Veltink et al.,1996; Kiani et al.,1997; Mathie, Celler, Lovell, & Coster, 2004), based on the particular mode of operation of the ambulatory system (clinical assessment or event monitoring), the specific hardware configuration, and the strategy upon which the development of the pattern recognition system is based.

Each node of the tree must have a classification algorithm associated to it. Only two approaches have been used to a greater extent in the design of these classifiers. In fixed-threshold classification, motion patterns are detected and discriminated by applying a threshold to the signal of interest. The threshold setting is critical, since intra- and inter-individual variations could easily lead to wrong classification.

In reference-pattern-based classification, the detection of motion patterns could be improved by developing a template pattern for each postural and activity condition and cross-correlating the signal with this (Veltink et al.,1996; Foerster & Fahrenberg, 2000). The precise placement of sensors is essential when thresholds are used for the classi-fication of motion.

Figure 5. Decision tree showing relationships between different movements of interest in a generic ambulatory monitor

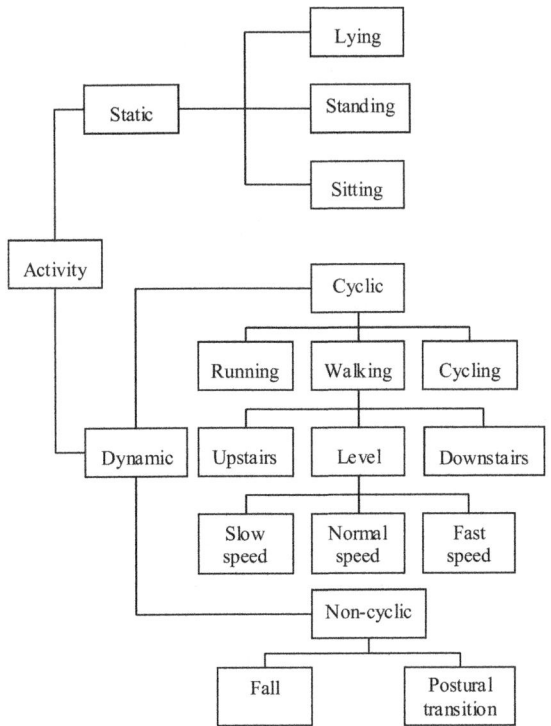

With a few exceptions, the behaviors most ambulatory monitors attempt to discriminate are standing, sitting, lying, postural transitions and walking. Standing, sitting, lying are instances of static activities, subtypes of which may be: "standing in stooped position, in upright position, lateral bending to the left, lateral bending to the right" (Kiani et al. 1997), "sitting upright or flexed," "standing upright or flexed" (Paraschiv-Ionescu et al., 2004), "standing lying on the left, on the right, supine or with back supported" (Foerster & Fahrenberg, 2000). Motion patterns may be non-cyclical (postural transitions) or cyclical (walking). Postural transitions such as "sit-to-stand, stand-to-sit, upright-to-lying, lying-to-upright" are included in the decision tree, among others, by Mathie et al. (2004) and Paraschiv-Ionescu et al. (2004). Walking is the most common example of cyclic dynamic activity, subtypes of which are "level walking at slow, normal, fast speed" (Kiani et al., 1997) and "walking upstairs, downstairs" (Bussmann et al. 1995; Foerster & Fahrenberg, 2000). Other dynamic activities may include cycling (Bussmann et al., 1995).

A classification that is based on individual reference patterns appears to be less susceptible than threshold-based classification, although careful sensor placement is critical for achieving good test-retest reliability.

Posture detection (sitting, standing, lying and their subtypes) (see Figure 5) requires that the orientation of a number of body segments (trunk, thigh, and so forth) relative to the gravitational field is computed, which involves necessarily the creation of a suitable sensor configuration arrangement. Crucial to the success of posture detection is then the ability to discriminate between gravitational and motion-related components of the raw acceleration signal. With some abuse of terminology, these components are commonly referred to as DC and AC components, respectively. Underlying this definition is the approach of low-pass filtering the raw signals for their separation, which is usually achieved by threshold-based tests for significant deviations of acceleration norm from gravity (Veltink et al.,1996).

Postural transitions are usually detected by threshold-based methods, and classified according to the nature of the static activities which precede and follow them. The process is claimed to be more robust when motion patterns are analyzed in specific frequency bands via wavelet-based multi-resolution analysis, which is also a better approach to deal with the discrimination between AC and DC components (Paraschiv-Ionescu, 2004).

Dynamic activity classification is based upon the observation that acceleration signals per cycle may differ in several aspects: morphology, statistical moments, and so forth, which calls for a suitable strategy upon which the development of the pattern recognition systems would be based.

The statistical approach to pattern recognition represents a pattern by features, viewed as points in suitably high-dimensional spaces (Figure 6). Machine-learning techniques are needed to turn raw data into useful contextual information, that is, information useful to provide knowledge about the state of the subject. To maximize the pattern recognition ability, these techniques would be able to generate and select data representations (features) with high discrimination ability. To perform classification, training patterns, captured from a large set of sampled movements and a range of conditions, can be used to determine the decision boundaries. There appears to be general consensus that a number of challenges exist in this regard. Different activities do not undergo abrupt changes, which makes their detection difficult. The variability in performing these activities by humans is generally high. The amount of information we can get from simple sensors is limited and noisy. Increasing the number of sensors, not necessarily inertial sensors, to achieve higher redundancy and robustness is not without practical (wearability, computational bottlenecks) or theoretical limits (curse of dimensionality, incurred by any machine learning approach we may favor). Numerous are the contact points between automatic recognition of activity as stated above and the research on methods for augmenting a computer with sensors which can make it aware of the context, as carried out in robotics and artificial intelligence.

Albeit identification of context without requiring external supervision seems to be better suited to become aware of context (Krause, Siewiorek, Smailagic, & Farringdon, 2003), most current approaches to the problem of automatic recognition of activity are based on using supervised machine learning techniques. The choice of good features is a fundamental step in statistical pattern recognition and a highly problem dependent task. A systematic approach to the search for features usable for analyzing daily-life

Figure 6. The general structure of a pattern recognition system

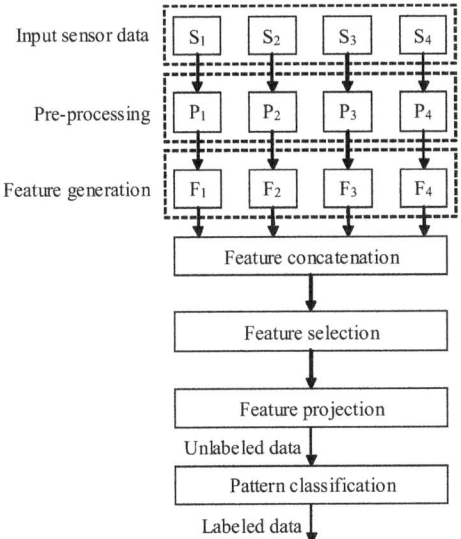

motor activities is illustrated in Kiani et al. (1997), where simple statistics such as cumulative sums, means, standard deviations, RMS values computed over sliding windows with selectable size and overlap are shown to be generally quite effective. Other possible choices include frequency-domain features (energy distribution, entropy, and so forth), and features that specifically measure correlations among signals from different sites and measurement directions — the latter ones are especially useful in the attempt to improve recognition of activities involving movements of multiple body parts (Bao & Intille, 2004).

The design of the clustering layer reflects different choices and approaches to the problem of feature selection and projection. Feature selection, either driven by an expert or by step-wise methods of selection, is an important step to deal properly with the issue of curse of dimensionality. Sometimes, the generation of additional features is costly (more sensors are needed), and their actual use may even be harmful to achieve optimality (Loosli, Canu, & Rakotomamonjy, 2003; Bao & Intille, 2004).

The ANN approach is perceived as a good tool to cluster noisy data from several sensors (Kiani et al., 1997; Van Laerhoven, Aidoo, & Lowette, 2001; Mäntyjärvi, Himberg & Seppänen, 2001). ANNs are used in (Mäntyjärvi et al., 2001). After raw accelerometer data are submitted to wavelet analysis for computation of the statistical power at selected scales of analysis, the principal component analysis (PCA) and the independent component analysis (ICA) are used for dimensionality reduction. Clustering is based on using a multilayer perceptron with back-propagation learning. The rationale for using

PCA and ICA is to find directions and scales where the sensor signals would be more informative for discrimination purposes. Van Laerhoven et al. (2001) discuss in depth advantages and disadvantages of using ANNs for clustering, in particular Kohonen self-organizing maps (KSOM). Their strong point is that activity contributes highly to context awareness, and adaptivity is indispensable in context awareness. Indisputably, the main disadvantage of many machine-learning algorithms is in their limited capability to properly handle the stability-plasticity dilemma, in the face of highly dynamic contexts (Van Laerhoven & Cakmakci, 2000). To overcome the limitations of KSOM in keeping up with the learning process over time, these authors consider the addition of a k-nearest neighbors clustering cascaded to the KSOM as a feasible way to stabilize its behavior.

Activity recognition on the selected features can be performed using any of different algorithms, including rule-based activity recognition, standard k-nearest neighbors search, naïve Bayes classifiers, and support vector machines (SVMs). Rule-based activity recognition seems to be good in capturing correlations between feature values that may lead to good recognition accuracy (Bao & Intille, 2004). Standard k-nearest neighbors search is shown to be a good work-horse in general problems of activity recognition (Van Laerhoven & Cakmakci, 2000; Bao & Intille, 2004). The problem with a naïve Bayes approach may be the excessive reliance on assumptions of conditional independence between different features in the Bayesian update (Golding & Lesh, 1999), and its greediness of training data to accurately model feature value distributions (Bao & Intille, 2004). In an attempt to overcome some limitations of neural network-based modeling, SVMs are increasingly recognized for their good performance on difficult classification tasks. Despite their emergence as a valuable tool for several applications in the biomedical field, including gait analysis (Begg & Kamruzzaman, 2005), only Loosli et al. (2003), to the best of our knowledge, have used them in the attempt to perform automatic recognition of activity. These authors claim that their implementation of SVMs is largely superior to the algorithms described in Van Laerhoven and Cakmakci (2000) and Van Laerhoven et al. (2001) over exactly the same datasets and selected features, although this claim has to be tempered by other findings of theirs, which show how SVMs are substantially equivalent to k-nearest neighbors clustering directly applied to the statistical features elaborated in Van Laerhoven and Cakmakci (2000).

An important step to improve upon the design of a pattern recognition system for analysis of daily-life motor activities is the addition of a supervision layer to introduce memory, so that individual movements are not classified in isolation (Mathie et al.,2004). Beside the classical approach to have a rule-based system to check sequences of classified movements and to correct those sequences which would be impossible, other approaches can be based on methods such as Markov modeling (Van Laerhoven & Cakmakci, 2000). The supervision layer on top of the classifier can supervise transitions from one context to another. A probabilistic finite state machine architecture is used where each context is represented by a state, and transitions are represented by arcs between states. A probability measure is built for each change of contexts, so every time a transition occurs, the supervision model checks if this is really likely (update of the probability). Observed state transitions that have a low probability to occur can thus be suspected to come from erroneous detections performed by the classification algorithms associated to the corresponding nodes of the decision tree.

Regarding a complex behavior as the juxtaposition of a sequence of elementary movements leads quite naturally to consider the hidden Markov model (HMM) analytical framework as a promising avenue of research in the field of automatic recognition of activity. An HMM is a doubly stochastic process with an underlying process of transitions between hidden states of the system and a process of emitting observable outputs, and, as such, it is ideal to implicitly model all of the many sources of spatio-temporal variability inherent in real movements. So far, HMMs have been used for representation and recognition of speech, handwriting and, of particular importance in the present context, gestures. Mäntyla, Mäntyjärvi, Seppänen, and Tuulari (2001) attempt to use hand gesture recognition, using cybergloves or accelerometer-based systems studied as a method of computer input for people with severe speech and motor impairment. Kallio, Kela and Mäntyjärvi (2003) develop a small wireless gesture-based input device based on accelerometers to provide methods to interact with different kinds of devices and environments. In both cases, trainable pattern recognition methods, namely HMMs, clustering and vector quantization are used to develop models for describing prototype gestures and to examine the gesture recognition performance. If we refer to gesture as a specific, intentional action by a human in which part of the body is moved in a predefined way indicating a specific event (Chambers, Venkatesh, West, & Bui, 2002), the way is paved to consider HMM-based gesture recognition a possible means to provide context knowledge, useful, in particular within occupational settings, not only to improve on standard clinical decision-making procedures (Uiterwal, Glerum, Busser, & van Lummel, 1998), but also to reduce the cognitive load on a worker during complex real world tasks (Lukowicz et al.,2004).

A final point is important to discuss. At first sight, it seems not difficult to achieve high values of accuracy, sensitivity and specificity in recognizing a relatively small set of postures and activities, although, sometimes, specific activities are resistant to robust classification and are frequently confused with one another, for instance level walking and upstairs walking (Veltink et al.,1996; Foerster & Fahrenberg, 2000). However, it should be pointed out that the highly specific system and methodologies used by each group make it difficult to directly compare different approaches. In most cases, these approaches are intended to meet the requirements of specific applications, and are tested in restricted environments and with small-size pools of subjects. Ironically, ambulatory systems intended to overcome the limitations of traditional laboratory-based systems for assessing motor performance are victims of their own validation being pursued, in the vast majority of cases, within laboratory environments which may artificially constrict and influence subject activity (Foerster, Smeja, & Fahrenberg, 1999). A thorough discussion about the need to use naturalistic data in training and testing ambulatory monitors is in Bao and Intille (2004). Needless to say, this need to cope with the idiosyncrasies of real-life worlds is precisely one of the driving factors behind the current interest in advanced methods for context-awareness and ambient intelligence among the artificial intelligence and robotic communities.

Functional Assessment of Specific Motor Disorders

Traditionally, accelerometers are applied to the quantitative analysis of tremor time series. The quantification of amplitude, frequency and occurrence time of tremor in patients affected with Parkinson's disease (PD) and its relation to posture and motion is described in Van Someren et al. (1998) and Foerster and Smeja (1999). The theory of spectral estimation is applied in Timmer, Lauk and Deuchl (1996) to assess whether a spectrum of accelerometric recordings exhibits multiple significant peaks and discuss different approaches to determine the amplitude and frequency of the tremor components from the spectrum. Another approach considers detection algorithms looking at tremor-related features in the time-domain, as in Van Someren et al. (1997), where the discrimination of tremor from intentional movements is based on the persistency of some features, such the temporal distances occurring between consecutive zero-crossings, the values of these periods of half waves, and the values of the peak-to-peak amplitudes in each half wave. Because of the time-varying nature of tremor-related components in accelerometric recordings, and the need for distinguishing the components that are due to voluntary movements and the components that are due to the movement disorder of interest, adaptive filtering techniques have also been considered, such as the Fourier Linear Combiner (Riviere, Rader, & Thakor, 1998) and the cascade learning architecture (Riviere & Khosla, 1997).

The assessment of postural stability is studied in Mayagoitia, Lötters, Veltink and Hermens (2002), where a tri-axis accelerometer-based system is developed for determining the long-term ability to maintain balance while standing. The acceleration measurement at the BCOM level allows one to introduce a number of performance parameters, similar to those that are extracted from force plates in classical posturographic studies. Two other approaches are interesting for their connection, in terms of the processing tools adopted, to the studies on assessment of physical activity (Cho & Kamen, 1998; Moe-Nilssen, 1998). More complex IMUs are those reported in Lee, Laprade and Fung (2001) and Wall and Weinberg (2003). Wall and Weinberg (2003) propose a balance prosthesis for postural control. In order to prevent falls in the balance impaired, body-tilt information is displayed to the subject via an array of tactile vibrators. The body-tilt information is obtained by integrating the information from a waist dual-axis accelerometer with the information from a single-axis gyro. Lee et al. (2001) deal with the problem of monitoring the lumbar spine motion, and to this aim they develop a portable system for three-dimensional motion analysis. A recent contribution towards the use of inertial sensing for the monitoring of specific motor disorders during postural transitions is illustrated in the work by Najafi, Aminian, Loew, Blanc and Robert (2002). They analyze the trunk tilt, corresponding to the angle between the vertical axis and anterior wall of the subject's thorax, and the trunk vertical displacement during stand-sit transfers, in the attempt to introduce good indicators of the risk of falls sustained by elderly people.

As for the assessment of gait, Smidt et al. (1971) apply Fourier series analysis to define a measure of smoothness of walking, which is found to provide an effective method for discriminating between normal gait patterns and gait patterns of subjects with gait defects. The approach by Sekine, Tamura, Togawa and Fukui (2000) and Sekine et al. (2002) is interesting for the connection established between wavelet properties and characteristics of motor fluctuations in normal and pathological conditions. In Sekine et al. (2000), waist 3D-accelerations are submitted to wavelet analysis in order to discriminate different types of walking in healthy subjects. Also in Sekine et al. (2002), the approach is further refined so that the distribution of wavelet coefficient power over selected scales of analysis is subject to a complexity analysis for the determination of a fractal dimension (the Hurst exponent) in different groups — healthy, elderly, subjects with PD. The Hurst exponent is shown to be significantly different among three types of walking (level/upstairs/downstairs) for individual subjects and show a great reproducibility. Moreover, the fractal dimensions are higher for elderly and PD patients as compared with healthy subjects.

The study of motor fluctuations in PD patients is reviewed in Keijsers, Hornstink and Gielen (2003). Their goal, in particular, is to assess levodopa-induced dyskinesia in these subjects. Dyskinesia are involuntary movements — jerky and characterized by sudden contractions followed by stretching, twisting and rotation — that usually occur after several years of using levodopa, a drug which is widely used to treat the symptoms of PD. The ambulatory monitor developed by Keijsers et al. (2003) is composed of several tri-axis accelerometes (at both upper arms, at both upper legs, at the wrist of the most dyskinetic side, and at the top of the sternum). Due to the complex, time-varying relations between voluntary movements and motor disorders, a simple supervised ANN is the selected classification technique exercised over a number of time- and frequency-domain features. The features with the highest discriminative ability are survivors of a pruning phase, which is part of the design procedure. At the end, the system is successfully trained to detect and assess the severity of dyskinesia, provided that the score of the severity of the symptoms is given by an experienced clinician.

Personal Navigation

The ability to locate the position/orientation of a person is of great importance in a number of applications, including electronic travel aids for the blind or visually impaired, integrated navigation systems for the dismounted infantry soldier, and, in general, context-aware applications in wearable and mobile computing.

The location-sensing techniques for personal navigation systems are based on either relative or absolute position measurements. Traditionally, inertial sensors have been employed for implementing dead-reckoning techniques, a variant of relative position measurements whose fundamental idea is to integrate incremental motion information over time. The problem of absolute positioning in outdoor environments is seemingly straightforward to solve by an externally referenced sensing technology such as GPS. However, the disadvantages with GPS — inability to work in indoors, unavailability of satellite signals in environments such as urban canyons, poor accuracy in relation to the needs of a specific application, inability to provide static heading information — suggest that dead-reckoning based on inertial sensing and GPS would have a better chance to work in practice (Ladetto, van Seeters, Sokolowski, Sagan, & Merminod, 2002). In the case of indoor environments, one approach for overcoming dead-reckoning limitations is to acquire location information by adding intelligence to the environment, so that it can supply location information to users via special infrastructures (Want, Hopper, Falcão, & Gibbons, 1992; Krumm, Williams, & Smith, 2002) — applications of externally referenced sensing techniques. The traditional computational techniques used to improve the performance of dead-reckoning systems aim at implementing either map matching or in-line sensor calibration procedures based on KFs which use external aids, such as GPS, in the attempt to estimate bias drifts of inertial sensors and compass disturbances.

The key problem in the design of a personal navigation system is to find a method to measure length and direction of displacement using step time as the basic unit of time, so as to determine the distance and heading from a known origin at an acceptable level of accuracy. Detecting step occurrences can be based on accelerometers or gyros, as stated above with regard to the problem of assessing temporal parameters of gait. A simple model-based approach to the problem of estimating step length hypothesizes that, once a method is available to determine step time, step length estimation can be based on cadence. Judd (1997) suggests that the step length could be estimated online based on a linear relationship between measured cadence and step length, whose validity, discussed in Ladetto et al. (2002), is however limited to level walking in open spaces. The approach by Ladetto et al. (2002) is an interesting example of so-called biokinematic navigation (Elwell, 1999), an inexpensive technique that matches an individual's gait to inertial measurements. Ladetto et al. (2002) exploit the relationship existing between walking speed and statistical features, such as the RMS values of waist 3D-accelerations, under the premise that the main goal of locomotion is to promote the BCOM displacement in space. The fact that step length is a time-varying process, with large environment-dependent variations, motivates the use of additional sensors for online model calibration, for instance GPS as in Perrin et al. (2000), Ladetto et al. (2002) and Jirawimut, Ptasinski, Garaj, Cecelia and Balachandran (2003) or to incorporate additional features of human walking dynamics into the algorithm (Vildjiounaite, Malm, Kaartinen, & Alahuhta, 2002; Lee & Mase, 2002). These authors note that, when the step length is longer, the accelerations in different parts of the legs tend to be higher. Hence, it may be hypothesized that the greater the vertical acceleration during a step, the greater the distance that has been traveled across the ground. This relationship can be exploited to adjust the pre-calculated step length and hence reduce the error associated with fixed values by using sensor look-up tables (Vildjiounaite et al.,2002) or a simple fuzzy-logic reasoning method (Lee & Mase, 2002).

Determining heading may require the implementation of gyro-compassing techniques. After INS alignment is performed, heading is estimated by gyro and magnetic compass data, provided both data are available. In contrast with a gyro, magnetic compass is not prone to drift, however its use is critical. Sometimes, serious inaccuracies in the magnetic compass readings are found which cannot be traced back to imperfect tilt compensation, but are likely due to perturbations in the magnetic field near or within the measurement space. The presence of magnetic disturbance can be checked by comparing the rate of change of the heading estimates from gyro and magnetic compass (Ladetto & Merminod, 2000). The conflicts in data interpretation existing about the existence and extent of turns indicated by gyro and magnetic compass warn not to consider magnetic compass reliable. Otherwise, its long-term low-frequency response can be combined with the high-frequency response of gyros. The use of GPS to dynamically recalibrate the heading error is appropriate in this context. In theory this combination should give the most accurate results. In particular, computed GPS heading can allow one to model the bias of the magnetic compass (Jirawimut et al.,2003) and the gyro bias and drift (Ladetto et al.,2002). In indoor applications, where GPS is unavailable, the performance of biokinematic navigation algorithms would be improved by map matching, as pointed out in Lee and Mase (2002) and Vildjiounaite et al. (2002). These authors show that the task of determining the person's location when the navigation space is constrained, such as within a building, can be greatly simplified by having some form of environment knowledge available to the navigation algorithm. This knowledge can be limited to indicate where some locations, or transition between locations, are in the map, such as the transition from one room to the next, the transition from level to upstairs, the presence of a corridor. When this knowledge is combined with an even relatively crude heading information, irregular walking styles or strong perturbations in the magnetic field can be accommodated and reasonably accurate estimates of person's location can still be obtained.

In our view, the important point is that there is something peculiar to the way humans move which makes personal navigation so different from vehicular navigation. Walking styles of humans imply frequent changes of speed and orientation, for reasons of stability, safety, and so forth, depending on the environment the activity is actually performed — the context. Because of this complexity, the prediction of the walked path from one point to another turns out to be de facto impossible. Moreover, inertial/magnetic sensors record movements depending on their actual placement on the body. Four examples point at the difficulties inherent with their use: changes of orientation that occur with minimal displacement changes (on-the-spot turning); changes of orientation not recorded, in spite that they occur over large angles (backward displacements, side-stepping); instantaneous changes of orientation that are recorded, in spite that the actual direction of displacement does not change (lateral bending of the trunk while walking straight ahead); steps which are not taken with normal rhythm or style, hence requiring special algorithms for their detection (side-stepping). Of course, these examples are not comprehensive of all situations taking place in practice. However, they point at the need that personal navigation systems would be capable of classifying "movements," and, for each classified movement, of reliably estimating length and direction of displacement in real-time conditions.

EMERGING TRENDS

In this chapter, the main computational approaches to use motion sensing based on inertial and magnetic sensors have been reviewed, in connection with a wealth of biomedical applications, including assessment of motor performance, automatic recognition of activity, and personal navigation.

Of utmost importance is the point that the answers to the problems raised by each application ought not to be considered in isolation, because of the relevant overlapping they have with each other:

1. Advanced signal processing and machine-learning techniques are used in the effort to identify signatures of human motion and perform motor pattern recognition;
2. Constraints in navigation space and dynamic features of unrestrained human motion concur to provide accurate long-term solutions to the navigation problem;
3. Topological information merged with an estimate of the traveled distance by dead-reckoning methods leads to improve the accuracy of methods for automatic recognition of activity.

The most promising avenue of research is in how artificial intelligence and biomechanics can reinforce each other, in the direction to pave a new way of inertial sensing that will not be hampered by the idiosyncrasies of current-generation sensing hardware. Impressive technological advances are about to turn wireless, wearable distributed multi-sensor systems into reality and further advances are also expected in the field of MEMS technologies. However, the capability of performing a robust, accurate real-time reconstruction of the trajectory of selected anatomical points, such as, for instance, the sacrum, the stern, the foot instep, by inertial and magnetic sensors is still the prerequisite for making body personal area networks attractive for long-term ambulatory monitoring of human subjects engaged in functional activities involving motion in unrestrained conditions. Interesting research is yet to be performed in regard to the signal processing algorithms the practitioners in the field will have to create in order to successfully cope with the difficult behavior of the motion sensors discussed in this chapter.

REFERENCES

Abbott, E., & Powell, D. (1999). Land-vehicle navigation using GPS. In *Proceedings of the IEEE, 87*(1), 145-162.

Aminian, K., Rezakhanlou, De Andres, Fritsch, Leyvraz, P.-F., & Robert, Ph. (1999). Temporal feature estimation during walking using miniature accelerometers: An analysis of gait improvement after hip arthroplasty. *Medical & Biological Engineering & Computing, 37*(6), 686-691.

Aminian, K., Robert, Ph., Jéquier, E., & Schutz, Y. (1995). Incline, speed, and distance assessment during unconstrained walking. *Medicine & Science in Sports & Exercise, 27*(1), 226-234.

Aminian, K., Robert Ph., Buchser, E. E., Rutschmann, B., Hayoz, D., & Depairon, M. (1999). Physical activity monitoring based on accelerometry: validation and comparison with video observation. *Medical & Biological Engineering & Computing, 37*(3), 304-308.

Aminian, K., Najafi, B., Büla, C., Leyvraz, P.-F., & Robert, Ph. (2002). Spatio-temporal parameters of gait measured by an ambulatory system using miniature gyroscopes. *Journal of Biomechanics, 35*(5), 689-699.

Auvinet, B., Chaleil D., & Barrey, E. (1999). Accelerometric gait analysis for use in hospital outpatients. *Rev. Rhum. (Engl. Ed.), 66*(7-9), 389-397.

Bachmann E. R., Yun, X., McKinney, D., McGhee, R. B., & Zyda, M. J. (2003, September 14-19). Design and implementation of MARG sensors for 3-DOF orientation measurement of rigid bodies. In *Proceedings of the IEEE International Conference on Robotics and Automation*, Taipei, Taiwan, (pp. 1171 - 1178).

Bao, L., & Intille, S. S. (2004). Activity recognition from user-annotated acceleration data. In A. Ferscha & F. Mattern (Eds.), *Proceedings of Pervasive 2004, LNCS* (Vol. 3001, pp. 1-17). Berlin Heidelberg: Springer-Verlag.

Begg, R., & Kamruzzaman, J. (2005). A machine learning approach for automated recognition of movement patterns using basic, kinetic and kinematic gait data. *Journal of Biomechanics, 38*(3), 401-408.

Bortz, J. E. (1971). A new mathematical formulation for strapdwon inertial navigation. *IEEE Trans. on Aerospace and Electronic Systems, 7*(1), 61-66.

Bouten, C. V. C., Koekkoek, K. T. M., Verduin, M., Kodde, R., & Janssen, J. D. (1997). A triaxial accelerometer and portable data processing unit for the assessment of daily physical activity. *IEEE Trans. on Biomedical Engineering, 44*(3), 136-147.

Bouten, C. V. C., Sauren, A. A. H. J., Verduin, M., & Janssen, J. D. (1997). Effects of placement and orientation of body-fixed accelerometers on the assessment of energy expenditure during walking. *Medical & Biological Engineering & Computing, 35*(1), 50-56.

Bussmann, J. B., Hartgerink, I., van der Woude, L. H., & Stam, H. J. (2000). Measuring physical strain during ambulation with accelerometry. *Medicine & Science in Sports & Exercise, 32*(2), 1462-1471.

Bussmann, J. B., Veltink, P. H., Koelma, F., van Lummel, R. C., & Stam, H. J. (1995). Ambulatory monitoring of mobility-related activities: The initial phase of the development of an activity monitor. *European Journal Physical Medicine Rehabilitation, 5*(1), 2-7.

Caruso, M. J. (1997). Applications of magnetoresistive sensors in navigation systems. *Sensor and Actuators, 1220*, 15-21.

Chambers, G. S., Venkatesh, S., West, G. A. W., & Bui, H. H. (2002). Segmentation of intentional human gestures for sports video annotation. In *Proceedings of the International Conference Pattern Recognition* (pp. 1082-1085).

Chen, K. Y., & Sun, M. (1997). Improving energy expenditure estimation by using a triaxial accelerometer. *Journal of Applied Physiology, 83*(6), 2112-2122.

Cho, C. Y., & Kamen, G. (1998). Detecting balance deficits in frequent fallers using clinical and quantitative evaluation tools. *Journal of the Americal Geriatrics Society, 46*(4), 426-430.

Curey, R. K., Ash, M. E., Thielman, L. O., & Barker, C. H. (2004, April 26-29). Proposed IEEE inertial systems terminology standard and other inertial sensor standards. *IEEE Position Location and Navigation Symposium, PLANS 2004*, Monterey, CA, (pp. 83-90).

Elwell, J. (1999). Inertial navigation for the urban warrior. In *Proceedings of the SPIE International Society of Optical Engineering* (Vol. 3709, pp. 196-204).

Evans, A. L., Duncan, G., & Gilchrist, W. (1991). Recording accelerations in body movements. *Medical & Biological Engineering & Computing, 29*(1), 102-104.

Foerster, F., & Smeja, M. (1999). Joint amplitude and frequency analysis of tremor activity. *Electromyography & Clinical Neurophysiology, 39*(1), 11-19.

Foerster, F., Smeja, M., & Fahrenberg, J. (1999). Detection of posture and motion by accelerometry: A validation in ambulatory monitoring. *Computers in Human Behavior, 15*, 571-583.

Foerster, F., & Fahrenberg, J. (2000). Motion patterns and posture: correctly assessed by calibrated accelerometers. *Behavior Research Methods, Instruments, & Computers, 32*(3), 450-457.

Foxlin, E. (2002). Motion tracking requirements and technologies. In K. Stanney (Ed.), *Handbook of virtual environments: Design, implementation, and applications.* Mahwah, NJ: Lawrence Erlbaum.

Gage, H. (1964). *Accelerographic analysis of human gait* (Paper No. 64-WA/HUF 8). Washington, DC: American Society for Mechanical Engineers.

Gebre-Egziabher, D., Elkaim, G. H., Powell, J. D., & Parkinson, B. W. (2000, March 13-16). A gyro-free quaternion-based attitude determination system suitable for implementation using low cost sensors. In *Proceedings of the IEEE Position, Location and Navigation Symposium PLANS 2000*, San Diego, CA, (pp. 185-192).

Giansanti, D., Macellari, V., Maccioni, G., & Cappozzo, A. (2003). Is it feasible to reconstruct body segment 3D position and orientation using accelerometric data? *IEEE Trans. on Biomedical Engineering, 50*(4), 476-483.

Golding, A., & Lesh, N. B. (1999, October 18-19). Indoor navigation using a diverse set of cheap, wearable sensors. In *Proceeding of the 3rd International Symposium on Wearable Computers*, San Francisico (pp. 29-36).

Ismail, A. R., & Asfour, S. S. (1999). Discrete wavelet transform: A tool in smoothing kinematic data. *Journal of Biomechanics, 32*(3), 360-368.

Kallio, S., Kela, J., & Mäntyjärvi, J. (2003, October 5-8). Online gesture recognition system for mobile interaction. In *Proceedings of the IEEE International Conference on Systems, Man and Cybernetics*, Washington, DC (Vol. 3, pp. 2070-2076).

Keijsers, N. L. W., Horstink, M. W. I. M., & Gielen, S. C. A. M. (2003). Online monitoring of dyskinesia in patients with Parkinson's disease. *IEEE Engineering in Medicine and Biology Magazine, 22*(3), 96-103.

Kemp, B., Janssen, A. J. M. W., & van der Kamp, B. (1998). Body position can be monitored in 3D using accelerometers and earth-magnetic field sensors. *Electroencephalography and Clinical Neurophysiology, 109*(6), 484-488.

Kiani, K., Snijders, C. J., & Gelsema, E. S. (1997). Computerized analysis of daily life motor activity for ambulatory monitoring. *Technology & Health Care, 5*(4), 307-318.

Krause, A., Siewiorek, D. P., Smailagic, A., & Farringdon, J. (2003, October 21-23). Unsupervised, dynamic identification of physiological and activity context in wearable computing. In *Proceedings of the 7th IEEE International Symposium Wearable Computers*, New York (pp. 88-97).

Krumm, J., Williams, L., & Smith, G. (2002). SmartMoveX on a graph - An inexpensive active badge tracker. In G. Borriello & L. E. Holmquist (Eds.), *UbiComp 2002, LNCS* (Vol. 2498, pp. 299-307). Berlin Heidelberg: Springer-Verlag.

Jirawimut, R., Ptasinski, P., Garaj, V., Cecelia, F., & Balachandran, W. (2003). A method for dead reckoning parameter correction in pedestrian navigation system. *IEEE Trans. on Instrumentation and Measurement, 52*(1), 209-215.

Judd, C. T. (1997, September). A personal dead reckoning module. In *Proceedings of the ION GPS '97*, Kansas City, MO, (pp. 47-51).

Ladetto, Q., & Merminod, B. (2002). In step with INS. *GPS World, 10*(10), 30-38.

Ladetto, Q., van Seeters, J., Sokolowski, S., Sagan, Z., & Merminod, B. (2002, October 14-16). Digital magnetic compass and gyroscope for dismounted soldier position and navigation. *NATO-RTO Meetings*, Istanbul, Turkey.

Lee, R. Y. W., Laprade, J., & Fung, E. H. K. (2003). A real-time gyroscopic system for three-dimensional measurement of lumbar spine motion. *Medical Engineering & Physics, 25*(10), 817-824.

Lee, S.-W., & Mase, K. (2002). Activity and location recognition using wearable sensors. *Pervasive Computing, 1*(3), 10-18.

Loosli, G., Canu, S., & Rakotomamonjy, A. (2003). *Détection des activités quotidiennes à l'aide des séparateurs à Vaste Marge*. RJCIA, France.

Lötters, J. C., Schipper, J., Veltink, P. H., Olthius, W., & Bergveld, P. (1998). Procedure for in-use calibration of triaxial accelerometers in medical applications. *Sensors and Actuators A, 68*(1-3), 221-228.

Luinge, H. J., & Veltink, P. H. (2004). Inclination measurement of human movement using a 3-D accelerometer with autocalibration. *IEEE Trans. on Neural Systems and Rehabilitation Engineering, 12*(1), 112-121.

Lukowicz, P., Ward, J. A., Junker, H., Stäger, M., Tröster, G., Atrash, A., & Starner, T. (2004). Recognizing workshop activity using body worn microphones and accelerometers. In A. Ferscha & F. Mattern (Eds.), *Pervasive 2004, LNCS* (Vol. 3001, pp. 18-32). Berlin Heidelberg: Springer-Verlag.

Maki, B. E. (1997). Gait changes in older adults: Predictors of falls or indicators of fear? *Journal of the American Geriatrics Society, 45*(3), 313-320.

Mayagoitia, R. E., Lötters, J. C., Veltink, P. H., & Hermens, H. (2002). Standing balance evaluation using a triaxial accelerometer. *Gait & Posture, 16*(1), 55-59.

Mayagoitia, R. E., Nene, A. V., & Veltink, P. H. (2002). Accelerometer and rate gyroscope measurement of kinematics: An inexpensive alternative to optical motion analysis systems. *Journal of Biomechanics, 35*(4), 537-542.

Mäntyla, V.-M., Mäntyjärvi, J., Seppänen, T., & Tuulari, E. (2000). Hand gesture recognition of a mobile device user. In *Proceedings of the International IEEE Conference on Multimedia and Expo*, New York (pp. 281-284).

Mäntyjärvi, J., Himberg, J., & Seppänen, T. (2001, October 7-10). Recognizing human motion with multiple acceleration sensors. 747-752. In *Proceedings of the IEEE International Conference Systems, Man and Cybernetics, 2*, Tucson, AZ, (pp. 747-752).

Mathie, M. J., Celler, B. G., Lovell, N. H., & Coster, A. C. F. (2003). Classification of basic daily movements using triaxial accelerometer. *Medical & Biological Engineering & Computing, 42*(5), 679-687.

Meijer, G. A. L., Westerkerp, K. R., Verhoeven, F. M. H., Koper, H. B. M., & ten Hoor, F. (1991). Methods to assess physical activity with special reference to motion sensors and accelerometers. *IEEE Trans. on Biomedical Engineering, 38*(3), 221-229.

Meyer, K., Applewhite, H. L., & Biocca, F. A. (1992). A survey of position trackers. *Presence: Teleoperators and virtual environments, 1*(1), 173-200.

Miyazaki, S. (1997). Long-term unrestrained measurement of stride length and walking velocity utilizing a piezoelectric gyroscope. *IEEE Trans. on Biomedical Engineering, 44*(8), 753-759.

Moe-Nilssen, R. (1998). Test-retest reliability of trunk accelerometry during standing and walking. *Archives Physical Medicine Rehabilitation, 79*(11), 1377-1385.

Morris, J. R. W. (1973). Accelerometry: A technique for the measurement of human body movements. *Journal of Biomechanics, 6*(6), 729-736.

Najafi, B., Aminian, K., Loew, F., Blanc, Y., & Robert, Ph. (2002). Measurement of stand-sit and sit-stand transitions using a miniature gyroscope and its application in fall risk evaluation in the elderly. *IEEE Trans. on Biomedical Engineering, 49*(8), 843-851.

Najafi, B., Aminian, K., Paraschiv-Ionescu, A., Loew, F., Büla, C. J., & Robert, Ph. (2003). Ambulatory system for human motion analysis using a kinematic sensor: monitoring of daily physical activity in the elderly. *IEEE Trans. on Biomedical Engineering, 50*(6), 711-723.

Padgaonkar, A. J., Krieger, K. W., & King, A. I. (1975). Measurement of angular acceleration of a rigid body using linear accelerometers. *ASME Journal of Applied Mechanics, 42*(3), 552-556.

Pappas, I. P. I., Popovic, M. R., Keller, T., Dietz, V., & Morari, M. (2001). A reliable gait phase detection system. *IEEE Trans. on Neural Systems and Rehabilitation Engineering, 9*(2), 113-125.

Paraschiv-Ionescu, A., Buchser, E. E., Rutschmann, B., Najafi, B., & Aminian, K. (2004). Ambulatory system for the quantitative and qualitativa analysis of gait and posture in chronic pain patients treated with spinal cord stimulation. *Gait & Posture, 20*(2), 113-125.

Perrin, O., Terrier, P., Ladetto, Q., Merminod, B., & Schutz, Y. (2000). Improvement of walking speed prediction by accelerometry and altimetry, validated by satellite positioning. *Medical & Biological Engineering & Computing, 38*(2), 164-168.

Randell, C., & Muller, H. (2000, October 31-November 3). Context awareness by analyzing accelerometer data. In B. MacIntyre & B. Iannucci (Eds.), *The Fourth International Symposium on Wearable Computers,* Arlington, VA (pp. 175-176).

Riviere, C. N., & Khosla, P. K. (1997, April 20-25). Augmenting the human-machine interface: improving manual accuracy. In *Proceedings of the IEEE International Conference on Robotics and Automation*, Albuquerque, NM (Vol. 4, pp. 2346-2350).

Riviere, C. N., Rader, R. S., & Thakor, N. V. (1998). Adaptive canceling of physiological tremor for improved precision in microsurgery. *IEEE Trans. Biomedical Engineering, 45*(7), 839-846.

Sabatini, A. M. (2005). Quaternion based strap-down integration method for applications of inertial sensing to gait analysis. *Medical & Biological Engineering & Computing, 43*(1), 94-101.

Sabatini, A. M., Martelloni, C., Scapellato, S., & Cavallo, F. (2004). Energy expenditure rate in level and uphill treadmill walking determined from empirical models and foot inertial sensing data. *Electronics Letters, 40*(2), 95-96.

Sabatini, A. M., Martelloni, C., Scapellato, S., & Cavallo, F. (2005). Assessment of walking features from foot inertial sensing. *IEEE Trans. on Biomedical Engineering, 52*(3), 486-494.

Sagawa, K., Sato, Y., & Innoka, H. (2000, October 8-11). Non-restricted measurement of walking distance. In *Proceedings of the IEEE International Conference Systems, Man, and Cybernetics*, Nashville, TN (Vol. 3, pp. 1847-1852).

Sekine, M., Tamura, T., Akay, M., Fujimoto, T., Togawa, T., & Fukui, Y. (2002). Discrimination of walking patterns using wavelet-based fractal analysis. *IEEE Trans. on Neural Systems and Rehabilitation Engineering, 10*(3), 188-196.

Sekine, M., Tamura, T., Togawa, T., & Fukui, Y. (2000). Classification of waist-acceleration signals in a continuous walking record. *Medical Engineering & Physics, 22*(4), 285-291.

Smeja, M., Foerster, F., Fuchs, G., Emmans, D., Hornig, A., & Fahrenberg, J. (1999). 24 hr assessment of tremor activity and posture in Parkinson's disease by multichannel accelerometry. *Journal of Psychophysiology, 13*, 245-256.

Smidt, G. L., Arora, J.S., & Johnston, R. C. (1971). Accelerographic analysis of several types of walking. *American Journal of Physical Medicine, 50*(6), 285-300.

Terrier, P., Aminian, K., & Schutz, Y. (2001). Can accelerometry accurately predict the energy cost of uphill/downhill walking? *Ergonomics, 44*(1), 48-62.

Terrier, P., Ladetto, Q., Merminod, B., & Schutz, Y. (2000). High-precision satellite positioning system as a new tool to study the biomechanics of human locomotion. *Journal of Biomechanics, 33*(12), 1717-1722.

Timmer, J., Lauk, M., & Deuschl, G. (1996) Quantitative analysis of tremor time series. *Electroencephalography and Clinical Neurophysiology, 101*(5), 461-468.

Tong, K., & Granat, M. H. (1999). A practical gait analysis system using gyroscopes. *Medical Engineering & Physics, 21*(2), 87-94.

Uiterwal, M., Glerum, E. B. C., Busser, H.J., & van Lummel, R. C. (1998). Ambulatory monitoring of physical activity in working situations, a validation study. *Journal of Medical Engineering and Technology, 22*, 168-172.

Van Laerhoven, K., Aidoo, K., & Lowette, S. (2001). Real-time analysis of data from many sensors with neural networks. In *Proceedings of the 5th International Symposium on Wearable Computers*, Zurich, Switzerland (pp. 115-123).

Van Laerhoven, K., & Cakmakci, O. (2000). What shall we teach our pants? In B. MacIntyre & B. Iannucci (Eds.), *The Fourth International Symposium on Wearable Computers* (pp. 77-83).

Van Someren, E. J., Vonk, B. F., Thussen, W. A., Speelman, J. D., Schuurman, P. R., Mirmiran, M., & Swaab, D. F. (1998). A new actigraph for long-term registration of the duration and intensity of tremor and movement. *IEEE Trans. on Biomedical Engineering, 45*(3), 386-395.

Veltink, P. H., Bussmann, H. B. J., de Vries, W., Martens, W. L. J., & van Lummel, R. C. (1996). Detection of static and dynamic activities using uniaxial accelerometers. *IEEE Trans. on Rehabilitation Engineering, 4*(4), 375-385.

Veltink, P. H., Slycke, P., Hemssems, J., Buschman, R., Bulstra, G., & Hermens, H. (2003). Three dimensional inertial sensing of foot movements for automatic tuning of a two-channel implantable drop-foot stimulator. *Medical Engineering & Physics, 25*(1), 21-28.

Verplaetse, C. (1996). Inertial proprioceptive devices: self-motion-sensing toys and tools. *IBM Systems Journal, 35*, 639-650.

Vildjiounaite E., Malm, E.-J., Kaartinen, J., & Alahuhta, P. (2002). Location estimation indoors by means of small computing power devices, accelerometers, magnetic sensors and map knowledge. In F. Mattern & M. Naghshineh (Eds.), *Pervasive 2002, LNCS* (Vol. 2414, p. 211-224). Berlin Heidelberg: Springer-Verlag.

Wall C., & Weinberg, M.S. (2003). Balance prostheses for postural control. *IEEE Engineering in Medicine and Biology Magazine, 2*, 84-90.

Want, R., Hopper, A., Falcao, V., & Gibbons, J. (1992). The active badge location system. *ACM Transactions on Information Systems, 10*, 91-102.

Weir, R. F. ff., & Childress, D. S. (1997). A portable, low-cost, real-time, clinical gait analysis system. *IEEE Trans. on Rehabilitation Engineering, 5*(4), 310-321.

Wertz, J. R. (Ed.). (1984). *Spacecraft attitude determination and control*. Dordrecht, The Netherlands: D. Reidel.

Willemsen, A. Th. M., Bloemhof, F., & Boom, H. B. K. (1990). Automatic stance-swing phase detection from accelerometer data for peroneal nerve stimulation. *IEEE Trans. on Biomedical Engineering, 37*(12), 1201-1208.

Willemsen, A. Th. M., Frigo C., & Boom, H. B. K. (1991). Lower extremity angle measurement with accelerometers - Error and sensitivity analysis. *IEEE Trans. on Biomedical Engineering, 38*(12), 1186-1193.

Willemsen, A. Th. M., van Alsté, J. A., & Boom, H. B. K. (1990). Real-time gait assessment utilizing a new way of accelerometry. *Journal of Biomechanics, 23*(8), 859-863.

Williamson, R., & Andrews, B. J. (2001). Detecting absolute human knee angle and angular velocity using accelerometers and rate gyroscopes. *Medical & Biological Engineering & Computing, 39*(3), 294-302.

You, S., Neumann, U., & Azuma, R. (1999). Orientation tracking for outdoor augmented reality registration. *IEEE Computer Graphics and Applications, 6*(6), 36-42.

Zijlstra, W., & Hof, A. L. (2003). Assessment of spatio-temporal parameters from trunk acceleration during human walking. *Gait & Posture, 18*(2), 1-10.

Chapter III

Monitoring Human Movement with Body-Fixed Sensors and its Clinical Applications

Kamiar Aminian, Ecole Polytechnique Fédérale de Lausanne (EPFL), Switzerland

ABSTRACT

In this chapter, first we outline the advantage of new technologies based on body-fixed sensors and particularly the possibility to perform field measurement, out of a laboratory and during the actual condition of the subject. The relevance of intelligent computing and its potential to enhance those features hidden in biomechanical signals are reviewed. An emphasis is made to show the results produced by these sensors when used alone and new possibilities offered when the information from different type of body fixed sensors are fused. In the second part, the relevance of body fixed sensors in medicine is presented by providing many clinical applications in orthopedics, Parkinson disease, physiology, pain management, and aging. Finally the chapter ends by emphasizing the potential of synergies between body fixed movement monitoring and other areas such as information technology which lead to the development of wearable body movement monitoring.

INTRODUCTION

Human movement analysis is an emerging field, which involves not only medical branches (orthopedics, physiology, neurology and sports), but even more technology and engineering sciences. Many investigators study and analyze human movement and particularly, gait, joint rotations and postural control in order to better understand motor function and describing motion ability alteration in term of health and disease. Standard technology for human motion capture is principally based on camera, magnetic and ultrasound systems which allow a complete 3D kinematics analysis of body segment but requires a dedicated laboratory. Ambulatory monitoring of body movement takes a different approach: collecting data from body-fixed sensors in the natural environment of the subject. It offers long-term monitoring, therefore providing complementary features, related both to the quantity (what activity, how often and for how long) and the quality (how) of the physical activities that are performed. Information provided by a set of complementary mobility-related parameters is essential for the objective assessment of a patient's functional ability.

Capturing human movement based on ambulatory technology is a relatively new field of research since it is directly related to the recent advances in miniature devices and sensors, new technologies for powerful microcontroller, high capacity memory and small power sources. Standard technologies provide body segment position relative to a fixed referential. Other kinematics such as velocity and acceleration are generally computed from the derivative of the positions. In contrast, the outputs of body fixed sensors are rather relative angles, segment acceleration or velocity. Finding 3D segment orientation, absolute angles and complete kinematics are a major difficulty when using body-fixed sensors. In this regard the use of intelligent computing such as bio-inspired algorithms is essential. Moreover, the current state of art in communication systems and wireless transmission has opened new possibilities for telemedicine and remote home care monitoring. Therefore, a new and promising field of research has been opened in gait and posture topics for in-field and outdoor measurement. This way many research questions that can not be elucidated in laboratory setting can now be studied based on these new technologies.

In this chapter, after a short review of standard technology, the advantage of body-fixed sensors and the relevance of intelligent computing for this technology are outlined. The major body-fixed sensors are presented and their uses for human motion capture are reported. An emphasis is made to show the results produced by these sensors when used alone and new possibilities offered when the information from different type of body fixed sensors are fused. Finally, the relevance of body-fixed sensors in medicine is presented by providing clinical applications in orthopedics, Parkinson disease, physiology, pain management, and aging. It is important to notice the potential of synergies between movement analysis based on ambulatory monitoring and other areas, such as nanotechnology, materials sciences, and information technology, which lead to the development of advanced mobile and ubiquitous body movement measurements such as wearable monitoring.

STANDARD ANALYSIS AND
THEIR LIMITATION

Movement analysis is mainly accomplished by laboratory techniques. Kinematics is measured using optical, electromagnetic and sonic technologies. Kinetics is estimated mainly by forceplate and electromyography.

Ultrasound emitter and receivers are used to record the body kinematic. The emitter sends out a burst of ultrasound, and the delay it takes for this burst in reaching the receivers is recorded (Kiss, Kocsis, & Knoll, 2004). Using this delay, the distances between the emitter and each receiver can be calculated from the sound velocity. Knowing the distance from three emitters, the coordinates of the receiver placed on body segment can be computed by triangulation (Château, Girard, Degueurce, & Denoix, 2003). Ultrasound motion capture is sensitive to perturbations such as wall refection, temperature change and air movement (wind) (Château et al., 2003). The working volume is limited to 2 to 4 meters and the use of several receivers is needed to allow complete motion capture (both side measurement). A sampling rate of 50Hz to 100Hz is reached with the current systems.

Electromagnetic motion capture consists of generating magnetic fields from transmitter with three mutually orthogonal coils. By supplying current to each one in turn, three different magnetic fields are alternately generated and this operation is repeated at a sampling rate of 50 to more than 100Hz depending on the performance of the system. For the set of three different magnetic fields generated in succession by the transmitter, a set of three such linear vectors can be determined (Kobayashi, Gransberg, Knutsson, & Nolen, 1997). Electromagnetic systems use 6 to 11 or more sensors per person to record body joint motion. Inverse kinematics is used to solve the angles for the various body joints, and compensate for the fact that the sensors are offset from the actual joint's center of rotation (Bodenheimer, Rose, Rosenthal, & Pella, 1997; Molet, Boulic, & Thalmann, 1996). Magnetic systems have very negative reactions to metal or magnetic fields in the environment and interferences from metal in floors, in walls, ceilings, structures, electrical devices are often present.

Optical motion capture has in recent years become an increasingly helpful tool in the area of human movement science, typically providing valuable information for assessing motor function pathologies. The number of companies providing motion capture systems targeting the areas of biomechanics, sports performance and gait analysis reflects their relevance in these domains. The technique consists of placing active or passive reflective markers on special parts of body segments. Cameras are placed around the subject so that each marker is visible by at least two of them and record markers position with a sampling rate of 50Hz to 200Hz (depending on the performance of the system). From the positions of the same marker viewed by at least two cameras, its actual 3D coordinates are computed. Optical motion capture systems are available from various providers, they allow the collection of information for illustrating and analyzing gait dynamics and studying the behavior of body limbs and joints during various motions, such as walking, running, limb raising, etc. Any discrepancies with respect to standard gait would indicate some type of dysfunction. The output enables the physician to detect motor function disorders, and to guide subsequent treatment. It also enables the physician to determine whether the disorder has been effectively corrected after treatment and/or surgery. A more complete description of the various clinical applications can be found in Richards (1999) and Ehara et al. (1997).

The level of detail that is available through motion capture is, however, open to discussion (Cappozzo, Catani, Leardini, Benedetti, & Della Croce, 1996; Cappozzo, Cappello, Della Croce, & Pensalfini, 1997; Chang, Su, Wu, & Wong, 1998). Because the markers are placed on the skin surface, and not directly on the bone and joints whose motion they are to identify, some systematic errors do occur. Furthermore, high quality systems can be very expensive and the use of markers tends to make them cumbersome. As a result, fielding these techniques typically requires a dedicated laboratory whose cost is often prohibitive, which has hindered the use of such measuring systems.

Forceplate is used both for gait and posture analysis. It is based on the measurement of ground reaction forces as the subject walks or stands on a platform equipped with force transducers (Barin, 1992; Winter, 1995, see also Chapter I). Standard forceplate consists of four triaxial force sensors measuring the three orthogonal components of the applied forces, the vertical torque, the anteriorposterior and mediolateral coordinates of the center of pressure (COP). Although posturography based on forceplate has proved to be useful for body sway evaluation and is considered as a standard method for balance assessment, there are many limitations for its use in clinical practice. First, for quantifying balance, the subject must remain in contact with the forceplate surface. Second, as it was noted, two forceplates are generally used to identify a separate load/unload mechanism for controlling the medio-lateral COP (Winter, Prince, Stergiou, & Powell, 1993). Third, the body is considered as a single-link inverted pendulum capable of rotating only about the ankle joints (Barin, 1992). Forceplate measurements provide an accurate estimate of the ground reaction force and the coordinates of center of pressure (COP). However, estimating the body sway angles from COP induces some significant error by assuming COP as a projection of center of mass (COM) and body behaving as a single-link inverted pendulum. It is a major misuse of the COP when referred to as sway (Winter, 1995). Gait analysis using forceplate provides COP trajectory and reaction forces needed for inverse dynamic (Winter, 1990). However, it is limited to a single gait cycle and does not provide any information related to stride-stride variability. Moreover, this technique requires the subjects to walk in a pre-defined specific path.

Electromyography (EMG) techniques provide detection and monitoring of electrical muscle activity by means of surface electrodes applied upon the belly of selected muscles (see Chapter I). EMG does not provide a direct measure of movement and a substantial number of electrodes are required for studying complex movements so that large amounts of data are to be processed. While EMG measurements have provided information regarding the latency and sequencing of active muscle responses, the relationship between specific muscular activation and the resultant outcome cannot be determined from the EMG because there are multiple muscles spanning each joint, and several muscles which act at more than one joint (Rietdyk, Patla, Winter, Ishac, & Little, 1999). In addition, the estimation of muscle force from EMG during dynamic contraction is difficult.

BODY-FIXED SENSORS AND NEW TECHNOLOGIES

Advantages of Body-Fixed Sensors

In order to overcome limitations of the standard technologies, some investigators have attached **sensors on body segments** to measure kinetic or kinematic features. Unlike the standard technology described above which need a dedicated controlled space, body-fixed sensors can be used just about anywhere. They are highly transportable and do not need any stationary units such as transmitters, receivers or cameras. Actually, body-fixed sensors (BFS) allow mobile and outdoor motion capture. All detected signals can be recorded by portable datalogger, allowing the subject to perform his/her activity in a real situation, at home, at work or at clinical center. There is no restriction in capturing volume and the data can be recorded for a long period depending on the performance and capacity of datalogger. In addition these sensors are much cheaper than sonic, magnetic and optical motion captures. They are easy to set up and use, and do not require highly skilled operators. Moreover, BFS allow a relatively higher sampling rate of recording than a standard system. A rate of several hundred hertz or higher can be reached with existing technologies. In addition, since there is no marker, signals can be continuously recorded without any lost trajectory due to marker hiding.

Body-Fixed Sensors and Intelligent Computing

BFS approach is very similar to the human sensory system used in postural control. Actually, the vestibular system acts as an inertial sensor: it detects linear and angular acceleration by two different receptors (otholits and semi-circular channels in internal ear). It can distinguish between inclination, linear and rotation acceleration (Berthoz, 1997). Signals from receptors are processed by integration or derivation and fused to detect and control head position. Other receptors in tendons and joints (mechano-receptors) measure force in joints and limb positions and in this manner, they provide the human body its own "body-fixed sensors." Signals from BFS can therefore be processed based on bio-inspired sensory processing, since nature has already found the solution to simplify the neurocomputation. For example, we do not have any receptors to directly measure the speed of movement. However, this speed is estimated by a derivation of position. This determines the existence of derivative function in CNS. Therefore the control of movement seems not continuous based on Bernstein theory, but discrete involving the internal physiological pacemaker. Llinás and Ribary (1993) claim that "we think at 40Hz and move at 10Hz" which means that the movement is controlled every 100ms. Like BFS, the brain uses multiple referentials instead of a unique referential (Euclidian) based on the task and sensory information. A hand will adapt its movement to the dimension and orientation of the object to grip (Jeannerod, 1991). BFS also use multiple references (relative to each member) while a camera-based system or other laboratory systems use a fixed referential. The triaxial accelerations detected by the semi-circular channels of internal ear not only need integrator and derivator neurons to estimate kinematics, but also involve tensor-like transformation between vectors.

Main Body-Fixed Sensors

Ambulatory systems using body-fixed sensors are used to measure human movement. In the following, the main types of body-fixed sensors are presented.

Electrogoniometer

An electrogoniometer is a sensor with attachments to the proximal and distal limb segments that spans a joint to be measured. The sensor operates on the assumption that the attachment surfaces move with (track) the midline of the limb segment onto which they are attached and thereby measures the actual angular change at the joint. Different kinds of electrogoniometers have been reported. A potentiometric electrogoniometer has a circular potentiometer as sensitive element (Kettelkamp, Johnson, Smidt, Chao, & Walker, 1970). Potentiometric electrogoniometer has been designed to study 3D wrist kinematics (Salvia et al., 2000) and 3D cervical spine motion ranges (Feipel, Rondelet, Le Pallec, & Rooze, 1999) movement. A more practical electrogoniometer includes a mobile and flexible part between two stiff surfaces attached on each moving segment (Figure 1). The flexible part allows free joint's motion. It can be used to measure an angle between two segments when the instantaneous center of rotation is not clearly defined or is moving during the rotation. Between the two end blocks, inside the protective spring, there is a composite cylinder which has a series of strain gauges mounted along its generatrix. As the angle between the two ends changes, the strain along the cylinder induces electrical resistive change which is measured through a voltage proportional to the angle. With this configuration, only angular displacements are measured (Martinet & Andr, 1994; Roduit, Besse, & Micalef, 1998). Flexible electrogoniometer are mostly used and commercialized (Myles, Rowe, Walker, & Nutton, 2002). These devices provide an output voltage proportional to the angular change between the two attachment surfaces. Unlike conventional potentiometric electrogoniometers, flexible

Figure 1. A biaxial flexibale electrogoniometer (Courtesy of Biometrics Ltd, Gwent UK)

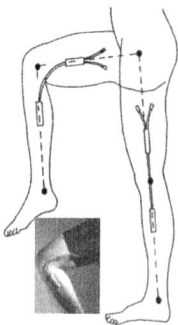

electrogoniometers do not suffer from any alignment problems with respect to the joint axis. The flexible electrogoniometers adapt better to all body parts and have been claimed to be more reliable with respect to the rigid conventional potentiometric goniometers (Tesio, Monzani, Gatti, & Franchignoni, 1995). The latter, however, are much cheaper. When considering flexible electrogoniometer for gait, their accuracy should be carefully evaluated by testing them on individuals of various statures. The error encountered for this device can range between 5 to 10 degrees (Shiratsu & Coury, 2003). Sensor attachment and the lack of robustness of the sensors are the major drawbacks of flexible electrogoniometer. Flexible electrogoniometer is used to evaluate proprioception before knee ligament surgery and after surgery. Testing consists of moving the joint into a predetermined amount of flexion, and the patient was told to remember the position. The knee was then passively returned to full extension, and the patient was asked to return the leg to the test position. The amount of discrepancy in degrees was then recorded. The use of electrogoniometer in ergonomy, lower limb asymmetry during walking, sit-stand lower limb patterns, studying the joint function has been reported.

Fiber-optic goniometers have been designed and commercialized more recently (www.measurand.com). The sensor uses fiber-optic sensing loops, each spaced along a flexible rod, which are treated to lose light proportional to bending. The rod is fixed at its ends on the rotated segments. Using an array of these sensors shape measurement system has been designed and evaluated (Yue, Li, Aissaoui, Lacoste, & Dansereau, 2004).

Earth-Magnetic Sensor

Measuring earth's magnetic field has been used to estimate body segment orientation relative to the magnetic-North direction (Kolen, Rhode, & Francis, 1993). Miniature and MEMS (MicroElectroMechanical Systems) electronic compass sensors use the magnetoresistive effect, which corresponds to a change of resistance under a magnetic induction (Roumenin, Dimitrov, & Ivanov, 2001). Kemp, Janssen and van der Kamp (1998) have shown the possibility to estimate the azimuth of the body position by using a magnetic compass. Magnetic compass has also been used in outdoor and mobile augmented reality applications, where accurately tracking the user's viewing orientation and position is crucial (Azuma et al., 2001). A disadvantage of the magnetic compass is its sensitivity to nearby mass of iron and local magnetic field. In addition, it needs to be calibrated for any change of location.

Plantar Pressure Sensor

Plantar pressure sensors attached to the sole are used to monitor temporal parameters during locomotion (Zhu, Wertsch, Harris, Loftsgaarden, & Price, 1991; Abu-Faraj, Harris, Abler, & Wertsch, 1997). Force sensing resistors (FSR) use the electrical property of resistance to measure the force (or pressure) applied to a sensor (Figure 2). A FSR is made up of two parts. The first is a resistive material applied to a film. The second is a set of digitating contacts applied to another film. The resistive material serves to make an electrical path between the two sets of conductors on the other film. When a force is applied to this sensor, a better connection is made between the contacts, hence the conductivity is increased. This technique generally presents satisfactory results for finding different phases of movement during normal walking (heel strike, swing and

Figure 2. The FSR sensor (above) with coated with alimentary silicon (below) (courtesy of Z. Pataky)

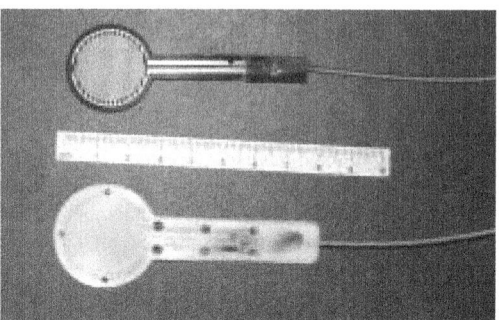

stance phase, heel and toe off). However, for pathological gait, many problems, such as shuffling, difficulty of appropriate positioning, connecting attachment, mechanical failure, and the subject's acceptance limit their applicability.

Based on FSR, Pataky, Faravel, da Silva and Assal (2000) have developed an ambulatory foot pressure device (AFPD, Figure 1) performing a relatively long-term measurement of foot pressure parameters: peak plantar pressure, foot-floor contact time and plantar the integral of the pressure over the time. The device is used for gait monitoring in diabetic patients with risk of ulcer. The system allows the data acquisition with the option of an acoustic signal for patients. This allows for possible changes in the pattern of ambulation to avoid high plantar pressure and risk of ulceration. The system is fully autonomous and battery operated for long-term continuous recording of plantar pressure measurement (up to 8 days).

The initial and terminal foot contact times during walking act as a reference phase in temporal gait parameters and as a means of distinguishing normal and pathologic gait. Using two FSR sensors attached respectively under the toe and heel, Hausdorff, Ladin and Wei (1995) have developed a footswitch system that provides accurate estimates of the start and end of stance phase for each gait cycle. Using forceplate as reference, the system can be used to estimate stance duration to within 3% of forceplate. Estimates of swing and stride duration also are within 5% of forceplate.

A Holter-type portable system has been developed which allows continuous recording of pressure data between the sole of the foot and the shoe during the performance of daily living activities. The system is based on resistive principle and peak plantar pressure, pressure-time integral as well as contact duration are determined for each of the 14 insole sensors. The system is capable of a relatively long-term recording — up to 8 hours (Abu-Faraj et al., 1997).

A portable, in-shoe pressure data-acquisition system (Paromed, DE) with two insole pressure transducers (16 piezoresistive microsensors each) was used to measure bilateral plantar pressure distribution and electromyography activity of the selected limb muscles. Sensors are linked to a datalogger fixed by a belt to the patient's back. The system consists of variable size insoles, which are approximately 3 mm in thickness and can be reused (Pertunnen, Anttila, Södergard, Merikanto, & Komi, 2004). Plantar pressure was also measured using capacitive sensors (Novel, DE). Each insole consists of 99 pressure sensors and plantar pressure distribution is recorded using a datalogger carried by the subject (Maluf & Mueller, 2003).

Accelerometer

Accelerometers have been used widely in movement recording (Mathie, Coster, Lovell, & Celler, 2004). They offer a practical and relatively low cost method for human movement analysis. An accelerometer consists of a mass, connected to a frame (accelerometer case) by a beam which can be represented by a dampened spring (Figure 3). The acceleration of the case of the sensor is transformed to the spring deformation, which is transduced to a change of electrical impedance (resistive or capacitive accelerometer) or charge generation (piezoelectric accelerometer). A single axis accelerometer measures the difference of acceleration (a) and gravity (g) along its sensitive axis given by the unit vector (n). The measured electrical signal (S) could thus be expressed as:

$$S = (a - g).n \tag{1}$$

Contrary to piezoelectric accelerometer, resistive and capacitive accelerometers are sensitive to DC as well as AC components of accelerations. Therefore, even in the absence of movement, they measure gravitational acceleration of body segment and more specifically its projection on the sensitive axis of the sensor. However, in most activities, body motion induces acceleration and a variable gravitational component due to the change of segment inclination with respect to the vertical axis. In these situations, estimating other parameters such as displacement or velocity as integral of acceleration provides important drift and errors due to segment inclination (DC component of

*Figure 3. An accelerometer consists of a mass, connected to a frame by a beam which can be represented by a dampened spring. A single axis accelerometer measures the difference of acceleration **a** and gravity **g** along its sensitive axis given by the unit vector **n**.*

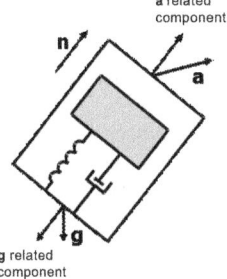

acceleration) and its variation (segment rotation). Both acceleration components are superimposed and their separation is necessary in order to properly analyze the movement. Moreover, the presence of drift and errors in velocity makes the estimation of displacement as the integral over time of velocity difficult.

The accelerations generated during human movement vary across the body and depend on the type of activity. Acceleration increases in amplitude from the head to the ankle. Mathie et al. (2004) provide an interesting review of the range of acceleration based on activity and body position and it frequency range. They reported that the maximum range of acceleration is reached during running at the ankle (12g). Trunk sway RMS acceleration (in sacrum) during standing range from 2.8mg up to 7.6mg and can reach 13mg during eyes closed (Moe-Nilssen, 1998) and even more on an unstable platform. We have recently found trunk acceleration ranges between 2 and 9g during falling. Although foot acceleration can reach a frequency of up to 60Hz, a spectral analysis of acceleration revealed that most components are below 16Hz. However, in order to be accurate in gait phase detection during walking (e.g., heel-strike and toe-off) a sampling rate of up to 200Hz is needed.

Using accelerometry in motion analysis has proven to be very promising and we have already reported many clinical applications of such a technique (Aminian et al., 1998; Aminian & Najafi, 2004). Gait parameters such as cadence were estimated from trunk accelerometry (Moe-Nilssen & Helbostad, 2004) and simple algorithms have been devised to detect some temporal events such as heel contact during walking (Zijlstra & Hof, 2003). The possibility to estimate temporal gait parameters from uniaxial accelerometer attached onto the thigh (Aminian et al., 1999) or two uniaxial accelerometers fixed on shank has been reported (Selles, Formanoy, Bussmann, Janssens, & Stam, 2005). Also, triaxial accelerometer on shank in conjunction with supervised machine learned rule has been investigated to detect gait phases (stance/swing) (Williamson & Andrews, 2000, September). Willemsen, van Alste and Boom (1990) developed a technique to measure human joint flexion-extension angle without the need for integration, which used four accelerometers on each segment. The system used two metal bars with eight accelerometers for measuring a single joint angle. As mentioned by the authors, the arrangement of the metal bars may be too encumbering for daily application (Williamson & Andrews, 2000).

Fyfe and Fyfe (2001) have proposed three accelerometers attached on the heel to measure the kinematics of the foot in the sagittal plane. They have used the double integral of angular acceleration to measure heel angle in respect to the ground, while the initial values were estimated during the stance phase and where the minimum foot velocity is zero.

As measurement of accelerometers fixed on a moving body depend on their location, many groups have tried to combine several accelerometers. Willemsen, Frigo and Boom (1991) used PVC brackets with four accelerometers to calculate one-degree of freedom angle joint. Giansanti, Macellari, Maccioni and Cappozzo (2003) investigated the possibility of using six or nine accelerometers to obtain the position and orientation of a body segment. Their simulations concluded that neither of the two systems was suitable for body position and orientation estimation.

Accelerometry has been used also as a new method for inverse dynamic during walking and running. Using four triaxial accelerometers attached on the trunk, hip force and moment during the stance phase were compared to standard method for inverse dynamic using force-plate and camera (van den Bogert, Read, & Nigg, 1996). Although accelerometry underestimated the joint force and moment by about 20%, it could be relevant in applications where the difference between right and left hip is considered.

Another important application of accelerometry is human body orientation detection and classification. Actually, at rest the accelerometer measures the inclination of body segment with respect to the vertical axis. By attaching these sensors on two or more segments of the body (e.g., trunk, thigh, shank), the body posture at rest (i.e., standing, sitting, lying) can be recognized (Aminian et al., 1999; Veltink, Bussmann, de Vries, Martens, & van Lummel., 1996; Ng, Sahakian, & Swiryn, 2000; Bussmann, van de Laar, Neeleman, & Stam, 2004). However, identifying body posture orientation from the acceleration of one segment, such as the trunk, is more challenging since the trunk orientation during sitting and standing is quite the same. Statistics of waist acceleration with neural network classifier have been used to estimate eight human activity states (lying on and back, sitting, standing, walking, running, upstairs and downstairs climbing) (Jonghun, Geehyuk, Wonbae, & Byoung-Ju, 2004). Other studies have demonstrated that level walking and climbing can be distinguished in the signals of a waist-fixed triaxial accelerometer (Sekine, Tamura, Togawa, & Fukui, 2000; Sekine et al., 2002; Tamura, Sekine, Ogawa, Togawa, & Fukui, 1997).

Trunk accelerometry is also used to evaluate postural control based on waist oscillation during standing (Moe-Nilssen & Helbostad, 2002; Mayagoitia, Joost, Lotters, Veltink, & Hermens, 2002). Interestingly, the vestibular system, which is the main component of postural control beside vision and proprioception, used biological transducers for both linear and angular acceleration sensing (otolith and semi-circular channels). In an original study, Baselli used up to 4 triaxial accelerometers attached on the head to estimate the acceleration that should be sensed by vestibular organ (Baselli et al., 2002).

Gyroscope

Gyroscopes or angular velocity sensors play an important role in the inertial sensing. The classical gyroscopes utilizing a conventional rotating wheel as well as precision fiber-optic and ring laser gyroscopes are usually heavy, large, and costly for human movement analysis. In contrast, micromachining can shrink the sensor size by orders of magnitude, reduces the fabrication costs significantly, and allows the electronics to be integrated on the same silicon chip. Almost all reported micromachined gyroscopes use vibrating mechanical elements to sense rotation (Yazdi, Ayazi, & Najafi, 1998). They have no rotating parts that require bearings, and hence they can be easily miniaturized and batch fabricated using micromachining techniques. All vibratory gyroscopes are based on the transfer of energy between two vibration modes of a structure caused by Coriolis acceleration. The Coriolis effect is an apparent force that arises in a rotating reference frame and is proportional to the angular rate of rotation. The amplitude of the Coriolis force (F_c) depends on the vibrating mass (m) and its velocity (v) and the angular velocity of the rotating frame ($w\partial$) by the following equation:

$$F_c = -2m \cdot v \otimes \omega \tag{2}$$

where $\otimes$ denotes the cross product.

The vibration force can be induced using piezoelectric, electromagnetic and electrostatic energy. The Coriolis force can be detected using piezoelectric, resistive or capacitive effect (Grétillat, 1998).

MEMS gyroscopes have been developed particularly for the automotive and navigation market as well as in consumer products, such as anti-jitters compensation for video cameras. The use of gyroscope for human motion capture is still new and under development, albeit gyroscopes can be considered a promising tool in this field. Actually, human motion consists mainly on rotations around joints. Unlike the accelerometer, there is no influence of gravity acceleration on the measured signal. The gyroscope can be attached to any part of any body segment as long as its axis is parallel to the measured axis. The angular rotation is still the same along this segment. Tong and Granat (1999) have shown that signals from different gyroscopes at different attachment sites of a segment are almost identical. Moreover, the angular rate signal is less noisy than acceleration since acceleration is the derivative of velocity and involves higher frequency components. Finally, rotation angles can be estimated from angular velocity by simple integration.

Since the actual nature of walking consists of lower limb rotation around joint articulation, the use of gyroscope has proven to be an alternative technique for gait analysis. By considering the shank and thigh as a unique segment, Miyazaki (1997) used a simple pendulum model to estimate speed of walking from a single gyroscope attached

Figure 4. Angular velocity of thigh, shank and foot segment in sagital plane measured by three uniaxial gyroscopes

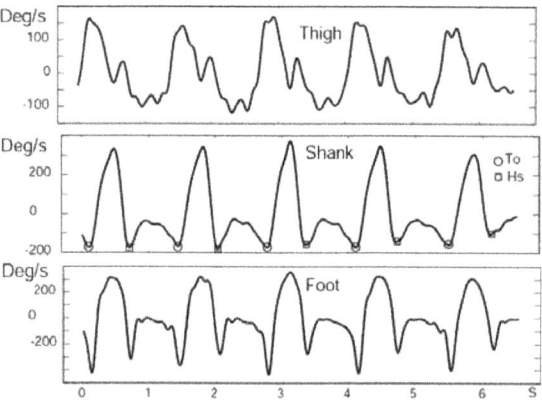

Hs and To show the time of heel strike (initial contact) and toe-off (terminal contact)

on the thigh. The system is very simple, but the accuracy is relatively low (15%). Aminian et al. (2002) proposed a model involving both shank and thigh, which consider a double pendulum model during the swing phase and the inverse double pendulum model during the stance phase. In this method a single gyroscope was placed on each shank and thigh. Wavelet transform was used to detect accurately heel strike and toe-off and to estimate stride length and velocity during walking. This method provides both temporal (stance, double stance and swing) and spatial (angle, stride and velocity) gait parameters. No significant error was observed for toe-off detection while a systemic delay (10ms) existed compared to the footswitch. Error for stride length and velocity were 0.07m and 0.06m/s. In addition this method provided a simple way to estimate the knee dynamic range of motion during walking. Sabatini, Martelloni, Scapellato and Cavallo (2005) provide an alternative method for estimating temporal gait parameters from a gyroscope attached on foot. The error for toe-off detection is slightly biased (35 ms), whilst the heel strike is not. Figure 4 illustrates typical angular velocities measured at thigh, shank and foot.

Recently, the possibility to detect walking upstairs periods using a gyroscope attached to the shank was provided (Coley, Najafi, Paraschiv-Ionescu, & Aminian, in press). The wavelet transform, in conjunction with a simple kinematics model, was used to detect the moments of toe-off, heel strike and foot flat, as well as the cycles corresponding to the stairs ascent. Walking upstairs was identified based on the pattern of angular velocity.

The use of angular velocity is particularly interesting in center of mass oscillation during quite stance. As suggested by Jeka, Kiemel, Creath, Horak, and Peterka (2004), velocity information is the most accurate form of sensory information used to stabilize posture during quiet stance. Adkin, Bloem and Allum (in press) proposed angular velocity of trunk as postural feedback and screening tool for balance disorders. Recently, Allum and Carpenter (2005) provided a comprehensive review of the gyroscopes in gait and balance assessment as well as some of its limitations: noise and drift, relative angular changes but not absolute position and sensor attachment. They show also the high potential that gyroscope has reached recently for use in biofeedback, or as an ambulatory device for long-term recording of balance and gait performance in both clinical and daily living conditions.

Body-Fixed Sensor Fusing

The above sensors considered separately are inexpensive and simple to use in movement detection and mobility measurement. However, in order to extract other interesting kinematics features such as angle, displacement and trajectory, complex and intelligent computing is necessary. Fusing the above sensors and using intelligent computing can provide a natural way to overcome this problem.

Examples of sensor fusing exist in our vestibular system. Semi-circular channel work as triaxial accelerometer which measures the rotation acceleration of the head based on the change of endolymphe speed. Otoliths in internal ear act as linear accelerometer but at a low frequency. By fusing semi-circular channels and otoliths and using mechanical and neuronal filtering of high acceleration, the inclination (gravity component), rotation and translation can be assessed by the brain. Even more, derivator neurons can provide the derivative of acceleration. Since derivation operation provides a positive phase shift in the signal it gives the possibility to anticipate and have fast reaction when confronting challenging movements, such as fall. Jerk, which is a derivative of acceleration, is an example of information in neural processing. Moreover, by using derivative in neural processing, non-linear motion control can be determined. Another example is the proprioception systems that fuse elongation and speed measured by neuromuscular spindle in muscle and force and its derivative detected by mechanoreceptor in tendon. The problem of how the nervous system fuses sensory information from multiple modalities remains unsolved. It is well established that the visual, vestibular, and somatosensory modalities provide position and rate (e.g., velocity, acceleration) information for estimation of body dynamics. However, it is unknown whether any particular property dominates when multisensory information is fused (Jeka et al., 2004). These biological sensors fusing are widely discussed by a very comprehensive and interesting book of Alain Berthoz, "Le sens du mouvement" (Berthoz, 1997).

Gyroscope-Accelerometer Combination

A simple example of sensor fusing is the combination of accelerometers and gyroscopes. Accelerometers and gyroscopes have complementary features. An accelerometer provides both a derivative of angular velocity and a body segment inclination during rest. Gyroscope estimates the derivative of the rotation and is not sensitive to gravity component (e.g., inclination). By considering a simple plane movement, combining a gyroscope with two accelerometers make possible a 2D estimation of body segment kinematics.

Let us consider the signals detected by such an inertial sensor placed on the body segment (Figure 5) The vertical (A_v) and forward (A_f) components of the body segment accelerations, $\vec{A}$, with reference to a fixed reference (room) are expressed in term of vertical (a_v) and forward (a_f) acceleration measured by the sensor as:

$$
A(t)
\begin{bmatrix} A_f(t) \\ A_v(t) \end{bmatrix}
\begin{bmatrix} \cos\ (t) & \sin\ (t) \\ \sin\ (t) & \cos\ (t) \end{bmatrix}
\begin{bmatrix} a_f(t) & g\sin\ (t) & 0 \\ a_v(t) & g\sin\ (t) & g \end{bmatrix}
$$

$$
\begin{bmatrix} a_f(t)\cos\ (t) & a_v(t)\sin\ (t) \\ a_f(t)\sin\ (t) & a_v(t)\cos\ (t) \end{bmatrix}
\tag{3}
$$

Figure 5. Body segment acceleration and rotation in a plane measured with a sensor

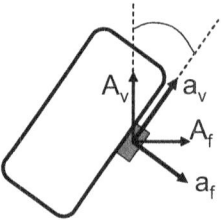

where θ *(t)* correspond to the orientation of the sensor with respect to the vertical axis at time *t* and g=9.81m/s².

θ *(t)* can be calculated from the medio-lateral angular velocity of the body segment θ *(t)* measured by the gyroscope:

$$\theta(t) = \int_t \omega(t)dt + \theta_0 \qquad (4)$$

with θ_0 initial value of the body segment angle with respect to the vertical axis.

Body segment velocity $\vec{V}(t)$ and displacement are obtained from the integral of the acceleration:

$$\vec{V}(t) = \int_t \vec{A}(t)dt + \vec{V}_0 \qquad (5)$$

$$\vec{R}(t) = \int_t \vec{V}(t)dt + \vec{R}_0 \qquad (6)$$

with $\vec{V}_0$ and $\vec{R}_0$ the initial conditions for velocity and displacement, respectively.

Therefore, from the above equations, the kinematics of the body segment in a sagittal plane can be estimated if the initial values for angle, velocity and displacement are known. By assuming some biomechanical hypothesis based on practical consideration, these values can be evaluated in many situations. For example, a period of rest is usually present before or after a movement. During any rest period acceleration variance and velocity $\vec{V}_0$ are almost zero, while the angle θ_0 can be estimated from the mean value of accelerations (a_{v0} and a_{f0}):

$$\theta_0 = \arctan(\frac{a_{f0}}{a_{v0}}) = \arcsin(\frac{a_{f0}}{g}) \qquad (7)$$

$\vec{R}_0$ has known constant values which can be estimated from the coordinate of the sensor's site at any rest standing period, albeit the variation of $\vec{R}(t)$ in many cases is valuable.

Several investigators have used similar methodology outlined above to measure human body motion. Sagawa and Satoh (2000) have used a 3D accelerometer and gyroscope on the toe to estimate the horizontal velocity and displacement of the toe. They reset the initial values for velocity and angle at each stance phase where horizontal velocity repeatedly returns to zero since the foot stands still on the ground during stance phase. Vertical displacement was calculated by integrating the change of the atmospheric pressure during ascending and descending detected by a pressure sensor. Najafi et al. (2002, 2003) have used a set of two accelerometers and gyroscope to measure the trunk tilt. The sensor attached on the chest includes two accelerometers with a gyroscope to determine the trunk rotation and translation in the sagittal plane. Sitting and standing positions are identified by recognizing the nature of sit-stand (respectively stand-sit)

Figure 6. Trunk movement during sit-stand and stand-sit transition. The time of transition is detected by the minimum of sin (θ).

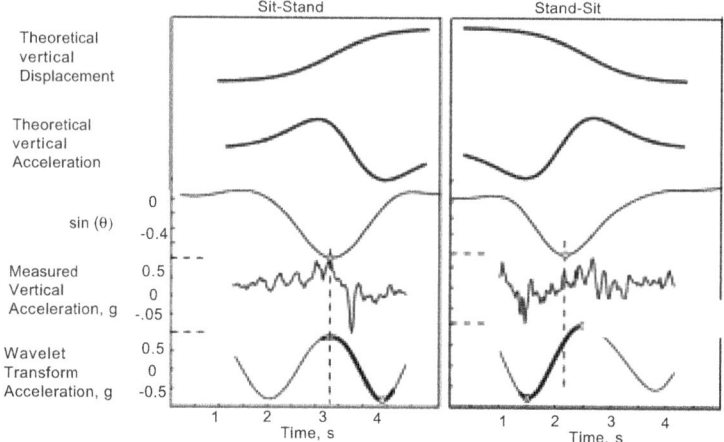

Using wavelet transform enhances the actual pattern of acceleration making distinction between sit-stand and stand-sit transition

transitions (Najafi & Aminian, 2002). Figure 6 shows the displacement of the trunk during a sit-stand transition (respectively stand-sit), as well as the trunk rotation and its theoretical acceleration. The detection of the negative peak of rotation from the gyroscope makes it possible to determine the time of postural transition. Pattern recognition of the vertical acceleration allows to classify the transition and to decide if the subject was in a standing or a sitting position. However, the measured acceleration is far from its theoretical pattern described in Figure 6. This error is mainly due to the fast variations (jerked movement) of the acceleration of the trunk at the time of transitions. To overcome this problem, a time-frequency analysis (wavelet transform) is necessary (Najafi et al., 2003). Figure 6 shows the measured vertical acceleration and how wavelet transform makes it possible to recognize the signature of the sit-stand and the stand-sit transition respectively.

Sabatini, Martelloni, Scapellato, and Cavallo (2005) placed an inertial sensor (two accelerometers and a gyroscope) on the foot to estimate sagittal kinematics of the foot during walking. He estimated initial value (θ_0) in (4) by estimating inclination during stance from the acceleration component and provided a simple way (a unique sensor site) to estimate incline and speed of walking.

Williamson and Andrews (2001) combined two sensor modules each comprising two accelerometers and a gyroscope. These are attached to the thigh and shank to estimate absolute human knee angle. The angular velocities were integrated with autonulling and auto-resetting based on the accelerometers' data. This system was restricted by the placement of the sensor, which had to fit the plane of motion and by the fact that motion should not occur in a horizontal plane. Usually, the human segments move in 3D, and using sensors which are only sensitive in one or two directions, implies making assumptions in the model, and losing accuracy.

Heyn, Mayagoitia, Nene, and Veltink (1996) showed that shank and thigh inclination angles can be measured with the need of signal integration with eight accelerometers as wells as two gyroscopes fixed on two rigid metal plate. They also found that using these metal plates was cumbersome. An elegant and real-time technique to accurately find joint angle by fusing gyroscope and accelerometers is presented recently by Dejnabadi, Jolles and Aminian (in press). With this technique, joint angles (e.g., knee) are found without the need for integration, so absolute angles can be obtained which are free from any source of drift. The model is based on estimating the acceleration of the joint center of rotation by placing a pair of virtual sensors on the adjacent segments at the center of rotation. An error of 1.3 degrees was obtained for absolute joint angle. The system is portable, easily mountable, and can be used for long term monitoring and gait analysis. Unlike electrogoniometers, the proposed system is more robust and also provides anterior-posterior rotations and linear accelerations of thigh and shank independently, which can further be used for a better estimation of lower limbs kinematics.

Accelerometer and gyroscope were used for inverse dynamic to estimate hip abduction moment during stance in walking and standing (Ziljstra & Bisseling, 2004). Trunk was considered as a two segments model. Data were collected from two modules attached on the upper and lower trunk. Each module included two triaxial accelerometers and a gyroscope measuring angular velocity in frontal plane. Thus, with this sensor configuration, linear and angular accelerations of these two segments are available. Comparing to standard technique for inverse dynamic, moderate to good correspondence has been found.

Figure 7. Autonomous Sensing Unit Recorder including two uniaxial accelerometers a gyroscope, datalogger, batteries and memory

The unit allows long-term movement monitoring (over a day or several days)

The relations (3-7) show also that the combination of two accelerometers with a gyroscope is a subtle way to estimate kinematics in a plane. We have recently proposed a new wireless and Autonomous Sensing Unit Recorder (ASUR) which includes sensors, datalogger, batteries and memory in a small box of 50gr, providing in this way a simple tool for long-term movement monitoring (Figure 7).

Kalman Filtering for Sensor Fusing

Both accelerometer and a gyroscope can measure inclination of a segment, but a gyroscope suffers from slow drift, and unknown initial inclination and an accelerometer is sensitive to dynamic accelerations. The gyroscope signal ω can be modeled as the superposition of three signals: a slowly varying offset b_ω, an angular velocity Ω, and a low white noise n_ω:

$$\omega = \Omega + b_\omega + n_\omega \tag{8}$$

Accelerometer signal (a) is composed of an acceleration (A) and gravity components (a_g), with superimposed measurement noise n_a:

$$a = A - a_g + n_a \tag{9}$$

Kalman filtering is seeking for a fusion method to make the best use of all the data available from different types of sensors. It is based on using information from several signals and a priori information about the behavior of the system under consideration in order to make a most-likely estimate of the system states.

Foxlin (1996) described the design of a Kalman filter to integrate the data from gyroscopes and inclinometers (gravity accelerometer). He used a complementary Kalman filter which operates only on the errors. One advantage of this structure is that it guarantees that the rapid dynamic response of the inertial system will not be compromised by the Kalman filter.

By studying mobile robot attitude estimation, Vaganay, Aldon, and Fournier (1993) provided a method in which gyroscope drift is compensated using two accelerometers. An extended Kalman filter was used to fuse the information from five inertial sensors: two accelerometers and three gyroscopes. The integration of angular rate is done outside of the Kalman filter, and is treated as part of a measurement system that provides gyroscopically determined measurements of pitch and roll.

Zhu and Zhou (2004) presented a real time motion-tracking system using tri-axis accelerometers, gyroscopes and earth magnetic sensors. Kalman-based fusion algorithm was applied to obtain dynamic orientations and further positions of segments of the subject's body. They showed that utilizing the Kalman filter to integrate the sensors could incorporate excellent dynamics of gyroscope and stable drift-free performance of gravity acceleration and magnetic field. Haid and Breitenbach (2004) presented a low cost inertial orientation tracking with Kalman filter using a gyroscope and a magnetic field sensor. They augmented the accuracy of an orientation detection system by estimating the drift of the gyroscope.

Rehbinder and Hu (2004) proposed a state estimation algorithm that fuses the data from gyroscopes and accelerometers to give long-term drift free attitude estimation. They combined two linear Kalman filters between which a trigger based switching takes place. Thus the kinematics representation used made it possible to construct a linear algorithm that was shown to give convergent estimates for this nonlinear problem.

Luinge and Veltink (2004) proposed a method for inclination measurement of human movement using a 3-D accelerometer. They designed a Kalman filter to estimate inclination from the signals of a triaxial accelerometer. The design was based on assumptions concerning the frequency content of the acceleration of the movement that was measured, the knowledge that the magnitude of gravity is 1g and taking into account a fluctuating sensor offset. They estimated inclination of trunk and pelvis and showed that their method was twice as accurate as an estimate obtained by low-pass filtering of the accelerometers. An important limitation of those methods comes from the accelerometer's non sensitivity of rotations around the vertical. One way of overcoming this problem is to add a triaxial earth magnetic sensor.

Bachmann, Yun, and McGhee (2003) employed this configuration to track the movement of human limbs. The gyroscopes were used for fast movement periods and magnetometers and accelerometers during slow periods. A quaternion-based Kalman filter was used to fuse the data from the nine sensors. They designed a constant-gain complementary filter to estimate the attitude of a rigid body. It was based on minimizing the error by adjusting the derivative of an estimated orientation quaternion using Gauss-Newton iteration. As a result of this approach, the measurement equations of the Kalman filter become linear, making it possible to estimate orientation in real time (Marins, Yun, Bachmann, McGhee, & Zydal, 2001). Lee, Laparde, and Fung (2003) used also such a combination to measure lumbar spine motion.

Other Body-Fixed Sensors Combination

In order to estimate foot angle during walking Pappas, Popovic, Keller, Dietz, and Morari (2001) combined gyroscope on the heel with three foot pressure sensors (FSR) placed respectively underneath the heel, the first and fourth heads of the metatarsal bones. The angular velocity of the heel was integrated to find angle in a sagittal plane. In order to avoid accumulation of drift errors in the integrated signal, the foot inclination (integrated gyro signal) was reset to zero during the stance-phase (i.e., foot-flat) when all three FSRs were loaded. Pappas et al. (2004) proposed also an instrumented insole for functional electrical stimulation (FES) whereby gait phases are identified based on FSR and gyroscope placed in insole. The extracted gait-phase signal was used in a finite state control scheme to time the electrical stimulation sequences.

Body-fixed sensor fusing is also used in pedestrian navigation. Earth magnetic sensor, gyroscope and air pressure sensor estimating altimeter are fused to estimate the trajectory of subject. In addition GPS receiver on subject allowed calibrating the body position during locomotion (Ladetto, 2002).

CLINICAL APPLICATIONS
OF BODY-FIXED SENSORS

Orthopedics

Outcome research is a relatively new field of interest in orthopedics (Keller, Rudicel, & Liang, 1993). The rapidly rising cost of healthcare with its financial impact on the individual and national economy, and deficiencies in clinical research methods such as a patient oriented evaluation that is a functional and quality-of-life assessment, have stimulated the emergence of this concept. The influence of surgical procedures on quality-of-life, subtle comparisons between two potentially efficacious treatments (e.g., two types of prosthesis) and the effect of the rehabilitation's program on patient function are some examples of the importance of outcome evaluation in orthopedics. The most common basis used for this evaluation is the scores derived from questionnaires and x-ray examinations. Although the questions on such evaluations are standardized, their answers are often subjective and the disparity between the patient and doctor's evaluations is significant (Lieberman et al., 1996; Anderson, 1972). Objective instruments usable in clinical practice with higher sensitivity and specificity in evaluating quality of life compared to traditional scoring systems is needed to enhance the surgeon's ability to assess the overall outcome in patients after joint replacement (Lieberman et al., 1997).

Walking is very easily accessible to a detailed clinical analysis and gait analysis is used often for outcome evaluation in orthopaedics. However, most standard quantitative approaches to gait analysis run the risk of biasing this activity (Doutrellot, 1992; Andriacchi, Hampton, Schultz, & Galante, 1979). Gait on a treadmill seems a relatively easy solution to implement, provided that the patient is able to walk on the treadmill. In addition, a number of studies have shown that the subject feels the rolling surface to be unstable. The afferent proprioception is thus modified. Moreover, the subject is immobile with respect to visual landmarks. As a result, sensorial afferent inputs are also modified. Finally, in order to provide meaningful collecting data, the subjects must be comparable among themselves. Therefore, they must all walk at a speed that is comfortable for them, that is at the speed where the energetic efficiency is at its optimum. In fact, the walking speed of a subject can vary between 50% and 150% of his average speed. This variation modifies several parameters such as energy consumption, articular amplitudes, points of impact and reaction forces on the ground (Walters, Lunsford, Perry, & Bird, 1988; Winter & Eng, 1988). Gait analysis using standard technologies (camera, forceplate, EMG) exist to assume the functional handicap of the patients. Their use in clinical practice, however has many limitations (e.g., time for instrumentation and analysis, cost, restricted volume) and require the subjects to walk in a pre-defined specific path, and assume that data measured from only a few steps are representative of usual gait performance. In this regard, the body fixed sensors method as a complementary method of gait analysis has many potential for outcome evaluation in joint degenerative disease.

Using accelerometry gait temporal parameters have been estimated in patients before and after hip arthroplasty. Measurements were taken on 12 hip osteoarthritic patients before and after operation — at three, six and nine months. A significant discrepancy between duration of initial and terminal double limb support, depending on affected hip, was observed. The contrast with their almost equal values for a population

Figure 8. Knee flexion-extension during a gait cycle obtained from gyroscope in comparison to the same angle measured by camera based motion capture

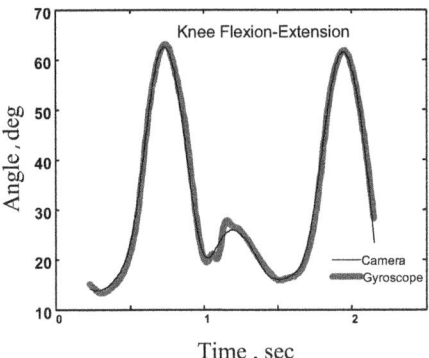

of normal subjects was paramount. A mean decrease of 88% of asymmetry of stance time and especially a mean decrease of 250% of asymmetry of double support time were observed nine months after the operation. This result underlines that the weight transfer phase is the most demanding task during a gait cycle and certainly a painful one for patients. Furthermore a tendency toward normality was also observed after arthroplasty (de Andrès, Aminian, Fritsch, Leyvraz, & Robert, 1998).

More recently, the validity of gyroscopes in outcome evaluation in orthopaedics has been tested by gait analysis in 11 patients with hip arthrosis (coxarthrosis), 8 patients with total hip prosthesis, and 9 healthy subjects. Camera-based system with reflective markers and forceplate were used for validation (Aminian et al., 2004). The estimated angles closely match the corresponding results of the camera-based system (r=0.97, standard error=4 deg). Figure 8 shows a typical knee angle estimated with the gyroscopes and obtained by camera-based system during walking. In addition, significant correlations were found between clinical Harris Hip Scores (HHS) (Harris, 1969) and range of flexion of the thigh (r=0.69, p<0.01). Moreover using principal components analysis, the possibility to classify patients into three groups (Control, with coxarthrosis and with hip prosthesis) has been reported (Dejnabadi et al., 2003).

Using three sites of attachment and six accelerometers, a "knee test" protocol has been proposed for motor function evaluation in orthopedics (van den Dikkenberg et al., 2002). Data are recorded as the patient performs a set of 29 tasks of everyday activities. "Knee score" is calculated by combining the weighted averages of movement features per task.

The functional ranges of movement of the knee during typical activity (walking at level, uphill and downhill, climbing, transfer task such as sit-stand) were investigated using flexible electrogoniometry in a group of patients with knee osteoarthritis before, and after total knee arthroplasty and in control subjects (Myles, Rowe, Walker, & Nutton, 2002). It is concluded that total knee arthroplasty gives rise to little improvement in knee motion during functional activities and that functional range of movement of the knee remains limited when compared to normal knee function for a minimum of 18 months following operation. Although knee arthroplasty produces a reduction in pain, following surgery the inability of patients to utilize the available range of motion at the knee during dynamic functional movement is the major finding of this study.

Using two inclinometers (low-pass accelerometer) on shank and thigh and electrogoniometers on knee, physical activity monitoring was performed before and after hip arthroplasty (Morlock et al.,2001). Main activity such as lying, sitting, standing, walking and stair climbing were identified by a subtle combination of the signals of the 3 sensors. The clinical HHS score correlated significantly with the number of stairs (r^2=0.26, p=0.003) and showed a positive tendency with the number of steps per day.

Parkinson

Parkinson's disease (PD) is by far the most common degenerative disease in the general population. Cardinal symptoms of PD include tremor, rigidity, akinesia, postural abnormalities and gait impairment. Advanced PD is thus typically characterized by severe, unpredictable and abrupt changes of the patient's motor function whereby OFF periods, characterized by the temporary loss of drugs' efficacy and the return of most parkinsonian symptoms, alternate, sometimes within minutes, with ON periods where the patients go back to "normal" state and a period during which the medication effects generate dyskinesia. The clinical assessment of these ON-OFF fluctuations, whose accuracy will determine the therapeutic interventions necessary to overcome them, is difficult and relies mostly on subjective historical data obtained from the patient itself or from relatives, and on standard clinical scales such as the UPDRS (Fahn et al.,1987) which are completed at the time of the patient's visit. To delineate precisely the temporal evolution of these complications, their characteristics and their severity, more objective instrumental methods are needed. Assessment of the PD disorders over time constitutes a major challenge for clinicians dealing with extrapyramidal patients, yet their accurate evaluation is the necessary prerequisite for appropriate therapeutic interventions to be undertaken. The development of body-fixed sensors allowing long-term ambulatory monitoring of activity and capable of classifying patterns of movement would have immediate clinical applications in the treatment of Parkinson's disease (Standaert, 2003).

Gait is a particular, semi-automatic motor task which is specifically sensitive to ON-OFF changes of parkinsonian state. Using 10 gyroscopes in 6 sensor sites (wrists, thighs and shanks) and using the Physiog datalogger systems with a sampling rate of 200Hz, Salarian et al. (2004) performed gait analysis on ten PD patients with sub-thalamic nucleus deep brain stimulation (STN-DBS), once STN-DBS was ON and once 180 minutes after turning it OFF. A group of ten age-matched normal subjects were also measured as controls. The results showed not only the change of spatio-temporal gait between PD patients and control, but provided significant change of gait parameters between ON and OFF periods (Table 1). Moreover stride-length, stride-velocity and range of shank rotation show a very good correlation with UPDRS gait sub-score.

Freezing is a phenomenon appearing during gait in PD patients. The feet are "glued" to the ground and the subject is unable to move forward. Using foot pressure sensors

Table 1. Measured gait parameters for patients during stim ON and stim OFF state and controls. p-values more than 0.05 were considered as not significant (NS) (Salarian et al., 2004)

Gait Parameters	Values for each group in mean±S.D.			p-value for the equivalence of mean of parameters			p-value for equivalence of mean of C.V,		
	Stim OFF	Stim ON	Control	ON/OFF paired	ON v.s. Control	OFF v.s. Control	ON/OFF paired	ON v.s. Control	OFF v.s. Control
Gait Cycle Time (s)	1.4± 0.6	1.2±0.2	1.0±0.1	N.S.	0.0312	0.0312	N.S.	0.0211	N.S.
Stance (%)	65.7± 8.6	61.5±4.5	59.4±1.2	0.0488	N.S.	0.0312	N.S.	0.0173	0.0211
Double Support (%)	31.4±17.1	23±9.1	18.7±2.5	0.0488	N.S.	0.0312	N.S.	N.S.	N.S.
Limp (%)	7.2±8.6	4.2±2.2	1.4±0.5	N.S.	0.0010	0.0006	N.S.	N.S.	N.S.
Stride Length (%h)	46.2±19.4	58.6±17.9	77.1±6.5	0.0020	0.0073	0.0004	N.S.	N.S.	0.0017
Stride Velocity (%h/s)	40.5±23.5	53.1±20.2	77.4±9.2	0.0039	0.0022	0.0003	0.0137	N.S.	N.S.
Range of Shank rotation (deg)	45.6±19.5	56.5±18.5	76±5.9	0.0020	0.0022	0.0002	0.0020	0.0113	0.0028
Range of Thigh rotation (deg)	28±8.5	34.4±8.6	34.4±11.8	0.0098	N.S.	N.S.	N.S.	0.0257	0.0028
Range of Knee rotation (deg)	39.4±13.7	45.4±15.8	60.4±7.9	0.0195	0.0113	0.0003	N.S.	N.S.	0.0140
Range of hand rotation, Pitch axis (deg)	8.4±5.1	17.8±12.6	20.2±6.4	N.S.	N.S.	0.0013	N.S.	N.S.	N.S.
Range of hand rotation, Roll axis (deg)	14±13.3	18.2±6.7	22.9±5	N.S.	N.S.	0.0028	N.S.	N.S.	N.S.
Range of hand rotation, Yaw axis (deg)	10.3±5.3	24.6±12.1	47.6±8.4	0.0039	0.0013	0.0002	N.S.	0.0452	N.S.
Peak Shank angular Velocity (deg/s)	225.2±103.5	275.4±110	386.3±40.1	0.0020	0.0058	0.0003	0.0020	N.S.	0.0452

and fractal analysis of the time series signal of foot pressure during freezing periods, Hausdorff, Balash, and Giladi (2003) found that the complex movement pattern is not random nor akinetic but has a fracal-like organizing mechanism which may reflect the activation of central control mechanism in response to freezing. Moreover, using the same sensor, an increase in the variability of spatiotemporal gait parameters in PD patient was reported (Hausdorff, Cudkowicz, Firtion, Wei, & Goldberger, 1998). In PD patients, the acceleration pattern of gait within each step (Sekine, Akay, Tamura, & Higashi, 2004) obtained by triaxial accelerometer on waist has also shown a fractal dimension which is higher than a healthy subject.

Dyskinesias is an involuntary movement which occur in PD when benefit form levodopa therapy are at the maximum. Objective assessment of dyskinesia is a challenging goal for many movement disorder specialists (Lang, 1995). Using a single gyroscope attached on the hand, Burkhard, Shale, Langston, and Tetrud (1999) have successfully quantified and characterized dyskinesia as well as underlined the high potential of gyroscopes for long-term monitoring of dyskinesia in PD patients. Keijsers, Horstink, and Gielen (2003, January) developed an ambulatory system based on 6 triaxial accelerometers attached at six different body positions to detect dyskinesia movement. A neural network was trained to distinguish dyskinesia from voluntary movements in daily life conditions. The input of the neural network consisted of temporal, statistical and spectral features of the acceleration data while the output was the severity score of dyskinesia evaluated in different body segment. The neural network was trained based on acceleration features and corresponding dyskinesia score derived from clinical evaluation (Keijsers, Horstink, & Gielen, 2003, February).

Tremor, another symptom of PD can be assessed by BFS on hand. Two uniaxial accelerometers (Ghika et al., 1993) and triaxial accelerometer (van Someren et al., 1998) have been used for the tremor recording. Ambulatory techniques with triaxial gyroscopes have also been reported for tremor monitoring, which have the advantage of providing tremor amplitude in degrees and therefore be independent of sensor position relative to the arm (Salarian et al., 2003).

Bradykinesia is the slowness of movement and the cause of impaired voluntary movement in PD patients. The possibility to quantify the level of bradykinesia during a test procedure was reported by using accelerometers on wrist (Dunnewold, Jacobi, & Can Hilten, 1997). Bradykinesia during spontaneous activity was investigated by using two triaxial gyroscopes attached on the wrist. The RMS value of angular velocity around each axis of rotation as well as the range of motion in degree for each motion have been estimated for a duration ranged from 25 to 40 minutes. These parameters showed a significant and high correlation with UPDRS subscore related to bradykinesia (r=0.83, p<0.0001) (Salarian et al., 2003). The same sensor configuration also provided sensitivity higher than 98% for tremor detection. The system is ambulatory and offers the possibility to record bradykinesia and tremor during long-term daily activity.

Aging

Older people make up a large and increasing percentage of the population. As people grow older they are increasingly at risk of falling and consequent injuries. Approximately 30% of people over 65 fall each year, and for those over 75 the rates are higher. Between 20% and 30% of those who fall suffer injuries that reduce mobility and independence and increase the risk of premature death (Todd & Skelton, 2004).

Motor function evaluation, fall risk estimation, fall prevention and fall detection are some main topics in aged population where body fixed sensors could provide significant contribution.

Nyberg and Gustafson (1995) reported that many falls in stroke patients occur during activities in which they change position (e.g., standing up, sitting down, or initiating walking). Pai and Rogers (1990) evaluated the effect of varying the speed of rising. Their studies have demonstrated that several conditions may influence the dynamics of movement of the whole body during sit-to-stand transfer, such as a subject's age and variance in the speed of rising.

Motor performance in elderly people has been evaluated using bi-axial accelerometers and a gyroscope attached as a single module on thorax. The duration of rising up from sitting or sitting down from a standing posture as well as the variability of this duration over repeated tasks have been considered to classify people with high and low risk of fall (Najafi et al., 2002). Trunk oscillation (sway) measured using biaxial gyroscopes (roll and pitch) has been used to identify a balance disorder and possibly to distinguish between different types of balance disorder (Allum et al., 2001). Larger trunk angular displacement in roll direction (medio-lateral) measured by gyroscope during walking has been used to distinguish elderly fallers from non-fallers (de Hoon et al., 2003). Using an ambulatory system with triaxial accelerometer on trunk, gait variability was estimated by an autocorrelation procedure (Moe-Nilssen & Helbostad, 2005). The authors reported that frail elderly showed lower mediolateral ($P = 0.015$), but higher vertical ($P = 0.015$) and anteroposterior ($P < 0.02$) trunk variability than the fit group. Trunk variability classified 80% of the subjects correctly into their respective group (sensitivity = 0.75, specificity = 0.85). By recording plantar pressure patterns detected by FSR sensors, Hausdorff et al. (1997) have also shown that gait inter-cycle variability

Table 2. Age-related decline of gait control under a dual-task condition (Beauchet et al., 2003)

	Walking alone	Walking while Backward counting	P-Value*
Young subjects (n = 12)			
Stride length CV (%)	2.3 ± 0.8	2.7 ± 1.2	0.308
Stride velocity CV (%)	3.2 ± 1.3	3.5 ± 1.8	0.638
Old subjects (n = 12)			
Stride length CV (%)	3.9 ± 1.6	10.2 ± 9.3	0.023
Stride velocity CV (%)	5.6 ± 2.2	12.5 ± 9.2	0.015

is significantly higher for faller that non fallers and young subjects. Moreover using the same sensors, Hausdorff et al. (2001) showed that long-range variability in gait pattern which can be expressed as the fractal dimension of gait variability decreased with aging. The degree of fractal dimension is strongly related to the degree of functional impairment.

Walking is a highly automated, rhythmic motor behavior that is mostly controlled by subcortical locomotor brain regions. Cognitive tasks such as backward counting can therefore interfere with gait regularity in older adults who need more attention. Gait analysis using gyroscope and "dual task" paradigm (backward counting during walking) was able to classify older and young subjects (Beauchet et al., 2003) (Table 2). Dual task increased stride-to-stride variability in older subjects. This strongly suggests the involvement of higher cortical regions for the motor control of gait in this population.

Several studies have attempted to provide a fall detector based on body-fixed accelerometers. Acceleration patterns during fall have been studied based on the peak detection of acceleration or using intelligent computing such as artificial neural network (Depeursinge, Krauss, & El-Khoury, 2001) in order to provide tele-alarm system. Although many investigations reported the possibility of fall detection with more or less effectiveness, there is no intelligent system with high sensitivity and specificity that can be used in every day life of elderly people. The main problem in the design of such a system is the difficulty of having enough actual patterns of fall obtained in real situations.

Pain

Pain is usually assessed by questionnaires and visual analogue scales (VAS) (Physical Rehabilitation Outcomes Measures, Finch, & Brooks, 2002). While these methods have been shown to be reliable, they have become generally accepted and are increasingly used. They remain essentially subjective and time-consuming. In addition since these questionnaires are based on self-reporting, their application is subjected to (but also restricted by) the cooperation of the patient. Pain behaviour may be characterized by changes in both, the quantity of activities and the quality of mobility. *Quantitative mobility parameters* could be the time spent in different postures (e.g., lying, sitting, standing, walking), the daily walking distance or the number of postural transitions (mobility intensity). *Qualitative mobility parameters* are more related to the way that an activity is performed for example, acceleration or deceleration of trunk during walking, gait speed, the cadence, the duration of sit-to-stand or stand-to-sit postural transition. In some cases the relation between the quantity of activity and pain is not very clear if the quality of activity is ignored in the analysis of data. This is due to the fact that activity levels depend generally on the environmental factors such as the work, the weather, the different days of the week and also, the personality structure of the patient. The provision of a reliable method for the measurement of physical activity would provide an instrument for the assessment of disability and many features of the health-related pain.

Gait feature changes have been reported by Moe-Nilssen, Ljunggren, and Torebjork (1999) where the intensity of horizontal acceleration (anterior-posterior and medio-lateral) has been significantly decreased during walking in patients with low back pain.

Pain related mobility was estimated in patients with complex regional pain syndrome type I (Stanton-Hicks et al.,1995) using body fixed accelerometers (Schasfoort et al., 2003). Physical mobility was detected using one sensor on the right thigh, two sensors on the sternum and two sensors on each forearm. The raw acceleration signals were stored digitally with a sample frequency of 32 Hz. Several outcome measures related to general mobility (time spent in different postures and body motions) and upper limb usage (intensity of upper limb activity during sitting and standing, their duration and the proportion of upper limb activity of one side relative to the other side) were compared between patients and controls. The results showed that the pain syndrome in the dominant upper limb had modest impact on general mobility while for upper limb usage outcome measures during sitting. There was a marked difference between painful patients and controls. Especially patients with dominant side involvement clearly showed less activity of their involved limb during sitting, indicated by significant differences for the mean intensity, percentage and proportion of upper limb activity.

Table 3. The variability between subjects regarding change from baseline of pain intensity, physical activity and gait parameters at 1, 3, 6 and 12 months after implantation of spinal cord stimulation system (Buchser et al., 2005)

CV(%)	1 month	3 months	6 months	*12 months*
Pain intensity	68.7	93.1	80.3	18.2
Total walking distance	191.7	145.6	121	140.6
Maximal walking distance	200.5	147.8	156.7	119.4
Lying	200	147.12	108.2	35
Walking and standing	180	220	103	72
Speed	82.6	81.2	88.7	77
Speed variability	337	161	41	38
Stride length	112	124	98	129
Stride length variability	110	197	75	96

Spinal cord stimulation (SCS) is increasingly used for the treatment of intractable pain syndromes due to vascular or neurogenic disorders. Although it is generally accepted that physical activity is decreased in chronic pain states, there seems to be no direct correlation between the intensity of pain and the restriction of physical activity. Using a combination of gyroscopes and accelerometers attached on chest, thigh and shank, changes in pain and spontaneous physical activity following SCS were evaluated under real life conditions (Paraschiv-Ionescu, Buchser, Rutschmann, Najafi, & Aminian, 2004). Five series of measurements were performed before the implant (baseline) and at one, three, six and 12 months after the implantation of a SCS system. Each of the five series was carried out during 7 hours per day and over 5 consecutive days, resulting in the collection of 175 hours corresponding to around 25 millions data samples on posture and gait for each patient. The pain was assessed using visual analog scale (VAS) (Buchser et al., 2005).

The authors showed that compared to baseline values, the physical activity increased consistently during the entire follow-up period. The time spent walking and standing was statistically increased after 6 months ($p<.001$) and the time spent lying decreased significantly ($p<.001$) at the same time. The average total walking distance increased up to 389% at 12 months, reaching statistical significance ($p<.05$) after 3 months. The stride length and the speed increased ($p<.01$) at all times. Table 3 illustrates the change of mobility parameters compared to the change in pain.

Energy Expenditure

The trunk and body segments movement requires acceleration and deceleration and these accelerations are, to a large extent, responsible for the energy expenditure that the movement requires. One of the aims of the accelerometry is to record these acceleration changes and therefore to evaluate energy expenditure. Accelerometry is widely used for temporal tracking of the frequency, intensity and duration of physical activities. A number of activity monitors have been commercialized with various characteristics. Some instigators have developed their own instruments (Bouten, Koekkoek, Verduin, Kodde, & Janssen, 1997). To determine energy expenditure, acceleration signals are processed to obtain the sum of the rectified and integrated acceleration from all three measurement directions. The integration time is set at 1 minute and the final output is expressed as "counts per minute" (Bouten, Verboeket-van de Venne, Werterterp, Verduin, & Janssen, 1996). The use of accelerometer in energy expenditure assessment has been widely discussed in a special issue of the journal *Medicine & Science in Sports & Exercises* (2000, September, supplement). The accuracy of the accelerometers for energy expenditure estimation depends on the activity being performed. Previous studies have demonstrated the usefulness of accelerometry for estimating energy expenditure of level walking (Balogun, Martin, & Clendenin, 1989; Haymes & Byrnes, 1993). However, accelerometers cannot measure with enough accuracy the energy cost of walking up or down a slope (Terrier, Aminian, & Schutz, 2001). Although use of additional processing techniques, such as artificial neural networks, have shown success in estimating the incline along which a subject is walking (Aminian, Robert, Jéquier, & Schutz, 1995), and this information may be used to provide improved estimation of energy expenditure. One of the main potential of energy expenditure knowledge is to study human obesity. Actually obesity occurs when energy intake exceeds energy expenditure. Low, moderate and high activity can be estimated based on intensity of the acceleration (Westerterp, 2001) and by adding inclinometer (gravity accelerometer) new insight between body postures and obesity has been reported (Levine et al., 2005).

FUTURE TRENDS: MOVEMENT CAPTURE AND INTELLIGENT COMPUTING IN WEARABLE TECHNOLOGIES

Ambulatory monitoring of human movement using body fixed sensors can provide satisfactory results over long-term monitoring reaching several days. However, these systems have some drawbacks. Sensors needed to be fixed on the limbs sometimes directly in the skin sometimes using a belt or cloth accessories. Special care is needed to avoid any inconvenience such as allergy or cumbersome situations. Moreover, the user (e.g., the physician) needs to spend a period of time- more or less short- for fixing the sensors on the body. Ambulatory systems need power and should be tested, powered and recharged regularly. Depending on the type of recording and device, subject should be trained for the use of the device or how to communicate with it, which is impossible with patients with severe handicaps. For all these reasons, ambulatory systems are not suitable when monitoring has to be accomplished over periods of several weeks or months, as is desirable in a number of clinical applications.

People have an increasing desire for *ubiquitous* access to information, anywhere, anyplace and anytime. For that, they need mobile and portable devices, but also adequate communication systems and software infrastructures. Totally unobtrusive devices that allow physicians to overcome the limitations of ambulatory systems and which can be worn by the subject have grown a lot of interest (Marculescu, Marculescu, Park, & Jayaraman, 2003). Wearable technology typically relies on wireless, miniature sensors enclosed in patches or bandages, or in items that can be worn, such as a ring or a shirt. The progress of technology in the development of miniature body fixed sensors open new possibilities in this new area. The design and implementation of sensors that are minimally obtrusive and reliably record movement beside other physiological signals provide valuable information to clinicians. The use of intelligent computing will be appropriate to extract relevant information from data recorded using multiple body-fixed sensors. Some of the potential of wearable technology in personal health monitoring, telerehabilitation and physical activity monitoring have been reviewed in (Jovanov et al., 2003; Winters, Wan, Yu, & Winters, 2003; Moy, Mentzer, & Reilly, 2003).

REFERENCES

Abu-Faraj, Z. O., Harris, G. F., Abler, J. H., & Wertsch, J. J. (1997). A holter-type, microprocessor-based, rehabilitation instrument for acquisition and storage of plantar pressure data. *J. Rehabil Res Dev, 34*, 187-194.

Adkin, A. L., Bloem, B. R., & Allum, J. H. J. (2005). Trunk sway measurements during stance and gait tasks in Parkinson's disease. *Gait & Posture, 22*(3), 240-249.

Allum, J.H., Adkin, A.L., Carpenter, M.G., Held-Ziolkowska, M., Honegger, F., & Pierchala, K. (2001). Trunk sway measures of postural stability during clinical balancetests: effects of a unilateral vestibular deficit. *Gait & Posture, 14,* 227-237.

Allum, J. H. J., & Carpenter, M. G. (2005). A speedy solution for balance and gait analysis: angular velocity measured at the centre of body mass. *Current Opinion in Neurology, 18,* 15-21.

Aminian, K., & Najafi, B. (2004). Capturing human motion using body fixed sensors: outdoor measurement and clinical applications. *Journal of Visualization and Computer Animation, 15,* 79-94.

Aminian, K., de Andrès, E., Rezakhanlou, K., Fritsch, C., Depairon, M., Schutz, Y., Leyvraz, P. F., & Robert, Ph. (1998). Motion analysis in clinical practice using ambulatory accelerometry. In *Proceedings of Captech '98, LNAI* (Vol. 1537, pp. 1-11). Springer-Verlag.

Aminian, K., Najafi, B., Bula, C., Leyvraz, P. F., & Robert, Ph. (2002). Spatio-temporal parameters of gait measured by an ambulatory system using miniature gyroscopes. *Journal of Biomechanics, 35*(5), 689-699.

Aminian, K., Rezakhanlou, K., de Andres, E., Fritsch, C., Leyvraz, P. F., & Robert, P. (1999). Temporal features estimation during walking using miniature accelerometers: an analysis of gait improvement after hip arthroplasty. *Medical & Biological Engineering & Computing, 37,* 686-691.

Aminian, K., Robert, Ph., Buchser, E. E., Rutschmann, B., Hayoz, D., & Depairon, M. (1999). Physical activity monitoring based on accelerometry: validation and comparison with video observation. *Medical & Biological Engineering & Computing, 37*(3), 1-5.

Aminian, K., Robert, Ph., Jéquier, E., & Schutz, Y. (1995). Incline, speed and distance assessment during unconstrained walking. *Medicine & Science in Sports & Exercise, 27*(2), 226-234.

Aminian et al. (2004). Evaluation of an ambulatory system for gait analysis in hip osteoarthritis and total replaced patients. *Gait & Posture, 20,* 102-107.

Anderson, G. (1972). Hip assessment: a comparison of nine different methods. *Journal of Bone & Joint Surgery, 54B*(4), 621-625.

Andriacchi, T. P., Hampton, S. J., Schultz, A.B., & Galante, J. O. (1979). Three dimensional coordinate data processing in human motion analysis. *Journal of Biomechanical Engineering, 101,* 279-283.

Azuma, R., Baillot, Y., Behringer, R., Feiner, S., Julier, S., & MacIntyre, B. (2001). Recent advances in augmented reality. *IEEE Computer Graphics and Applications, 21*(6), 34-47.

Bachmann, E. R., Yun, X., & McGhee, R. B. (2003). Sourceless tracking of human posture using small inertial/magnetic sensors. *IEEE International Symposium on Computational Intelligence in Robotics and Automation,* 822-829.

Balogun, J. A., Martin, D. A., & Clendenin, M. A. (1989). Calorimetric validation of the Caltrac accelerometer during level walking. *Journal of the American Physical Therapy Association, 69*(6), 501-509.

Barin, K. (1992). Dynamic posturography. *IEEE Engineering in Medicine and Biology, 11*(4), 52-56.

Baselli, G., Legnani, G., Franco, P., Brognoli, F., Marras, A., Quaranta, F., & Zappa, B. (2002). Assessment of inertial and gravitational inputs to the vestibular system. *Journal of Biomechanics 34*(6), 821-826.

Beauchet, O., Najafi, B., Dubost, V., Aminian, K., Mourey, F., & Kressig, R.W. (2003). Age-related decline of gait control under a dual-task condition. *Journal of the American Geriatrics Society, 51*(8), 1187-1188.

Berthoz, A. (1997). *Le sens du mouvement.* Paris: Odile Jacob.

Bodenheimer, B., Rose, C., Rosenthal, S., & Pella, J. (1997). The process of motion capture: dealing with the data. In D. Thalmann & M. van de Panne (Eds.), *Computer Animation and Simulation '97, Eurographics Animation Workshop* (pp. 3-18). New York: Springer.

Bouten, C. V. C., Koekkoek, K. T. M., Verduin, M., Kodde, R., & Janssen, J. D. (1997). A triaxial accelerometer and portable data processing unit for the assessment of daily physical activity. *IEEE Transactions on Biomedical Engineering, 44*(3), 136-147.

Bouten, C. V. C., Verboeket-van de Venne, W. P. H. G., Westerterp, K. R., Verduin, M. & Janssen, J. D. (1996). *Journal of Applied Physiology, 81,* 1019-1026.

Buchser, E., Paraschiv-Ionescu, A., Durrer, A., Depierraz, B., Aminian, K., Najafi, B., & Rutschmann, B. (2005). Improved physical activity in patients treated for chronic pain by spinal cord stimulation. *Neuromodulation, 8,* 40-48.

Burkhard, P. R., Shale, H., Langston, J. W., & Tetrud, J. W. (1999). Quantification of dyskinesia in Parkinson's disease: validation of a novel instrumental method. *Movement Disorders, 14*(5), 754-763.

Bussmann, J. B. J., van de Laar, Y. M., Neeleman, M. P., & Stam, H. J. (1998). Ambulatory accelerometry to quantify motor behavior in patients after failed back surgery: A validation study. *Pain, 74,* 153-161.

Cappozzo, A., Cappello, A., Della Croce, U., & Pensalfini, F. (1997). Surface-marker cluster design criteria for 3-D bone movement reconstruction. *IEEE Transactions on Biomedical Engineering, 44,* 1165-1174.

Cappozzo, A., Catani, F., Leardini, A., Benedetti, M. G., & Della Croce, U. (1996). Position and orientation in space of bones during movement: Experimental artefact. *Clinical Biomechanics, 11*(2), 90-100.

Chang, W. L., Su, F. C., Wu, H. W., & Wong, C. Y. (1998). Motion analysis of scapula with combined skeleton and skin-based marker system. *3rd World Congress of Biomechanics*, Sapporo, Japan (pp. 2-8).

Château, H., Girard, D., Degueurce, C., & Denoix, J. M. (2003). Analyse des contraintes méthodologiques liées à l'utilisation d'un système d'analyse cinématique tridimensionnelle fondé sur le principe de la triangulation ultrasonore. *Innovation et technologie en biologie et médecine, ITBM-RBM, 24,* 69-78.

Coley, B., Najafi, B., Paraschiv-Ionescu, A., & Aminian, K. (2005). Stair climbing detection during daily physical activity using a miniature gyroscope. *Gait & Posture, 22,* 287-294.

de Andrès, E., Aminian, K., Fritsch, C., Leyvraz, P. F., & Robert, Ph. (1998). Interest of gait analysis in hip and knee. In *Proceeding of the 2nd Mediterranean Congress of Physical Medicine and Rehabilitation,* Valencia, Spain (p. 347).

de Hoon, E. W., Allum, J. H., Carpenter, M. G., Salis, C., Bloem, B. R., Conzelmann, M., & Bischoff, H. A. (2003). Quantitative assessment of the stops walking while talking test in the elderly. *Archives of Physical Medicine and Rehabilitation, 84*(6), 838-842.

Dejnabadi, H., Jolles, B.M., & Aminian, K. (2005). A new approach to accurate measurement of uniaxial joint angles based on a combination of accelrometers and gyroscopes. *IEEE Transactions on Biomedical Engineering, 52*(8), 1478-1484.

Dejnabadi, H., Jolles, B. M., Najafi, B., Trevisan, C., Marinoni, E. C., & Aminian, K. (2003, March 23-27). A robust gait parameterization technique for hip arthroplasty outcome evaluation. In S. R. Lord & H. B. Menz (Eds.), *Proceedings of the ISPGR, Posture and gait throughout the lifespan*, Sydney, Australia, (p. 52).

Depeursinge, Y., Krauss, J., & El-Khoury, M. (2001). Device for monitoring the activity of a person and/or detecting a fall, in particular with a view to providing help in the event of an incident hazardous to life or limb. *Brevet US, 6,201,47. CSEM - Centre Suisse d'Electronique et de Microtechnique S.A.*

Doutrellot, J. P. (1992). Analyse critique de l'utilisation du tapis roulant dans l'étude biomécanique de la locomotion. *Mémoire de DEA. Sciences et Techniques appliquées au handicap et à la réadaptation. Université de Bourgogne.*

Dunnewold, R. J. W., Jacobi, C. E., & Can Hilten, J. J. (1997). Quantitative assessment of bradykinesia in patients with Parkinson's disease. *Journal of Neuroscience Methods, 74,* 107-112.

Ehara, Y., Fujimoto, H., Miyazaki, S., Mochimaru, M., Tanaka, S., & Yamamoto, S. (1997). Comparison of the performance of 3D camera systems .2. *Gait & Posture, 5*(3), 251-255.

Fahn, S., Elton, R. L., & members of the UPDRS Development Committee. (1987). Unified Parkinson's disease rating scale. In S. Fahn, C. D. Marsden, D. Calne, & M. Goldstein (Eds.), *Recent developments in Parkinson's disease* (Vol. II, pp. 153-163). Florhan Park: Mc Millan Health Care Information.

Feipel, V., Rondelet, B., Le Pallec, J. P., & Rooze, M. (1999). Normal global motion of the cervical spine: an electrogoniometric study. *Clinical Biomechanics, 14*(7), 462-470.

Foxlin E. (1996). Inertial head-tracking sensor fusion by a complementary separate-bias Kalman filter. In *IEEE Proceedings of VRAIS* (pp. 185-195).

Fyfe, K. R., & Fyfe, K. W. (2001). Motion analysis system. *European patent, EP1066793.*

Ghika, J., Wiegner, A. W., Fang, J. J., Davies, L., Young, R. R., & Growdon, J. H. (1993). Portable system for quantifying motor abnormalities in Parkinson's disease. *IEEE Transactions on Biomedical Engineering, 40*(3), 276-283.

Giansanti, D., Macellari, V., Maccioni, G., & Cappozzo, A. (2003). Is it feasible to reconstruct body segment 3-D position and orientation using accelerometric data. *IEEE Transactions on Biomedical Engineering, 50*(4), 476-483.

Grétillat, F. (1998). *Silicon micromachined vibrating gyroscopes with piezoresistive detection and electromagnetic excitation.* Dissertation, University of Neuchâtel.

Haid, M., & Breitenbach, J. (2004). Low cost inertial orientation tracking with Kalman filter. *Applied Mathematics and Computation, 153,* 567-575.

Harris, W. H. (1969). Traumatic arthritis of the hip after dislocation and acetabular fractures: treatment by mold arthroplasty. *Journal of Bone and Joint Surgery, 51A,* 737-755.

Hausdorff, J. M., Ashkenazy, Y., Peng, C. K., Ivanov, P. C., Stanley, H. E., & Goldberger, A. L. (2001). When human walking becomes random walking: fractal analysis and modeling of gait rhythm fluctuations. *Physica A., 302*(1-4), 138-147.

Hausdorff, J. M., Balash, Y., & Giladi, N. (2003). Time series analysis of leg movements during freezing of gait in Parkinson's desease: akinesia, rhyme or reason? *Physica A: Statistical mechanics and its applications, 321*(3-4), 565-570.

Hausdorff, J. M., Cudkowicz, M. E., Firtion, R., Wei, J. Y., & Goldberger, A. L. (1998). Gait variability and basal ganglia disorders: Stride-to stride variations of gait cycle timing in Parkinson's disease and Huntington's disease. *Movement disorders, 13*(3), 428-437.

Hausdorff, J. M., Edelberg, H. K., Mitchell, S. L., Goldberger, A. L., & Wei, J. Y. (1997). Increased gait unsteadiness in community-dwelling elderly fallers. *Archives of Physical Medicine and Rehabilitation, 78,* 278-283.

Hausdorff, J. M., Ladin, Z., & Wei, J. Y. (1995). Footswitch system for measurement of the temporal parameters of gait. *Journal of Biomechanics, 28*(3), 347-351.

Haymes, E. M., & Byrnes, W. C. (1993). Walking and running energy expenditure estimated by Caltrac and indirect calorimetry. *Medicine & Science in Sports & Exercise, 25,* 1365-1369.

Heyn, A., Mayagoitia, R. E., Nene, A. V., & Veltink, P. H. (1996, October 31-November 3). The kinematics of the swing phase obtained from accelerometer and gyroscope measurements. In *Proceedings of the 18th Annual Conference of the IEEE Engineering in Medical Biology Society*, Amsterdam (Vol. 2, pp. 463-464).

Jeannerod, M. (1991). *A neurophysiological model for the directional coding of reaching movements*. In J. Paillard (Ed.), *Brain and space* (pp. 49-69). Oxford, UK: Oxford University.

Jeka, J., Kiemel, T., Creath, R., Horak, F., & Peterka, R. (2004). Controlling human upright posture: Velocity information is more accurate than position or acceleration. *Journal of Neurophysiology, 92,* 2368-2379.

Jonghun, B., Geehyuk, L., Wonbae, P., & BJ, Yun (2004). Accelerometer Signal Processing for User Activity. *Lecture Notes in Computer Science, 3215,* 610-617.

Jovanov, E., Lords, A. O., Raskovic, D., Cox, P. G., Adhami, R., & Andrasik, F. (2003). Stress monitoring using a distributed wireless intelligent sensor system. *IEEE Engineering in Medicine and Biology Magazine, 22*(3), 49-55.

Keijsers, N. L., Horstink, M. W., & Gielen, S. C. (2003, January). Movement parameters that distinguish between voluntary movements and levodopa-induced diskinesia in Parkinson's disease. *Human Movement Science, 22*(1), 67-89.

Keijsers, N. L., Horstink, M. W., & Gielen, S. C. (2003, February). Automatic assessment of levodopa-induced diskinesias in daily life by neural networks. *Movement Disorders, 18*(1), 70-80.

Keller, R. B., Rudicel, S. A., & Liang, M. H. (1993). Outcomes research in orthopaedics. *Journal of Bone & Joint Surgery, 75A,* 1562-1574.

Kemp, B., Janssen, A. J., & van der Kamp, B. (1998). Body position can be monitored in 3D using miniature accelerometers and earth-magnetic field sensors. *Electroencephalography and Clinical Neurophysiology, 109,* 484-488.

Kettelkamp, D. B., Johnson, R. J., Smidt, G. L., Chao, E. Y. & Walker, M. (1970). An electrogoniometric study of knee motion in normal gait. *Journal of Bone and Joint Surgery, 52-A,* 775-790.

Kiss, R. M., Kocsis, L., & Knoll, Z. (2004). Joint kinematics and spatial-temporal parameters of gait measured by an ultrasound-based system. *Medical Engineering & Physics, 26*(7), 611-620.

Kobayashi, K., Gransberg, L., Knutsson, E., & Nolen, P. (1997). A new system for three-dimensional gait recording using electromagnetic tracking. *Gait & Posture, 6*(1), 63-75.

Kolen, P. T., Rhode, J. P., & Francis, P. R. (1993). Absolute angle measurement using the earth-field-referenced Hall effect sensors. *Journal of Biomechanics, 26*(3), 265-270.

Ladetto, Q. (2002). *Capteurs et algorithmes pour la localisation autonome en mode pédestre.* Thesis EPFL 2710.

Lang, A. (1995). Clinical rating scales and videotape analysis. In W. Koller & G. Paulson (Eds.), *Therapy of Parkinson's disease* (pp. 21-46). New York: Marcel Dekker.

Lee, R. Y. W., Laparde, J., & Fung, E. H. K. (2003). A real-time gyroscopic system for three-dimensional measurement of lumbar spine motion. *Medical Engineering & Physics, 25,* 817-824.

Levine et al. (2005). Interindividual variation in posture allocation: Possible rote in human obesity. *Science 307*(5709), 584-586.

Lieberman, J. R., Dorey, F., Shekelle, P., Schumaker, L., Thomas, B. J., Kilgus, D. J., & Finerman, G. A. (1996). Differences between patients and physicians evaluations of outcome after total hip arthroplasty. *Journal of Bone & Joint Surgery, 78*(A), 835-838.

Lieberman, J. R., Dorey, F., Shekelle, P., Schumacher, L., Kilgus, D. J., Thomas, B. J., & Finerman, G. A. (1997). Outcome after total hip arthroplasty. Comparison of a traditional disease-specific and a quality-of-life measurement outcome. *Journal of Arthroplasty, 12*(6), 639-645.

Llinás, R., & Ribary, U. (1993). Coherent 40Hz oscillation characterizes dream state in humans. *Proceedings of the National Academy of Sciences, 90,* 2078-2081.

Luinge, H. J., & Veltink, P. H. (2004). Inclination measurement of human movement using a 3-D accelerometer with autocalibration. *IEEE Transactions on Neural Systems and Rehabilitation Engineering, 12*(1).

Maluf, K. S., & Mueller, M. J. (2003). Comparison of physical activity and cumulative plantar tissue stress among subjects with and without diabetes mellitus and a history of recurrent plantar ulcers. *Clinical Biomechanics, 18,* 567-575.

Marculescu, D., Marculescu, R., Park, S., & Jayaraman, S. (2003). Ready to ware. *IEEE Spectrum, Vol. 40*(10), 28-32.

Marins, J. L., Yun, X., Bachmann, E. R., McGhee, R. B., & Zyda, M. (2001, October 29-November 3). An extended Kalman filter for quaternion-based orientation estimation using MARG sensors. In *Proceedings of the International Conference on Intelligent Robots and Systems,* Maui, Hawaii (pp. 2003-2011).

Martinet, N., & Andr, J. M. (1994). Analyse cinématique: approche goniométrique. In J. Pelissier & V. Brun (Eds.), *La marche humaine et sa pathologie* (No 27, pp. 75-82). Collection de pathologie locomotrice. Paris: Masson.

Mathie, M. J., Coster, A. C. F., Lovell, N. J., & Celler, B. G. (2004). Accelerometry: providing an integrated, practical method for long-term, ambulatory monitoring of human movement. *Physiological Measurement, 25,* R1-R20.

Mayagoitia, R. E., Joost, C., Lotters, J. C., Veltink, P. H., & Hermens, H. (2002). Standing balance evaluation using a triaxial accelerometer. *Gait & Posture, 16,* 55-59.

www.measurand.com (2005).

Medicine & Science in Sports & Exercise (2000). *Supplement 32*(8), S439-S516.

Miyazaki, S. (1997). Long-term unrestrained measurement of stride length and walking velocity utilizing a piezoelectric gyroscope. *IEEE Transactions on Biomedical Engineering, 44*(8), 753-759.

Moe-Nilssen, R. (1998). Test-retest reliability of trunk accelerometry during standing and walking. *Archives of Physical Medicine and Rehabilitation, 79*, 1377-1385.

Moe-Nilssen, R., & Helbostad, J. L. (2002). Trunk accelerometry as a measure of balance control during quiet standing. *Gait & Posture, 16(1)*, 60-68.

Moe-Nilssen, R., & Helbostad, J. L. (2004). Estimation of gait cycle characteristics by trunk accelerometry. *Journal of Biomechanics, 37*(1), 121-126.

Moe-Nilssen, R., & Helbostad, J. L. (2005). Interstride trunk acceleration variability but not step width variability can differentiate between fit and frail older adults. *Gait & Posture, 21*, 164-170.

Moe-Nilssen, R., Jorunn, L., & Helbostad, J. L. (2002). Trunk accelerometry as a measure of balance control during quiet standing. *Gait & Posture, 16*, 60-68.

Moe-Nilssen, R., Ljunggren, A. E., & Torebjork, E. (1999). Dynamic adjustments of walking behavior dependent on noxious input in experimental low back pain. *Pain, 83*, 477-485.

Molet, T., Boulic, R., & Thalmann, D. (1996). A real-time anatomical converter for human motion capture. *Proceedings of the 7th Eurographics Workshop on Animation and Simulation* (pp. 79-94). Wien: Springer-Verlag.

Morlock, M., Schneider, E., Bluhm, A., Vollmer, M., Bergmann, G., Müller, V. & Honl, M. (2001). Duration and frequency of every day activities in total hip patients, *Journal of Biomechanics, 34(7)*, 873-881.

Moy, M. L., Mentzer, S. J., & Reilly, J. J. (2003). Ambulatory monitoring of cumulative free-living activity. *IEEE Engineering in Medicine and Biology Magazine*, 89-95.

Myles, C. M., Rowe, P. J., Walker, C. R. C., & Nutton, R. W. (2002). Knee joint functional range of movement prior to and following total knee arthroplasty measured using flexible electrogoniometry. *Gait & Posture, 16*, 46-54.

Najafi, B., & Aminian, K. (2002). Body movement monitoring system and method. *Patent no EP1195139.*

Najafi, B., Aminian, K., Loew, F., Blanc, Y., & Robert, Ph. (2002). Measurement of stand-sit and sit-stand transitions using a miniature gyroscope and its application in fall risk evaluation in the elderly. *IEEE Transactions on Biomedical Engineering, 49*(8), 843-851.

Najafi, B., Aminian, K., Paraschiv-Ionescu, A., Loew, F., Büla, C., & Robert, Ph. (2003). Ambulatory system for human motion analysis using a kinematic sensor: monitoring of daily physical activity in elderly. *IEEE Transactions on Biomedical Engineering, 50*(6), 711-723.

Ng, J., Sahakian, A. V., & Swiryn, S. (2000). Sensing and documentation of body position during ambulatory ECG monitoring. *IEEE Computers in Cardiology, 27*, 77-80.

Nyberg, L., & Gustafson, Y. (1995). Patients falls in stroke rehabilitation. A challenge to rehabilitation strategies. *Stroke, 26*, 832-842.

Pai, Y. C., & Rogers, M. W. (1990). Control of body mass transfer as a function of speed of ascent in sit to stand. *Medicine & Science in Sports & Exercise, 22*, 378-384.

Pappas, I. P. I., Keller, T., Mangold, S., Popovic, M.R., Dietz, V., & Morari, M. (2004). A reliable gyroscope-based gait-phase detection sensor embedded in a shoe insole. *IEEE Sensors Journal, 4*(2), 268-274.

Pappas, I. P. I., Popovic, M.R., Keller, T., Dietz, V., & Morari, M. (2001). A reliable gait phase detection system. *IEEE Transactions on neural systems and rehabilitation engineering, [see also IEEE Transactions on Rehabilitation Engineering], 9*(2), 113-125.

Paraschiv-Ionescu, A., Buchser, E. E., RutschmannB., Najafi, B., Aminian, K. (2004). Ambulatory system for the quantitative and qualitative analysis of gait and posture in chronic pain patients treated with spinal cord stimulation. *Gait & Posture, 20*, 113-125.

Pataky, Z., Faravel, L., da Silva, J., & Assal, J. P. (2000). A new ambulatory foot pressure device for patients with sensory impairment. A system for continuous measurement of plantar pressure and a feed-back alarm. *Journal of Biomechanics, 33*, 1135-1138.

Perttunen, J. R., Anttila, E., Södergård, J., Merikanto, J., & Komi, P. V. (2004). Gait asymmetry in patients with limb length discrepancy. *Scandinavian Journal of Medicine and Science in Sports, 14*(1), 49.

Finch, E., Brooks, D., Stratford, P. W., Mayo, N. E. (2002). *Physical rehabilitation outcomes measures: A guide to enhanced clinical decision-making*. Lippincott Williams & Wilkins, 244-245.

Rehbinder, H., & Hu, X. (2004). Drift-free attitude estimation for accelerated rigid bodies. *Automatica, 40*, 653-659.

Richards J. G. (1999). The measurement of human motion: A comparison of commercially available systems. *Human Movement Science, 18*, 589-602.

Rietdyk, S., Patla, A. E., Winter, D. A., Ishac, M. G., & Little, C. E. (1999). Balance recovery from medio-lateral perturbations of the upper body during standing. *Journal of Biomechanics, 32*, 1149-1158.

Roduit, R., Besse, P. A., & Micalef, J. P. (1998). Flexible Angular sensor, *IEEE Transactions on Instrumentation and Measurement, 47*(4), 1020-1022.

Roumenin, Ch., Dimitrov, K., & Ivanov, A. (2001). Integrated vector sensor and magnetic compass using a novel 3D Hall structure. *Sensors and Actuators A: Physical, 92*(1-3), 119-122.

Sabatini, A. M., Martelloni, C., Scapellato, S. & Cavallo, F. (2005). Assessment of walking features from foot inertial sensing. *IEEE Transactions on Biomedical Engineering, 52*(3), 486-494.

Sagawa, K., & Satoh, Y. (2000). Non-restricted measurement of walking distance, Systems, Man, and Cybernetics. *IEEE International Conference, 3*, 1847-1852.

Salarian, A., Russmann, H., Vingerhoets, F., Dehollain, C., Blanc, Y., Burkhard, P. R., & Aminian, K. (2004). Gait assessment in Parkinson's disease: Toward an ambulatory system for long-term monitoring. *IEEE Transactions on Biomedical Engineering, 51*(8), 1434-1443.

Salarian, A., Russmann, H., Vingerhoets, F. J., Burkhard, P. R., Blanc, Y., Dehollain, C., & Aminian, K. (2003, April 24-26). An ambulatory system to quantify bradykinesia and tremor in Parkinson's disease. In *Proceedings of 4th IEEE International Conference On Information Technology Applications in Biomedicine. New Solutions for New Challenges* (pp. 35-38).

Salvia et al. (2000). Analysis of helical axes, pivot and envelope in active wrist circumduction. *Clinical Biomechanics, 15*(2), 103-111.

Schasfoort, F. C., Bussmann, J. B., Zandbergen, A. M., & Stam, H. J. (2003). Impact of upper limb complex regional pain syndrome type 1 on everyday life measured with a novel upper limb-activity monitor. *Pain, 101*, 79-88.

Sekine, M., Akay, M., Tamura, T., & Higashi, Y. (2004). Fractal dynamics of body motion in patients with Parkinson's disease. *Journal of Neural Engineering, 1*, 8-15.

Sekine, M., Tamura, T., Akay, M., Fujimoto, T., Togawa, T., & Fukui, Y. (2002). Discrimination of walking patterns using wavelet-based fractal analysis. *IEEE Transactions on Neural Systems and Rehabilitation Engineering, 10*, 188-96.

Sekine, M., Tamura, T., Togawa, T., & Fukui, Y. (2000). Classification of waist-acceleration signals in a continuous walking record. *Medical Engineering & Physics, 22*(4), 285-291.

Selles, R. W., Formanoy, M. A. G., Bussmann, J. B. J., Janssens, P. J., & Stam, H. J. (2005). Automated estimation of initial and terminal contact timing using accelerometers; Development and validation in transtibial amputees and controls. *IEEE Transactions on Rehabilitation Engineering, 13*(1), 81-88.

Shiratsu, A., & Coury, H. J. C. G. (2003). *Clinical Biomechanics, 18*, 682-684.

Standaert, D. G., (2003). Wearable technology's applications in Parkinson's disease. *IEEE Engineering in Medicine and Biology Magazine*, 25-26.

Stanton-Hicks, M., Janig, W., Hassenbusch, S., Haddox, J. D., Boas, R., & Wilson, P. (1995). Reflex sympathetic dystrophy: Changing concepts and taxonomy. *Pain, 63*, 127-133.

Tamura, T., Sekine, M., Ogawa, M., Togawa, T., & Fukui, Y. (1997). Classification of acceleration waveforms during walking by wavelet transform. *Methods of Information in Medicine, 36*(4-5), 356-369.

Terrier, P., Aminian, K., & Schutz, Y. (2001). Can accelerometry accurately predict the energy cost of uphill/downhill walking? *Ergonomics, 44*(1), 48-62.

Tesio, L., Monzani, M., Gatti, R., & Franchignoni, F. (1995). Flexible electrogoniometers: kinesiological advantages with respect to potentiometric goniometers, *Clinical Biomechanics, 10*(5), 275-277.

Todd, C., & Skelton, D. (2004). *What are the main risk factors for falls among older people and what are the most effective interventions to prevent these falls?* Synthesis of the Health Evidence Network (HEN). World Health Organization Regional Office for Europe, Copenhagen.

Tong, K., & Granat, M. H. (1999). A practical gait analysis system using gyroscopes. *Medical Engineering and Physics, 21*, 87-94.

Vaganay, J., Aldon, M. J., & Fournier, A. (1993). Mobile robot attitude estimation by fusion of internal data. In *Proceedings of IEEE International Conference on Robotics and Automation* (pp. 277-282).

van den Bogert, A. J., Read, L., & Nigg, B. (1996). A method for inverse dynamic analysis using accelerometry. *Journal of Biomechanics, 29*, 949-954.

Van den Dikkenberg et al. (2002). Measuring functional abilities of patients with knee problems: Rationale and construction of the DynaPort knee test. *Knee surgery sports traumatology arthroscopy, 10*(4), 204-212.

Van Someren, E. J., Vonk, B. F., Thijssen, W. A., Speelman, J. D., Schuurman, P. R., Mirmiran, M., & Swaab, D. F. (1998). A new actigraph for long-term registration of the duration and intensity of tremor and movement. *IEEE Transactions on Biomedical Engineering, 45*(3), 86-395.

Veltink, P. H., Bussmann, H. B. J., de Vries, W., Martens, W. L. J., & van Lummel, R. C. (1996). Detection of static and dynamic activities using uniaxial accelerometers. *IEEE Transactions on Biomedical Engineering, 4*(4), 375-85.

Walters, R. L., Lunsford, B. R., Perry, J., & Bird, R. (1988). Energy-speed relationship of walking: Standard tables. *Journal of Orthopaedic Research, 6*, 215-222.

Westerterp, K. R. (2001). Pattern and intensity of physical activity. *Nature 410 (6828),* 539.

Willemsen, A. T., Frigo, C., & Boom, H. B. (1991). Lower extremity angle measurement with accelerometers - Error and sensitivity analysis. *IEEE Transactions on Biomedical Engineering, 38*(12), 1186-1193.

Willemsen, A. T., van Alste, J. A, & Boom, H. B. (1990). Real-time gait assessment utilizing a new way of accelerometry. *Journal of Biomechanics, 23*(8), 859-863.

Williamson, R., & Andrews, B. J. (2000). Gait event detection for FES using accelerometers and supervised machine learning. *IEEE Transactions on Rehabilitation Engineering, 8*(3), 312-319.

Williamson, R., & Andrews, B. J. (2000, September). Sensor systems for lower limb functional electrical stimulation (FES) control. *Medical Engineering & Physics, 22*, 313-325.

Williamson, R., & Andrews, B.J. (2001). Detecting absolute human knee angle and angular velocity using accelerometers and rate gyroscopes. *Medical and Biological Engineering and Computing, 39*, 294-302.

Winter, D. A. (1990). *Biomechanics and motor control of human movement* (3rd ed.). Nes York: Wiley-Interscience Publishings.

Winter, D.A. (1995). Human balance and posture control during standing and walking. *Gait & Posture, 3*, 193-214.

Winter, D. A., & Eng, P. (1988). Energy generation and absorption at the ankle and knee during, fast natural and slow cadences. *Clinical Orthopaedics and Related Research, 175*, 147-154.

Winter, D. A., Prince, F., Stergiou, P., & Powell, C. (1993). Medial-lateral and anterior-posterior motor responses associated with centre of pressure changes in quiet standing. *Neuroscience Research Communications, 12*, 141-148.

Winters, J. M., Wan, Y., & Winters Jill M., (2003). Wearable sensors and telerehabilitation integrating intelligent telerehabilitation assistants with a model for optimizing home therapy. *IEEE Engineering in Medicine and Biology Magazine,* 56- 65.

Yazdi, N., Ayazi, F., & Najafi, K. (1998). Micromachined inertial sensors. In *Proceedings of the IEEE, 86*(8), 1640-1659.

Yue, Li, Aissaoui, R., Lacoste, M., & Dansereau, J. (2004). Development and evaluation of a new body-seat interface shape measurement system. *IEEE Transactions on Biomedical Engineering, 51*(11), 2040-2050.

Zhu, H. S., Wertsch, J. J., Harris, G. F., Loftsgaarden, J. D., & Price, M. B. (1991). Foot pressure distribution during walking and shuffling. *Archives of Physical Medicine and Rehabilitation, 72*, 390-397.

Zhu, R., & Zhou, Z. (2004). A real-time articulated human motion tracking using tri-axis inertial/magnetic sensors package. *IEEE Transactions on Neural Systems and Rehabilitation Engineering, 12*(2), 295-302.

Zijlstra, W., & Bisseling, R. (2004). Estimation of hip abduction moment based on body fixed sensors. *Clinical Biomechanics, 19*, 819-827.

Zijlstra, W., & Hof, A. L. (2003). Assessment of spatio-temporal gait parameters from trunk accelerations during human walking. *Gait & Posture, 18*(2), 1-10.

Chapter IV

Computational Intelligence Techniques

Bharat Sundaram, The University of Melbourne, Australia

Marimuthu Palaniswami, The University of Melbourne, Australia

Alistair Shilton, The University of Melbourne, Australia

Rezaul Begg, Victoria University, Australia

ABSTRACT

Computational intelligence (CI) encompasses approaches primarily based on artificial neural networks, fuzzy logic rules, evolutionary algorithms, support vector machines and also approaches that combine two or more techniques (hybrid). These methods have been applied to solve many complex and diverse problems. Recent years have seen many new developments in CI techniques and, consequently, this has led to many applications in a variety of areas including engineering, finance, social and biomedical. In particular, CI techniques are increasingly being used in biomedical and human movement areas because of the complexity of the biological systems. The main objective of this chapter is to provide a brief description of the major computational intelligence techniques for pattern recognition and modelling tasks that often appear in biomedical, health and human movement research.

INTRODUCTION

Computational intelligence is a branch of the study of artificial intelligence. Computational intelligence research aims to use learning, adaptive, or evolutionary algorithms to create programs that are, in some sense, intelligent. Computational intelligence research either explicitly rejects statistical methods, or tacitly ignores statistics.

Computational intelligence, as the name suggests, relies on number crunching. The field has developed enormously due to quantum jumps in computational power over the last two decades. The problems however, solved by computational intelligence techniques viz. search, optimization, adaptation and learning are age old. So, to understand computational intelligence, we must have a perspective of the other techniques that researchers have used to solve the same problems. These include statistical and syntactic approaches to solve the same problems.

The fundamental research question is: How to create a machine that can store information (not mere data) and interpret the learnt information in a useful manner? Add to this, the further requirement that the machine needs to be able to update its information database based on novel data and do this optimally. This objective takes different forms in different problems. For instance, in pattern recognition, the machine needs to be able to represent particular patterns, classify them, retrieve particular patterns if required, mark a novel pattern as previously unknown and generate a representation for it and so

Figure 1. Overview of problems in computational intelligence and machine learning

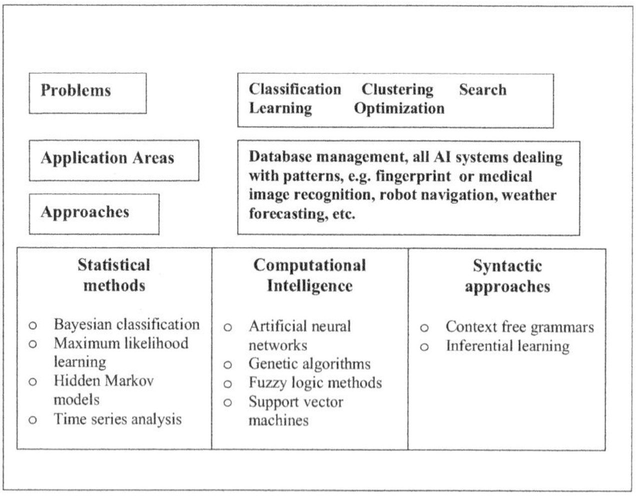

Problems	Classification Clustering Search Learning Optimization		
Application Areas	Database management, all AI systems dealing with patterns, e.g. fingerprint or medical image recognition, robot navigation, weather forecasting, etc.		
Approaches			
Statistical methods	**Computational Intelligence**	**Syntactic approaches**
o Bayesian classification o Maximum likelihood learning o Hidden Markov models o Time series analysis	o Artificial neural networks o Genetic algorithms o Fuzzy logic methods o Support vector machines	o Context free grammars o Inferential learning

on. In this particular sense, all research problems in pattern recognition, classification, information retrieval, data mining, authentication etc are, in essence, pattern representation problems. If one can program a machine to "understand" data and extract reproducible information out of it, then, in principle, the machine is behaving "intelligently." Statistical approaches had a first shot at achieving this objective and were later overtaken by computational intelligence techniques, due in large part to the revolution in number crunching abilities of modern day computers.

In this chapter, we will cover the computational intelligence techniques along with an historical perspective and important landmarks in computational intelligence research. In order that our coverage is complete, we will cover the statistical techniques to solve problems for which computational intelligence (CI) techniques were later devised.

STATISTICAL APPROACHES

Statistical approaches include all learning methods in which the learning machine converts available data into statistical models. Different statistical constructs achieve this objective in different ways. We will cover the important ones in this section.

Bayesian Learning

Bayes' theorem gives an expression for the a-posteriori probability $P(A|B)$ of an event A given an event B in terms of the a-priori unconditional probability of event A.

Consider a hypothesis set H. Given training data D, we can find the most probable hypothesis $h_{MAP} \in H$ such that, $h_{MAP} = \max_{h \in H} P(h/D)$. From Bayes' theorem:

$$h_{MAP} = \max_{h \in H} \frac{P(D/h)P(h)}{P(D)} = P(D/h)P(h) \qquad (1)$$

This choice of h is the natural maximum probable hypothesis and is called *maximum a-posteriori hypothesis* (MAP).

Bayes' principle can be extended to the case when each hypothesis in H outputs a vector $v \in V$. Thus, given a new instance of v, what is its probability of occurrence. The answer is not merely $P(v/h_{MAP})$ but is given by $\sum_{h \in H} P(v/h)P(h/D)$.

This idea forms the basis of hidden Markov models (HMMs) when H is the set of hidden states and V is the set of all possible outputs in a particular state.

In summary, a Bayesian network is defined by a graphical structure, a family of conditional probability distributions, and their parameters, which together specify a joint distribution over a set of random variables of interest. Bayesian methods have been employed since the 1960's and the interest has been steadily growing especially in the fields of economics (Zellner, 1984) and pattern processing (Duda & Hart, 1973). HMMs have been at the forefront of speech recognition as well as speaker recognition for a long time and Bayesian methods have also had a large bearing on image reconstruction algorithms.

SUPPORT VECTOR MACHINES

The starting point for support vector machines (SVMs) is in binary pattern recognition that involves constructing a decision rule to classify vectors into one of two classes based on a training set of vectors whose classification is known a-priori. Support vector machines (Cortes & Vapnik, 1995; Guyon, Boser, & Vapnik, 1993; Vapnik, 1995) do this by implicitly mapping the training data into a higher-dimensional feature space. A hyperplane (decision surface) is then constructed in this feature space that bisects the two categories and maximizes the margin of separation between itself and those points lying nearest to it (called the support vectors). This decision surface can then be used as a basis for classifying vectors of unknown classification.

The main advantages of the SVM approach are:

- SVMs implement a form of structural risk minimization (Vapnik, 1995) — They attempt to find a compromise between the minimization of empirical risk and the prevention of over fitting.
- The problem is a convex quadratic programming (QP) problem. So there are no non-global minima, and the problem is readily solvable using quadratic programming techniques.
- The resulting classifier can be specified completely in terms of its support vectors and kernel function type.

Support vector regressors (Drucker, Burges, Kaufman, Smola, & Vapnik, 1997; Vapnik, Golowich, & Smola, 1997; Smola & Scholkopf, 1998) are a class of non-linear regressors inspired by Vapnik's SV methods for pattern classification. Like Vapnik's method, SVRs first implicitly map all data into a (usually) higher dimensional feature space. In this feature space, the SVR attempts to construct a linear function of position that mimics the relationship between input (position in feature space) and output observed in the training data by minimizing a measure of the empirical risk. To prevent over fitting, a regularization term is included to bias the result toward functions with smaller gradient in feature space.

Two major advantages that SVRs have over competing methods (unregularised least-squares methods, for example) are sparseness and simplicity (Smola & Scholkopf, 1998; Burges, 1998). SVRs are able to give accurate results based only on a sparse subset of the complete training set, making them ideal for problems with large training sets. Moreover, such results are achievable without excessive algorithmic complexity, and use of the kernel trick makes the dual form of the SVR problem particularly simple.

SVR methods may be broken into ε-SVR (Drucker, Burges, Kaufman, Smola, & Vapnik, 1997; Vapnik, Golowich, & Smola, 1997) and ε-SVR methods (Scholkopf, Bartlett, Smola, & Williamson, 1999; Scholkopf & Smola, 1998), both of which require a-priori selection of certain parameters. Of particular interest is the ε (or υ in υ-SVR methods) parameter, which controls the sensitivity of the SVR to presence of noise in the training data. In both cases, this parameter controls the threshold ε (directly for ε-SVR, indirectly for υ-SVR) of insensitivity of the cost function to noise through use of Vapnik's ε-insensitive loss function.

The standard ε-SVR approach is associated with a simple dual problem, but unfortunately selection of ε requires knowledge of the noise present in the training data (and its variance in particular) which may not be available (Smola, Murata, Scholkopf, & Muller, 1998). Conversely, the standard υ-SVR method has a more complex dual form, but has the advantage that selection of υ requires less knowledge of the noise process (Smola, Murata, Scholkopf, & Muller, 1998) (only the form of the noise is required, not the variance). Thus both forms have certain difficulties associated with them.

Yet another approach is that of Suykens' least-squares SVR (Suykens, Van Gestel, De Brabanter, De Moor, & Vandewalle, 2002), which uses the normal least-squares cost function with an added regularization term inspired by Vapnik's original SV method. The two main advantages of this approach are the simplicity of the resulting dual cost function, which is even simpler than ε-SVR; and having one less constant to choose a-priori. The disadvantages include loss of sparsity and robustness in the solution. These problems may be ameliorated somewhat through use of a weighted LS-SVR scheme (Suykens, Van Gestel, De Brabanter, De Moor, & Vandewalle, 2002). However, while this method is noticeably superior when extreme outliers are present in the training data, in our experience the performance of the weighted LS-SVR may not be significantly better than the standard LS-SVR if such outliers are not present.

Usually, support vector machines are trained using a batch model. Under this model, all training data is given a priori and training is performed in one batch. If more training data is later obtained, or parameters are modified, the SVM must be re-trained from scratch. But if only a small amount of data is to be added to a large training set (assuming that the problem is well posed) then it will, most likely have only a minimal effect on the decision surface. Re-solving the problem from scratch seems computationally wasteful.

An alternative is to warm-start the solution process by using the old solution as a starting point to find a new solution. This approach is at the heart of active set optimization methods (Coleman & Hubert, 1989; Fletcher, 1981) and, in fact, incremental learning is a natural extension of these methods. While many papers have been published on SVM training, relatively few have considered the problem of incremental training. For a description of SVM classification, the reader is referred to Chapter VIII.

ARTIFICIAL NEURAL NETWORKS

Historical Perspective

The earliest work in neural computing goes back to the 1940's when McCulloch and Pitts introduced the first neural network computing model. In the 1950's, Rosenblatt's work resulted in a two-layer network, the perceptron, which was capable of learning certain classifications by adjusting connection weights. Although the perceptron was successful in classifying certain patterns, it had a number of limitations. The perceptron was not able to solve the classic XOR (exclusive or) problem. Such limitations led to the decline of the field of neural networks. However, the perceptron had laid foundations for later work in neural computing.

Neuroscience was influential in the development of neural networks, but psychologists and engineers also contributed to the progress of neural network simulations.

Rosenblatt (1958) stirred considerable interest and activity in the field when he designed and developed the perceptron. The perceptron had three layers with the middle layer known as the association layer. This system could learn to connect or associate a given input to a random output unit.

Another system was the ADALINE (adaptive linear element), which was developed in 1960 by Widrow and Hoff (of Stanford University). The ADALINE was an analogue electronic device made from simple components. The method used for learning was different to that of the Perceptron, it employed the least-mean-squares (LMS) learning rule.

In 1969 Minsky and Papert wrote a book titled *Perceptrons* (MIT Press, Cambridge) in which they generalized the limitations of single-layer perceptrons to multilayered systems. The significant result of their book was to eliminate funding for research with neural network simulations. The conclusions supported the disenchantment of researchers in the field. As a result, considerable prejudice against this field was activated.

Although public interest and available funding were minimal, several researchers continued working to develop neuromorphically based computational methods for problems such as pattern recognition.

During this period several paradigms were generated which modern work continues to enhance. Grossberg's (Grossberg & Carpenter, 1988) influence founded a school of thought which explores resonating algorithms. They developed the ART (adaptive resonance theory) networks based on biologically plausible models. Anderson and Kohonen developed associative techniques independent of each other. Klopf (1972) developed a basis for learning in artificial neurons based on a biological principle for neuronal learning called heterostasis.

Werbos (1974) developed and used the *back-propagation* learning method. However, several years passed before this approach was popularized. Back-propagation nets are probably the most well known and widely applied of the neural networks today. We shall cover it in greater detail in the following section.

Amari Shun-Ichi (1967) was involved with theoretical developments. He published a paper which established a mathematical theory for a learning basis (error-correction method) dealing with adaptive pattern classification. While Fukushima Kunihiko developed a step-wise trained multilayered neural network for interpretation of handwritten characters. The original network was published in 1975 and was called the cognitron.

Significant progress has been made in the field of neural networks. Advancement beyond current commercial applications appears to be possible, and research is advancing the field on many fronts. Neural network embedded chips are emerging and applications to complex problems developing. Clearly, today is a period of transition for neural network technology.

We now proceed to the technical details of a standard neural network. A neural network consists of four main parts:

1. Processing units u_j, where each u_j has a certain activation level $a_j(t)$ at any point in time.
2. Weighted interconnections between the various processing units which determine how the activation of one unit leads to input for another unit.

Figure 2. Schematic of a neural processing unit or perceptron

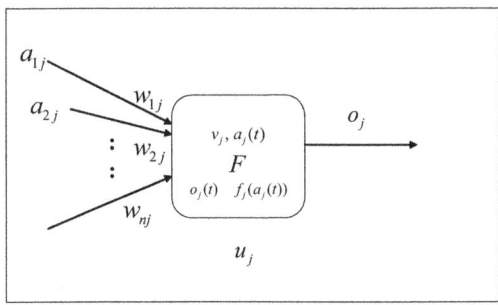

3.   An activation rule which acts on the set of input signals at a unit to produce a new output signal, or activation.
4.   Optionally, a learning rule that specifies how to adjust the weights for a given input/output pair.

A processing unit u_j takes a number of input signals, say a_{1j}, a_{2j}, ..., a_{nj} with corresponding weights w_{1j}, w_{2j}, ..., w_{nj} respectively. The net input to u_j is given by:

$$v_j = \sum w_{ij} a_{ij}$$

The new state of activation of u_j is given by:

$$a_j(t+1) = F(a_j(t), v_j)$$

where F is the activation rule and $a_j(t)$ is the activation of u_j at time t. The output signal o_j of unit u_j is a function of the new state of activation of u_j:

$$o_j(t+1) = f_j(a_j(t+1))$$

One of the most important features of a neural network is its ability to adapt to new environments. Therefore, learning algorithms are critical to the study of neural networks.

The Activation Rule F

Many types of activation functions have been used in ANNs. A linear activation function simply multiplies the net input of the network by some constant to produce the output. The threshold function is an important activation function, which was used in

Figure 3. The single variable sigmoid function

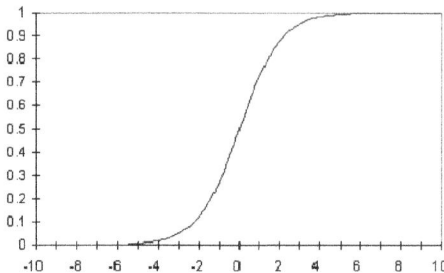

early work in neural network models. This function sets the output of the neuron equal to one if the net input exceeds a given threshold. Otherwise, the neuron's output is set at zero. The logistic function or sigmoid function is also commonly used as an activation function and is described by the following equation and is plotted in Figure 3.

$$F(x) = \frac{1}{1 + e^{-x}}$$

There are two important features of the sigmoid function that should be noted. The first is that the function is highly non-linear. For large values of activation, the output of the neuron is restricted by the activation function. The second important feature is that the sigmoid function is continuous and continuously differentiable. Both this feature and the non-linearity feature have important implications in the network learning procedure. One more convenient feature of the sigmoid function is that its first derivative can be expressed in terms of the function itself, that is, $F'(x) = F(x)(1 - F(x))$. This property is very useful when deriving the back-propagation learning rule.

The Multi-Layer Perceptron (MLP)

The real capabilities of the perceptron can be realized by connecting a large number of them together to form a network just like the neurons form a network in biological systems. Such a computational counterpart, however, is arranged in layers with interconnections between the layers. The generic structure of a MLP is shown in Figure 4.

The MLP in Figure 4 has one hidden layer and is called a 3-layer MLP. Each perceptron in the hidden layer is connected with every other perceptron in the input and output layer. This is a fully connected MLP although the connections are only in the forward (left-to-right) direction. Hence, this network is called the feed-forward MLP. The activation function used throughout is usually the sigmoid function with a bias (referred

Figure 4. 3-layer fully connected MLP

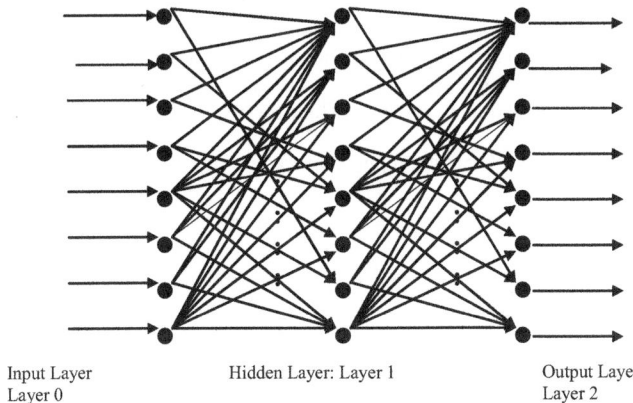

Input Layer   Hidden Layer: Layer 1   Output Layer
Layer 0   Layer 2

to as activation threshold), that is, $F_{ij}(x;\theta_{ij}) = \dfrac{1}{1+e^{-(x-\theta_{ij})}}$. Each layer can have a different number of perceptrons and there can also be more than one hidden layer. Thus, in a general MLP, there are n layers $(0, 1...n-1)$, with the input and output layers indexed as 0 and $n-1$ respectively. Each layer has $(N_0, N_1...N_{n-1})$ nodes respectively. For such a fully connected MLP, the connection weight between the ith node of the $(k-1)^{th}$ layer and the jth node of the kth layer is denoted by w^k_{ji}. If the dimension of the input vector is p, then $N_0 = p$ and if the dimension of the required output is q, $N_{n-1} = q$. The task that remains is to find the optimal weights and bias values for a given problem.

A Basic Digit Recognizer Using MLP

Our aim is to build a 3-layered MLP to recognize digits. The digits are in the form of 256×256-sized grayscale images and at the output we expect the network to recognize the correct digit. We choose to input as the normalized sum of grayscale values of disjoint 16×16 pixel blocks. Since there are 256 such blocks in each input image, $N_0 = p = 256$ and $N_2 = q = 10$, one for each of the digits from 0 to 9. In some cases, we would want the network to signal a flag indicating its inability to recognize any digit. In such a scenario, we would have 1 additional perceptron at the output for indicating such a flag. Now, we have a large number of such images called the *training set* that we can use to train the network to learn the weights and a separate set of images called the *test set* to test the network later. The output is in the form of probability (i.e., a number between *0* and *1*). Given an input, the perceptrons at the output will hold some value between *0* and *1*. We take the perceptron with the maximum value to denote the recognized digit. In order to learn the weights and the activation thresholds for each perceptron, we present the back-propagation method which is a classical method employed in training the MLP.

Back-Propagation in MLP

First, let us see how the output is computed given a 256 dimensional normalized input vector. The input to the ith perceptron in the input layer is $x_i^0 = w_i^0 I_i$ where $I = [I_0, I_1,...,I_{p-1}]$ is the input vector generated from the input image. The output of the first layer is given as $v_i^0 = F(x_i; \theta_i^0)$ where θ_i^0 is the bias for the denoted perceptron. This output appears as weighted input to the next layer. Thus for the 2nd layer, the input is

$$x_j^1 = \sum_{i=1}^{N_0=256} w_{ji}^1 v_i^0 \text{ for } j = 0,1,...N_1 - 1 \text{ and the output of this layer is } v_j^1 = F(x_j; \theta_j^1) \text{ which is fed}$$

as input to the 3rd layer. Thus the input at the 3rd layer is $x_k^2 = \sum_{i=1}^{N_1} w_{kj}^1 v_j^0$ for $k = 0,1,...N_2 - 1$

and the output is $v_k^2 = F(x_k; \theta_k^2)$. Let us say, the train set of images contains M train images. For each such image, the network generates an output vector $y_{net}^m = [v_0^2, v_1^2,...v_{q-1}^2]$ for $m = 0,1,...M - 1$. For the train set, we know the correct output a-priori. So, we formulate the correct output vector which will have a 1 in the ith entry if the digit is i and 0 otherwise. Thus the correct output vector for the digit 0 would be [1, 0, 0, 0, 0, 0, 0, 0, 0, 0] a.s.o. The general desired output vector can be written as $y_d^m = [y_0^m, y_1^m,...y_{q-1}^m]$ for $m = 0,1,...M - 1$.

We can thus compute the error between the output given by the network and the actual output for each pattern in the training set as some function of the desired and actual outputs. This error function $e^m = E(y_d^m, y_{net}^m)$ is usually chosen as the sum of squares or sum of absolute value of error in each component. For using the Delta Rule, explained in the next section, the sum of squares taken over the entire training set and scaled by the size of the training set (MSE) is used as the error function. Thus, moving forward in the network gives us an output and an error value for each input pattern. The MSE is used to update the weights and bias terms using what is called the Delta Rule, as we shall see in the next section.

The Delta Rule

Initially, the weights and bias terms are randomly set and the outputs computed. Once the error is computed at the output layer, the update rule for the weights is given as:

$$\Delta w_{ji}^l = \eta \times \delta_j^l \times v_j^l$$

where η is the learning rate and is the most important parameter for the user to control the performance of the network. δ_j^l is the delta coefficient and is called the local gradient of the error. It is a measure of how much the error at the output is due to the particular weight w_{ji}^l.

1.   If node j is an output node, then δ_j^l is the product of $F'(x_j^l)$ and the difference in the jth component between desired output and the network output, that is, $\delta_j^2 = F'(x_j^2) \times (y_j^m - v_j^2)$ for each m, where x_j^2 is the total input to node j.

2.   If node j is a hidden node, then δ_j is the product of $F'(x'_j)$ and the weighted sum of the δ's computed for the nodes in the next hidden or output layer that are connected to node j, obviously weighted by the connection weights to the next layer (i.e., the next forward layer).

The important point to note is that the weights are updated beginning from the output. The errors are propagated from the output to the input. Hence, the algorithm is called back-propagation. The back-propagation algorithm has been generalized for updating the bias terms as well as updating networks with more than one hidden layer (Rumelhart, Hinton, & Williams, 1986; Hush & Horne, 1993).

In the derivation of the above technique, the change Δw in the weights is proportional to the partial derivatives of the error function. Thus, the change in weights will go to zero or in other words, the algorithm will converge when the partial derivatives go to zero. Thus, the algorithm climbs down a hill in the multi-dimensional search space of weights versus error and settles at the bottom which is why is referred to as the **gradient-descent technique.** The back-propagation algorithm developed so far requires that the weight changes be proportional to the derivative of the error. The larger the learning rate η, the larger the weight changes on each epoch, the quicker the network learns. However, the size of the learning rate can also influence whether the network achieves a stable solution. If the learning rate gets too large, then the weight changes no longer approximate a gradient descent procedure. (True gradient descent requires infinitesimal steps). Oscillation of the weights is often the result.

Ideally, we would like to use the largest learning rate possible without triggering oscillation. This would offer the most rapid learning. One method that has been proposed is a slight modification of the back-propagation algorithm so that it includes a momentum term. Applied to back-propagation, the concept of momentum is that previous changes in the weights should influence the current direction of movement in weight space. This concept is implemented by the revised weight-update rule:

$$\Delta w'_{ji}(n+1) = \eta \times \delta'_j \times v'_j + \mu \times \Delta w'_{ji}(n)$$

where μ is called the momentum parameter. The index n denotes the current epoch. Due to the momentum term, if the weights start moving in a particular direction, they tend to continue moving in that direction, thereby overriding the tendency to oscillate in case the learning rate gets too large. To grasp the advantage of this approach, imagine a ball rolling down a hill when it encounters a ditch. To prevent the ball from getting stuck in the ditch, it must come down with enough momentum. This is exactly what the momentum term seeks to achieve. For a detailed analysis of back-propagation algorithms with momentum, refer to Phansalkar and Sastry (1994).

Back-propagation and gradient descent have been applied extensively, almost exhaustively in practical applications of neural networks particularly in pattern recognition, data mining, search and function estimation. These techniques are a starting point for any discussion on neural networks. These algorithms have been extended to incremental forms in which the user does not need to scan through the entire training to update the weights (Meeden, 1996). In the derivation of the back-propagation algorithm, the error is taken over the complete pass of the training set (epoch). For small learning rates, the weight update can be done after each training pattern, but as the learning rate grows larger, one must adhere to the strict rules of the algorithm and update the weights only at the end of each epoch. This is unsuitable for learning new data which is why the incremental versions of the algorithm have gained much importance. A lot of research has gone into speeding up the convergence of the back-propagation algorithm that has enhanced its application in real-time control environments (Jiang, McCorkell, & Zmood, 1995; Yan, Rad, Wong, & Chan, 1996; Hunt, Sbarbaro, Zbikowski, & Galwtrop, 1992). From here, we move to the equally famous neural network architecture: the radial basis function network.

The Radial Basis Function Network (RBFN)

Figure 5 shows a typical RBFN architecture with a single output. The arrows show connection weights. The activation function (called basis function in the RBFN context) is Gaussian. Each perceptron in the hidden layer is characterized by its center which is a member of the input space (the mean of the Gaussian). For each given input vector, the Euclidian distance from the center of each perceptron is computed. The output of the hidden layer is computed by applying the basis function to each of the Euclidian distances computed. The output is a weighted sum of the hidden layer outputs along with a unity bias. In case there is more than 1 output, each output is computed as a linear combination of the outputs of the hidden layer. Thus, the RBFN differs from the MLP in the form of the output. The output is a linear combination of the outputs of the hidden layer. The only source of non-linearity is from the basis function of the hidden layer. The training of the RBFN requires prior knowledge about the centers and widths (mean and variance) of the Gaussian bases. Once the centers are known, the connection weights can be evaluated using a host of search techniques including back-propagation and gradient descent. The novelty lies in computing the optimal centers. To understand the computation of centers and center widths, let us take a closer look at the functionality of the RBFN hidden perceptrons. At the centers, the Euclidian distance is computed and transformed by the Gaussian basis. Due to the structure of the Gaussian basis, the output is high if the input is close to the center and decreases as the input moves away from the center. Since the output is a linear combination of the hidden nodes' output, the output

Figure 5. Typical RBFN architecture

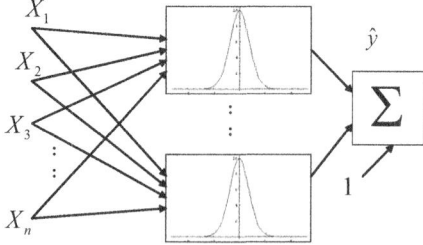

of the RBFN is dominated by that hidden node whose center is relatively closer to the input (in the Euclidian sense). Thus, the RBFN classifies the input as belonging to a particular cluster characterized by the center of the cluster that is the same as the center of the hidden layer perceptron representing that cluster. The RBFN performs clustering of the input data. Hence, the cluster heads can be found by the many clustering techniques available in literature. The common strategy is to create a pool of candidate hidden perceptrons using K-means clustering and then selecting the best hidden perceptrons using orthogonal least squares, Maximum correlation, local error maximization or some such method. The distinguishing feature here is that the learning of the centers is not supervised learning (i.e., the learning is not dictated by any error function of the desired and actual output of the network). Computing the center widths is a more involved problem. In some applications, the center widths are set by the user as a constant throughout the network and become a user-defined network parameter. There are obvious disadvantages to such a heuristic setting. For a more involved approach, please refer (Benoudjit & Verleysen, 2003, October). Once the centers and their widths are set, the weights can be updated using back-propagation or any other supervised learning method. It must be however mentioned here that there exist supervised learning rules for training the RBFN, including the centers and center widths, but the two-stage training described above is more in tune with the functionality of RBFNs.

Concisely put, RBFNs are easier to train than MLP networks simply because they have lesser non–linearity. RBFNs have been used in most applications that use MLP networks including Face recognition (De Silva, De Silva, & Ranganath, 2003), biomedical applications (Tumer, Ramanujam, Ghosh, & Richards-Kortum, 1998; Merzagora, Bracchil, Cerutti, Rossi, Bianchi, & Gaggiani, 2004), bioinformatics (Ibrikci, Brandt, Wang, & Acikkar, 2002), non-linear signal processing (Kassam & Inhyok, 1993) and non-linear control (Behera, Gopal, & Chaudhury, 1995).

NEURAL NETWORKS WITH ASSOCIATIVE MEMORY

The two networks that we have yet considered, MLP and RBFN, have a representation of the input in parameters internal to the network. The MLP stores the input information in its weights and the RBFN, in the centers and weights. The outputs of these networks are some abstraction of the input that the user interprets as correct or incorrect depending on a ready-made desired output value. This approach, although widely used, begs a pertinent question: is it possible for a network to store input information in terms of the input itself and retrieve that stored information in its original form? This question can be rephrased as, "Can the network associate input patterns as such with one or more of its internal states?" This is the basic principle of any network having associative memory. The MLP and RBFN do have associative memory in some abstract sense, since they store input information in their weights. But this concept of associating a particular input with a particular state of the network is brought out explicitly in another neural network by Hopfield (1982) called the Hopfield Neural net. This network is able to store patterns in its internal state in a manner similar to the human brain (which was the inspiration for Hopfield to come up with the network design)

The Hopfield net (Hopfield, 1982) has only one layer of units. These units play a triple role as input, output, and processing units. For a schematic, see Figure 6. The units are globally interconnected and every unit is thus connected to every other unit, including itself. The total input x_i of the ith unit is given by the linear combination:

$$x_i = \sum_{j=1}^{N} w_{ij} v_j + b_i$$

where, v_j is the output of the jth unit, w_{ij} is the connection weight from unit j to unit i and b_i is the bias term for the ith unit. N here, is the total number of Hopfield units also called Hopfield neurons, since they were designed as a model for human memory. Now, let us see how this network can be programmed to remember patterns.

Say, we want the network in Figure 6 to learn, remember and recall a set of M patterns. These could be images, in which case we could take each individual pixel to represent the input to each neuron, thus making the total number of neurons equal to the number of pixels.

Now, let us perform the following operations on the network:

1.   Fix the weights such that they are symmetric, that is, $w_{ij} = w_{ji}$ and $w_{ii} = 0$ for all i. To achieve this, set the weights as:

$$w_{ij} = \left. \begin{array}{ll} \dfrac{1}{N} \displaystyle\sum_{m=0}^{M} p_i^m p_j^m & \text{for } i \neq j \\ 0 & \text{for } i = j \end{array} \right\}$$

Figure 6. Fully connected Hopfield net architecture

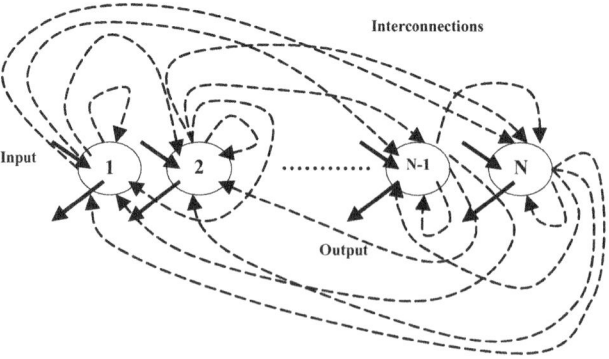

In Hopfield's original work, the self-connection was set to zero for stability of the solution but later, other researchers (Matsuda, 1993) showed that this is not necessary and in fact, the network, without the self-connection term, will have unstable solutions.

2. Once the weights are set, the **sign function** is chosen as transfer function for the neurons. A test input pattern is given at the input of each neuron and the network evolution begins. The input to each node is ±1, since the sign function is used as the transfer function, the output is also ±1.

3. There are many strategies for evolving the network. The most common strategy is to randomly choose a neuron and update its output. This process is called asynchronous update. This random selection and update is continued until the output of the network remains constant.

Given an input pattern, a concept of energy is associated with it, *w.r.t* the Hopfield network. It is defined as:

$$E = -\frac{1}{2}\sum_{i=1}^{N}\sum_{j=1}^{N} w_{ij}x_i x_j$$

where x_i is the ith component in the input vector and is fed as input to the ith neuron. Similarly, the energy associated with a particular state of the Hopfield network (i.e., its output state) is defined as:

$$E_{net} = -\frac{1}{2}\sum_{i=1}^{N}\sum_{j=1}^{N} w_{ij}v_i v_j$$

where v_i is the output of the ith neuron. The Hopfield network can be shown to evolve and finally settle down in such a state so as to minimize the energy of the network. At each step of the random update, the Hopfield network changes its state in such a manner so as to reduce E_{net}. These minimum energy states are called the attractors of the Hopfield network. It is basically these attractor patterns that are stored, recalled and memorized by the network. Empirical results suggest that the capacity of the Hopfield network is not very large. The number of patterns that a given net can store and recall is around 15% of the total neurons in the net.

The Hopfield network, though simple and useful, has one particular caveat related to its evolution. There is a high possibility that the energy minimization of the net is not complete before the net stabilizes. It is likely that the network settles in a local energy minima and the corresponding state is not really a stable state. To avoid such discrepancies, the evolution strategy is made probabilistic. One of the most famous probabilistic evolution techniques used to evolve the Hopfield net is called simulated annealing. A Hopfield neural network, evolved using simulated annealing, is called the Boltzmann machine. It was originally proposed by Geoffrey Hinton and Terry Sejnowski. In the Boltzmann machine, a probabilistic step is introduced before computing the output of a neuron from its input after it is selected for asynchronous update. One more feature of the Boltzmann machine is that the states are binary (*0 or 1*). The deterministic output of a neuron selected for update is given by:

$$v_i = \sum_{j=1}^{N} w_{ij} v_j$$

where v_j is the state of the jth neuron. The change in the network energy (defined in the same manner as that for the Hopfield net) due to a change in the state of one neuron can be shown to be:

$$\Delta E_i = \sum_{j} w_{ij} v_j$$

Instead of using the Sign function to update the state, the new state is given by the following function:

$$P(v_i(t+1) = 1) = \frac{1}{1 + e^{-\Delta E_i/T}}$$

where T is called the temperature of the network. Initially, T is set to a large value so that it dominates over ΔE_i. Thus, there is a lot of randomness in the network. The state can change or remain the same with roughly equal probability. Over time, T is gradually reduced and the neurons' transfer function begins to govern the evolution. This process is similar to the forming of crystals through annealing. The matrix is heated to a high temperature so that all atoms have high kinetic energy. Then, as they are slowly cooled down, they get time to arrange themselves in regular patterns. This avoids defects in the final crystal formed. Similarly, due to a spurious initial value of the states of the network, the network may converge to a pseudo-stable state (i.e., local minimum in the multidimensional energy space). Due to the presence of the temperature term, this scenario is discouraged. Since the network has a high initial temperature, all the states are almost equally unstable and have a strong tendency to keep with every update. Hence, it is unlikely that the network will "*settle*" into a local minimum. Eventually, as the temperature decays, the network has been through a large number of spurious pseudo-stable states and it is more likely that any state that still remains stable is actually a stable state of the network.

The Boltzmann machine is impractical in real–time applications since the probabilistic learning process takes a long time. Yet, it is a very good theoretical exercise to help understand how concepts from diverse fields can be brought together to yield novel architectures. Boltzmann machines have many off-line applications in pattern recognition (Ma, 1995), speech processing (Trehern, Jack, & Laver, 1986) and can be used in most scenarios where Hopfield nets are used.

FEEDBACK IN NEURAL NETWORKS

The Hopfield neural network and Boltzman machine discussed in the previous section had a self–connection weight in each neuron. We mentioned that this self-connection was initially considered undesirable in Hopfield networks, but later proven to be useful to obtain stable attractors. Such self-connections cause the output of a

Figure 7. 3-layer recurrent MLP

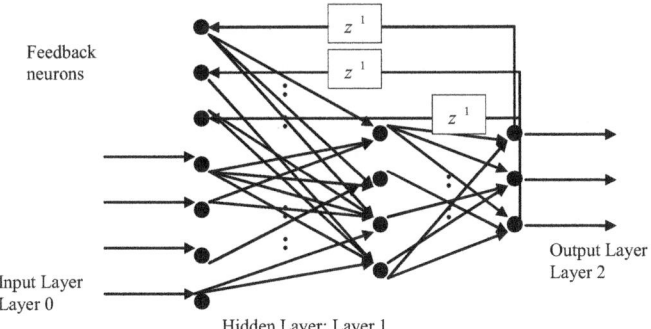

Input Layer
Layer 0

Hidden Layer: Layer 1

neuron to affect its future state evolution and are the essence of feedback in neural networks. A similar concept of feedback can be applied to the MLP and this was considered the natural extension to the MLP for applications in which the input is time varying (e.g., speech recognition). Neural networks with feedback are called **recurrent neural networks** (RNNs). To convert the standard MLP into a RNN, simply connect the output of each neuron in the output layer with one neuron in the input layer with a time delay. Extra neurons in the input (without the feedback connection) serve as the input neurons and the input vector is applied only to these neurons. For a schematic, refer Figure 7.

Training of this network is done using a neat trick. The feedback connection is broken and the network redrawn as a feed-forward network, but with the feedback neurons taking the output at the previous time stamp as their current input. Thus, the network is rolled over versus time, as some neurons in the input layer having time varying inputs. Still, at a given time, the network can be treated as a feed-forward network and the weights can be updated using the back-propagation rule. The corresponding algorithm is called back-propagation through time (BPTT) and has been well established in literature (Sato, 1990; Werbos, 1990). RNNs along with BPTT have been applied in many areas including speech recognition, control, robotics, and gait synthesis among many others. The major drawback with such networks is the computationally expensive training time.

Before we end our discussion on ANNs, we need to look at one more important class of networks called self-organizing maps.

SELF-ORGANIZING MAPS (SOM)

In feed-forward neural networks, the output is a transformation of the input via the neuron weights. In feedback neural networks, the output is fed back to the input and temporal processing takes place. In self-organizing maps, the neurons compete locally to learn the input and end up developing into specific detectors of input patterns. The learning in such networks is unsupervised and competitive and the neurons self-organize to form the final network structure. It follows that such networks can be used to learn complex functions and also be used to generate functional groups of neurons (i.e., groups of network neurons that respond to a particular feature in the input). Let us study such networks with reference to an example. Consider a 2-D lattice of neurons with randomly initialized weight vectors. This lattice is expected to learn a stochastic source that produces input vectors $x(t)$. The learning proceeds as follows:

1. For each $x(t)$, the neuron that has weight vector closest to $x(t)$ is computed. The closeness could be defined by any distance measure. (Euclidian distance is used in many cases.)
2. The neuron at minimum distance is called the *winner*. A spatial neighborhood $N_w(t)$ around the neuron is demarcated and only neurons within the neighborhood will take part in any update due to $x(t)$.
3. The simplest weight update rule that is used to learn the input is $w_i(t+1) = w_i(t) + \eta(t)(x(t) - w_i(t))$, where w_i is the weight of the neuron being updated and η is the learning rate.
4. Note that the neighborhood function as well as the learning rate is a function of time. Both these terms decrease with time. Initially, the neighborhood is taken to include a large number of neighboring neurons to ensure a rough global order in the lattice. With time, as the network becomes more structured, shrinking the neighborhood results in good spatial resolution. The learning rate is analogous to the learning rate term using in MLP in which we saw that a large learning rate after training is complete, drives the network into oscillation. Due to a similar reason and to avoid over-learning, the learning rate is progressively decreased with time.
5. In addition, the learning rate can be different for different neurons within the neighborhood itself. Usually, a Gaussian learning rate is used within the neighborhood which looks like:

$$\alpha_i(t) = \alpha_0(t)\exp(-\|r_i - r_w\|^2 / \sigma^2(t))$$

where r is the position vector of the neuron in the spatial lattice. Again, note that α and σ are decaying functions of time.

Once the SOM is trained for a sufficient time, it is able to map the input features into an internal representation stored in the spatial location and weights of the neurons. When a test input is given to the SOM, it excites only those neurons whose weights are close to the input in the Euclidian sense. Thus, the SOM can be made to respond to particular kinds of inputs as well as particular features within the input. Armed with an additional output vector, in addition to the weight vector, the SOM can be used to model complex functions whose explicit functional form is unknown. Such a self-organized map can retrieve the output in a single computation step and for this very reason SOMs find uses in real-time applications, such as robot navigation and control (Behera, Chaudhury, & Gopal, 1998).

Comments

This section of ANNs is by no means a comprehensive discussion of neural networks. The volume of work and literature in this field is just too dense to fit into a single book, let alone a single section! However, we have covered the basic types of neural networks that have shaped neural network research through the last 50 years. Each of individual networks discussed viz. MLP, RBFN, SOM, RNN, Hopfield net and Boltzmann machine have developed extensively from their humble beginnings presented in this section. Along the way, many new networks have been invented. Cellular Automata is one such field that has grown out of the research on SOMs. *Neural network inversion* in another technique that has gained in prominence since the 1990's. With increasing computational power, researchers are expected to come up with more intricate designs for neural networks. In conclusion, it can be safely said that ANNs will be at the forefront of computational intelligence research and applications in the years to come.

FUZZY LOGIC

The Starting Point: Linguistic Variable

Fuzzy logic is primarily concerned with quantifying and reasoning about vague or fuzzy terms that appear in our natural language. In fuzzy logic, these fuzzy terms are referred to as linguistic variables (also called fuzzy variables). For example, in the statement *"The size of the box is large,"* the *size* is a linguistic variable and its value is *large*. We call the range of possible values of a linguistic variable the universe of discourse. The underlying power of fuzzy set theory is that it uses linguistic variables, rather than quantitative variables, to represent imprecise concepts.

Fuzzy Set and Membership

Let us begin by considering conventional set theory that has a digital world view. For instance, conventional set theory allows the statement S: *"the size of the box is large"* to have only one of two truth values (true or false), usually written 1 or 0. This can be expressed as a binary membership function $m_s(x)$. Mathematically, for a crisp S, we have $m_s(x) = 1$ if $x \in S$ and $m_s(x) = 0$ if $x \notin S$. This binary membership function cannot express attributes like *very large, very small, medium size, etc.* Hence, we proceed to a natural generalization of conventional set theory by allowing variables to have degrees of truth. This is precisely fuzzy set theory. A fuzzy set F defined on a universe of discourse U is characterized by a membership function $m_F(x)$ which takes on values in the interval [0, 1]. Figure 8 clearly illustrates the membership functions for a crisp set and for a fuzzy set. Viewing a set as a function capable of defining partial membership is the basis of fuzzy set theory. In engineering applications of fuzzy logic, the most commonly used shapes for membership functions are triangular, trapezoidal, piece-wise linear and Gaussian.

Until very recently, the user chose membership functions arbitrarily, based on the user's experience. Hence, the membership functions for two users could be quite different. The number of membership functions is dependent on the users. Greater resolution is achieved by using more membership functions at the price of greater

Figure 8. Binary and fuzzy membership functions

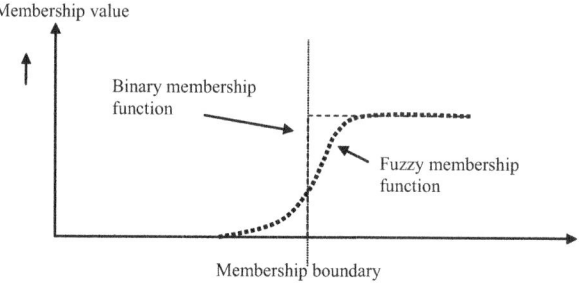

computational complexity. In fuzzy set theory, we can go further, as in our natural language, to use a linguistic hedge or modifier to modify the meaning of a term or, more generally, of a fuzzy set. For example, if *large box* is a fuzzy set, then *very large box, more-or-less large box, extremely large box and not-so-large box* are examples of hedges that are applied to this fuzzy set. Hedges can be viewed as operators that can act on a fuzzy set's membership function to modify it. For instance, let us look at the:

(a)   **Concentration operator** which basically adds the linguistic modifier *"very"* — the concentration operation has the effect of further reducing the membership values of those elements that have smaller membership values. This operation is given as:

$$m_{conS}(x) = m_S^2(x);$$

(b)   **Dilation operator** which can be viewed as adding the linguistic modifier *"some-what."* The dilation operation dilates the fuzzy elements by increasing the membership value of those elements with small membership values more than those elements with high membership values. This operation is given as:

$$m_{dilS}(x) = \sqrt{m_S(x)}$$

Fuzzy Logic and Fuzzy Operation

Conventional logic uses "and" (intersection), "or" (union) and "not" (complement). Combinations of these operators can describe any relationship but, in fuzzy logic, union, intersection and complement are defined in terms of their membership functions. Let fuzzy sets A and B be described by their membership functions. Then we have:

(a)   **Fuzzy union**: the union of A and B, denoted by $A \vee B$ is defined by $m_{A \vee B}(x) = m_A(x) \vee m_B(x) = \max(m_A(x), m_B(x));$

(b)   **Fuzzy intersection**: the intersection of A and B, denoted by $A \wedge B$, is defined by $m_{A \wedge B}(x) = m_A(x) \wedge m_B(x) = \min(m_A(x), m_B(x));$ and

(c)   **Fuzzy complement**: The fuzzy complement of B, denoted by $\sim B$, is defined by $m_{\sim B}(x) = 1 - m_B(x)$.

A fuzzy relationship R from Set A to Set B can be visualized as a fuzzy graph and can be characterized by the two variable membership function $m_R(x, y)$ which satisfies the composition rule as follows: $m_B(y) = \max(\min(m_R(x, y), m_A(x)))$. In summary, fuzzy logic may be viewed as an extension of multi-valued conventional logic. Such extension and generalization is very significant. The fact that fuzzy logic deals with approximate rather than precise modes of reasoning implies that, in general, the chains of reasoning in fuzzy logic are short in length, and rigor does not play as important a role as it does in classical logical systems. Tersely put, in fuzzy logic, everything, including truth, is a matter of degree.

Fuzzy Logic Systems

Although fuzzy logic deals with continuous truth values, in the real world we deal with concrete truth values. For instance, given a box, it is of a fixed size. Hence, the input to fuzzy systems is crisp and the real world expects a crisp output. Thus a fuzzy logic system maps crisp inputs into crisp outputs. It contains four major components: fuzzifier, rules, inference engine and defuzzifier.

Fuzzification

Fuzzification is the process of making a crisp quantity fuzzy. The fuzzification interface involves the following functions:

(a)   measures the value of input variables.
(b)   performs a scale mapping that transfers the range of values of input variables into corresponding universes of discourse.
(c)   performs the function of fuzzification that converts input data into suitable linguistic values which may be viewed as labels of fuzzy sets. Non-singleton fuzzification provides a means for handling uncertainties totally within the framework of fuzzy logic systems. A non-singleton fuzzifier is one for which $m_A(x') = 1$ for some fixed x' and $m_A(x)$ decreases from unity as x moves away from x'. In non-singleton fuzzification, x' is mapped into a fuzzy number (i.e., a fuzzy membership function is associated with it). Examples of such membership functions are the Gaussian and triangular. The broader these functions, the greater is the uncertainty about x'.

Fuzzy Rules

A fuzzy system is characterized by a set of linguistic statements based on expert knowledge. The knowledge base comprises of knowledge of the application domain and the attendant control goals. It consists of a "database" and a "rule base." The database provides necessary definitions which are used to define linguistic control rules and fuzzy data manipulation. Generally the design of fuzzy controllers is based on the operator's understanding of the process behavior instead of its detailed mathematical model. The main advantage of this approach is that it is easy to implement "rules of thumb" experiences and heuristics. These rules are often expressed using syntax of the form: If

<fuzzyproposition>, then <fuzzy proposition >, where the fuzzy propositions are of the form, *"x is Y"* or *"x is* not *Y, "x* being a scalar variable and *Y* being a fuzzy set associated with that variable. This rule establishes a relationship or association between the two propositions. Fuzzy logic systems store rules as fuzzy associations (i.e., for the rule IF *A* THEN *B*, where *A* and *B* are fuzzy sets, a fuzzy logic system stores the association *(A,B)* in a matrix *M)*. The fuzzy associative matrix *M* maps fuzzy set *A* to fuzzy set *B*. This fuzzy association or fuzzy rules is called a fuzzy associative memory (FAM).

Fuzzy Inference

Fuzzy inference is the kernel in a fuzzy logic system. It has the capability of simulating human decision making based on fuzzy concepts and of inferring fuzzy control actions employing fuzzy implication and the rules of inference in fuzzy logic. In the fuzzy inference engine, fuzzy logic principles are used to combine fuzzy "IF-THEN" rules from the fuzzy rule base into a mapping from fuzzy input sets to fuzzy output sets.

(a) **Max-min inference**: In max-min inference the implication operator used is min, that is, $m_{ij} = truth\ (a_i \rightarrow b_j) = \min\ (a_i \rightarrow b_j)$, where $a_i \rightarrow b_j$ is a fuzzy implication which effectively means *if* a_i *then* b_j. Given two fuzzy sets *A* and *B,* this equation can be used to form the matrix *M.*

(b) **Max-product inference**: Max-product inference uses the standard product as the implication operator when forming the components of *M*: $m_{ij} = a_i b_j$. Following the calculation of this matrix, max-min composition is used to determine the induced matrix *B'* from some subset vector *A'*.

A myriad of inference methods exist for different applications like sum-product inference, Mamdami's inference (Jamshidi, Vadiee, & Ross, 1993) to name a few.

Defuzzification

This is the process of converting fuzzy outputs into crisp outputs. Again this is a necessary step because the real world requires a specific response. For instance, say the output of a fuzzy logic controller is *"increase excitation of the pendulum slightly."* This cannot be fed to the pendulum because it expects a finite fixed increase in its excitation. Hence, the fuzzy value *slightly* must be converted to the corresponding numerical value which is where defuzzification comes in.

The defuzzification interface performs the following functions: (a) a scale mapping which converts the range of values of output variables into corresponding universes of discourse; and (b) defuzzification, which yields a nonfuzzy control action from an inferred fuzzy control action. Defuzzifier produces a crisp output for the fuzzy logic system from the fuzzy set that is the output of the inference block. Many defuzzifiers have been proposed in the literature (Mendil & Benmahammed, 1998; Pelayo, Rojas, Ortega, & Prieto, 1993; Chen, Huang, & Liu, 1997) and they are largely specific to a particular application. In the control domain that forms a major part of the application area of fuzzy logic, one obvious criterion for the choice of a defuzzfier is computational simplicity. This criterion has led to the following candidates for defuzzifiers:

(a) **Maximum defuzzifier**: This defuzzifier examines the fuzzy set B and chooses as its output the value of y for which it is a maximum.

(b) **Mean of maxima defuzzifier**: This defuzzifier examines the fuzzy set B and first determines the values of y for which it is a maximum. It then computes the mean of these values as its output.

(c) **Centroid defuzzifier**: This defuzzifier determines the centre of gravity of the final fuzzy control space and uses this value as the output of the fuzzy logic system. The resultant output is sensitive to all the rules executed. By and large, process control applications use centroid, while information-based applications like risk evaluation and terrain analysis use the composite maximum. Efficient defuzzifier development is an ongoing research problem.

GENETIC ALGORITHMS

"Survival of the fittest" is the theme that has spawned an entire research area called genetic algorithms (GA). In nature, we see this theme repeated over and over again. Can this theme be applied to Computational methods? It only needs an alternate way of looking at the problems in computational intelligence, in particular, optimization. In order to optimize a given quantity, w.r.t a set of unknown parameters X, we need to minimize a particular cost $C(X)$. Now, if one defines a concept called *"fitness"* for a given X, then, given X_1 and X_2, X_2 have a greater *fitness* than X_1 if $C(X_2) < C(X_1)$. So, in given population of X, we can evolve the population using Darwinian principles so that after a few evolutions, only the fittest X survives. All that remains is expressing every element of X in a manner that aids easy simulation of the evolution process within the population. In GA terminology, each X is called a chromosome and each parameter in X is called a gene. Borrowing further from the biological theory of chromosomes, the GA evolves through reproduction of chromosomes, their crossover and the occasional mutation. Before we go into the specifics of GAs let us take a look at how this research area developed into its present form.

Historical Perspective

"Evolutionary computation was definitely in the air in the formative days of the electronic computer" (Mitchell, 1996, p. 2). As early as 1962, researchers such as Box, Friedman, Bledsoe and Bremermann had all independently developed evolution-inspired algorithms for function optimization and machine learning, but all this research did not attract too much attention. A more successful development in this area came in 1965, when Rechenberg developed a technique called *evolution strategy*. In this technique, there was no population or crossover; one parent was mutated to produce one offspring, and the better of the two was kept and became the parent for the next round of mutation (Haupt & Haupt, 1998). Later versions introduced the idea of a population. Evolution strategies are still employed today by engineers and scientists, especially in Germany.

In 1966, the next signification development in the field occurred when Fogel, Owens and Walsh introduced in America a technique they called *evolutionary programming*. In this method, candidate solutions to problems were represented as simple finite-state machines. Like Rechenberg's evolution strategy, their algorithm worked by randomly mutating one of these simulated machines and keeping the better of the two (Mitchell, 1996, p. 2; Goldberg, 1989, p.105). Crossover was a feature that was lacking in both these methods.

Earlier, in 1962, John Holland had explicitly proposed crossover but its potential remained unexplored. The seminal work in the field of genetic algorithms came in 1975, with the publication of the book *Adaptation in Natural and Artificial Systems* (Holland & John, 1975). This book was the first to systematically and rigorously present the concept of adaptive digital systems using mutation, selection and crossover, thus simulating the process of biological evolution as a problem-solving strategy. The book also attempted to put genetic algorithms on a firm theoretical footing by introducing the notion of schemata (Mitchell, 1996, p. 3; Haupt & Haupt, 1998, p. 147). That same year, Kenneth De Jong's important dissertation established the potential of GAs by showing that they could perform well on a wide variety of test functions, including noisy, discontinuous, and multimodal search landscapes (Goldberg, 1989, p. 107).

Thus evolutionary computation had been set on a firm theoretical footing, ready to be applied. By the early- to mid-1980's, genetic algorithms were being applied to a broad range of subjects, from abstract mathematical problems like bin-packing and graph coloring to tangible engineering issues such as pipeline flow control, pattern recognition and classification, and structural optimization (Goldberg, 1989, p. 128).

At first, these applications were mainly theoretical. However, as research continued to proliferate, genetic algorithms migrated into the commercial sector, their rise fueled by the exponential growth of computing power and the development of the Internet. Today, evolutionary computation is a thriving field, and genetic algorithms are "solving problems of everyday interest" (Haupt & Haupt, 1998, p. 147) in areas of study as diverse as stock market prediction and portfolio planning, aerospace engineering, microchip design, biochemistry and molecular biology, and scheduling at airports and on assembly lines.

THE GENETIC ALGORITHM

Encoding the Chromosomes

Figure 10 describes a flowchart of a typical GA. The first and most important step is encoding the parameters into a single parameter vector. The obvious choice is binary encoding because, after all, we want to simulate the GA on a computer. This is where the ingenuity of the user is most required. A good encoding ensures faster convergence of the GA and a lesser chance that it will get stuck in local optima. This step is also typical to the area of application of the GA which ranges from robot navigation (Geisler & Manikas, 2002) to optimal antenna array generation (Haupt, 1995). One programming problem that haunts all GA users is encoding of non-integer parameters (in practice, most genes are floating point values). In most cases, a finite number of bits are allocated to each gene in the chromosome and the expected range of values for that gene is covered using the allocated number of bits. Optimal bit allocation to each parameter is a research problem in itself and is highly domain- and problem-specific.

The Genetic Operators: Selection, Crossover and Mutation

All these three operators take part in the reproduction step of the GA to generate the next generation. Depending on the user and the application, some operators may not be used at all while others may be used sparingly (i.e., not in every generation). **Selection** is usually done by placing a threshold on the minimum allowable fitness and is omitted in many instances of GAs. The trade-off is using this operator is one of speed versus the risk of falling into local optima. At the end of the selection step, the number of chromosomes is again the same as that in the parent population. Other selection strategies include the *roulette wheel selection* which is described below and *tournament selection*:

1. Evaluate fitness for all chromosomes in the parent population. Normalize the fitness values so that the sum of all fitness values of the population equals unity.
2. Define accumulated normalized fitness of a single chromosome x as the sum of fitness values of all chromosomes up to x.
3. Pick a random number N between 0 and 1.
4. The selected individual is the first one whose accumulated normalized fitness is greater than N.

Crossover is the step in which genetic information is exchanged between two parent chromosomes that have been chosen by the selection operator or otherwise to give two novel offsprings. Given two-parent chromosomes, the probability of using the crossover operator is usually governed by a single parameter in the GA called *crossover probability*. Crossover is implemented as one-point crossover, multi-point crossover and non-uniform crossover which are illustrated in Figure 9. It is clear that non-uniform crossover gives offsprings of different lengths which is mostly undesirable. The point of crossover is also governed by a probability parameter in the GA simulation. There are two more methods of Crossover called arithmetic crossover and heuristic crossover that have been reported in literature (Davis, 1991; Holland & John, 1975).

Mutation, like in natural evolution, occurs rarely in GAs. Mutation is a necessary operator, especially in GAs where the user has minimal a-priori information about the global optimum. Mutation is implemented by flipping a selected bit in a selected chromosome. The probability of mutation in a given chromosome (immediately after it has been produced by crossover), is an adjustable parameter of the GA. Once selected for mutation, a random number is used to choose the bit to flip within the chromosome. The probability of mutation can be smaller than 0.01. Other methods of implementing mutation include boundary mutation, uniform mutation and non-uniform mutation (Holland, 1975; Davis, 1991). The purpose of mutation is to maintain genetic diversity in the population with the sole aim of preventing local optima.

Comments

GAs are often viewed as a global optimization method, although convergence to a global optimum is only guaranteed in a weak probabilistic sense. However, one of the strengths of GAs is that they perform well in search spaces where there may be multiple

Figure 9. Crossover strategies in GAs

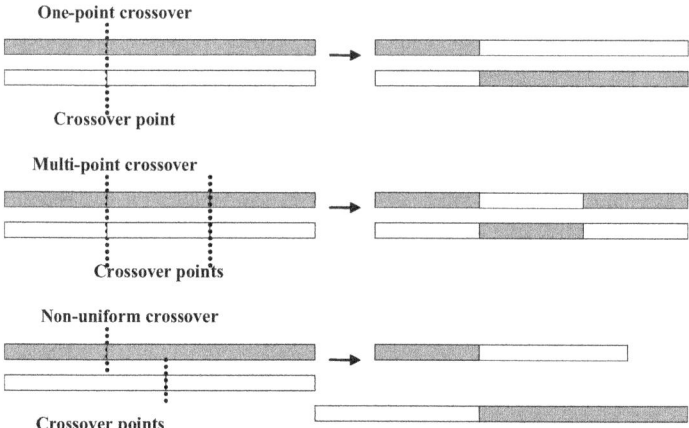

local optima. GAs tend not to get stuck on a local minima since all the operators have enough randomness to ensure a thorough search. Ironically, herein lies one of the major cons of GAs. The search mechanism becomes exhaustive although thorough, and consequently the convergence becomes impractical. "Due to the randomness of the GA operation, it is difficult to predict its performance, a factor that is crucial for hard-deadline, real-time applications" (Tang, Man, Kwong, & He, 1996, p. 35). In the same work, though, the authors show that GAs are better than classical optimization techniques for multi-objective optimization which covers a large class of practical problems in industry and economics. It is for this reason that GAs are used in applications such as resource allocation especially in circuit partitioning (Nan, Li, & Kou, 2004). The power of evolutionary computation has touched virtually any field one cares to name, and new uses continue to be discovered by ongoing research. The most amazing fact however still remains that the random chance of variation, coupled with the law of selection (*natural selection as expounded by Darwin*), is a problem-solving technique of immense power and nearly unlimited application.

GAs have been proposed for use with neural networks and fuzzy logic systems. These systems, also called AI-hybrid systems will be introduced in the next section

HYBRID APPROACHES

In this chapter, we have discussed the main techniques in computational intelligence viz. artificial neural networks, genetic algorithms and fuzzy logic. These three techniques address different problems in computational intelligence. Broadly, ANNs can be termed as information storage networks. They store information given in the form of test data in their weights. Thus, they are used in applications which require content-

Figure 10. Flowchart of a typical GA

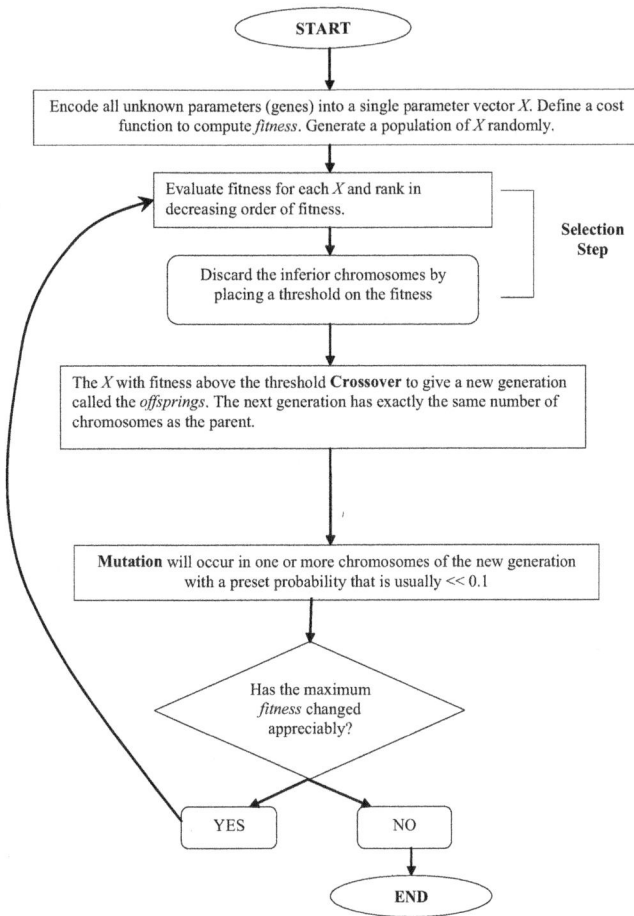

based discrimination of the data. In a similar vein, fuzzy systems can be termed as non-deterministic decision makers. They are consequently used in many expert systems and operate at the highest levels of an AI architecture. Genetic algorithms serve as an intelligent search and optimization technique. An AI system has all these requirements viz. information storage and retrieval, decision making, search and optimization. Thus, in practical AI systems, these techniques occur in combination or at different levels of the system hierarchy. Yet, there are cases in which one of these techniques is used in the other. For example, GAs being an optimization tool, are used to optimize the membership functions of the fuzzy logic system (Tang, Man, & Chan, 1994). Fuzzy logic systems usually act as expert systems in controller design. In older systems, the fuzzy rules were hand-coded based on human experience with non-fuzzy controllers. If such a system failed to perform, then the rules were tinkered with. In recent years, the approach has shifted to learning the fuzzy rules using another computational intelligence technique, usually GAs. The basic idea is to use the GA to generate a population of fuzzy rules and evaluate the performance of the controller. A performance-based fitness function improves the merit of the fuzzy rules with each generation. Such a system is described in detail in Chiang, Chung, and Lin (1997).

ANNs and GAs have been combined by using GAs to optimize the parameters of an ANN. For instance, when the Gradient Descent Algorithm is used to modify the weights of an ANN, it needs to have a specific network topology, in particular, the number of hidden nodes and the number of connection weights. These values, if preset by hand are not usually optimal (Husbands & Mill, 1991). Hence GAs are used to optimize the ANN architecture prior to learning of the weights. For a typical example of the use of GAs in developing ANN architecture refer to Leung, Lam, Ling, and Tam) (2003). The major challenge in combining GAs with ANNs or fuzzy logic systems, again, lies in encoding the chromosomes (i.e., designing the parameter vector and cost function).

Another hybrid approach is to use both GAs and a back-propagation neural network to solve the same optimization problem. Since GAs have a better chance of getting to the global optimum and ANNs are faster, we can combine the best of both worlds by first getting close to the global optimum using the GA and then using the ANN to improve our result (Janson & Frenzel, 1993).

In conclusion, it can be safely said that the research area of hybrid approaches begs to be investigated further. The major challenges lie in designing scalable algorithms for combining two or more of these computational intelligence techniques. A grand unification of these three cornerstones of computational intelligence would definitely help in the development of a new generation of intelligent machines.

SUMMARY

In this chapter, we reviewed the gamut of techniques that fall under the broad category of computational intelligence. The review is not exhaustive. It is meant to serve as a starting point for the chapters ahead. It gives a feel of the tools at the hands of a researcher who plans to use computational intelligence techniques. For a description of real systems using computational intelligence techniques the reader is expected to refer the particular papers that have been included focusing on various aspects of biomedical and human movement science areas.

REFERENCES

Amari, S.-I. (1967). A theory of adaptive pattern classifiers. *IEEE Transactions on Electronic Computers*, 16, 299-307.

Behera, L., Chaudhury, S., & Gopal, M. (1998, March). Application of self-organizing neural networks in robot tracking control. In *IEE Proceedings on Control Theory and Applications, 145*(2), 135–140.

Behera, L., Gopal, M., & Chaudhury, S. (1995, November). Inversion of RBF networks and applications to adaptive control of nonlinear systems. In *IEE Proceedings on Control Theory and Applications, 142*(6), 617–624.

Behera, L., & Sundaram, B. (2004). Stochastic filtering and speech enhancement using a recurrent quantum neural network. In *Proceedings of the ICISIP 2004* (pp. 165–170).

Benoudjit, N., & Verleysen, M. (2003, October). On the kernel widths in radial-basis function networks. *Neural Processing Letters, 18*(2), 139– 54.

Burges, C. J. C. (1998). A tutorial on support vector machines for pattern recognition. *Knowledge Discovery and Data Mining, 2*(2), 121–167.

Carpenter, G.A., & Grossberg, S. (1988). The ART of adaptive pattern recognition by a self-organizing neural network. *Computer: Special Issue on Artificial Neural Systems,* 21, 77-88.

Chen, C.-Y., Huang, C.-Y., & Liu, B.-D. (1997, October). Current-mode defuzzifier circuit to realize the centroid strategy. In *IEE Proceedings on Circuits, Devices and Systems, 144*(5), 265–271.

Chiang, C.-K., Chung, H.-Y., & Lin J.-J. (1997, August). A self-learning fuzzy logic controller using genetic algorithms with reinforcements. *IEEE Transactions on Fuzzy Systems, 5*(3), 460–467.

Coleman, T. F., & Hubert, L. A. (1989). A direct active set method for large sparse quadratic programs with simple bounds. *Mathematical Programming, 45*, 373-406.

Cortes, C., & Vapnik, V. (1995). Support vector networks. *Machine Learning, 20*(3), 273-297.

Dalton, J., & Deshmane, A. (1991, April). Artificial neural networks. *Potentials, IEEE, 10*(2), 33-36.

Davis, L. (1991). Handbook of genetic algorithms. In V. N. Reinhold (Ed.). New York.

De Silva, C. R., De Silva, L. C., & Ranganath, S. (2003, December). A new radial basis function network classifier for holistic recognition of universal facial expressions. In *The Fourth Pacific Rim Conference on Multimedia and Proceedings of the 2003 Joint Conference of the Fourth International Conference on Information, Communications and Signal Processing* (Vol. 2, pp. 1206-1210).

Drucker, H., Burges, C. J. C., Kaufman, L., Smola, A., & Vapnik, V. (1997). Support vector regression machines. In M. C. Mozer, M. I. Jordan, & T. Petsche (Eds.), *Advances in neural information processing systems* (Vol. 9, p. 155). Cambridge, MA: MIT.

Duda, R., & Hart, P. (1973). *Pattern classification and scene analysis*. New York: John Wiley & Sons.

Fletcher, R. (1981). *Practical methods of optimization, Vol 2: Constrained optimization*. NJ: John Wiley & Sons.

Geisler, T., & Manikas, T. W. (2002). Autonomous robot navigation system using a novel value encoded genetic algorithm. *The 2002 45ᵗʰ Midwest Symposium on Circuits and Systems (MWSCAS-2002)* (Vol. 3, pp. 45-48).

Goldberg, D. E. (1989). *Genetic algorithms.* Reading, MA: Addison-Wesley.

Goser, K., Hilleringmann, U., Rueckert, U., & Schumacher, K. (1989, December). VLSI technologies for artificial neural networks. *IEEE Micro, 9*(6), 28-44.

Guyon, I., Boser, B., & Vapnik, V. (1993). Automatic capacity tuning of very large VC-dimension classifiers. In S. Hanson, J. Cowan, & C. Giles (Eds.), *Advances in neural information processing systems* (Vol. 5, pp. 147-155), Morgan Kaufmann.

Hassoun, M. H. (1996, June). Fundamentals of artificial neural networks. In *Proceedings of the IEEE, 84*(6), 906.

Haupt, R. L. (1995, April). An introduction to genetic algorithms for electromagnetics. *IEEE Antennas and Propagation Magazine, 37*(2), 7-15.

Haupt, S. E., & Haupt, R. L. (1998, March). Optimizing complex systems. In *Proceedings of the IEEE Aerospace Conference* (Vol. 4, pp. 241-247).

Holland, J. H., & John, H. (1975). *Adaptation in natural and artificial systems.* Ann Arbor, MI: University of Michigan.

Hopfield, J. J. (1982). Neural networks and physical systems with emergent collective abilities. In *Proceedings of the National Academy of Sciences, USA.* (pp. 2554-2558).

Hsu, Y.-Y., & Yang, C.-C. (1994, May). A hybrid artificial neural network-dynamic programming approach for feeder capacitor scheduling. *IEEE Transactions on Power Systems, 9*(2), 1069-1075.

Hunt, K. J., Sbarbaro, D., Zbikowski, R., & Galwtrop, P. J. (1992). Neural networks for control system - A survey. *Automatica, 28*(6), 1083-1113.

Husbands, P., & Mill, F. (1991). Simulated co-evolution as the mechanism for emergent planning and scheduling. In *Proceedings of the Fourth Int. Conf. on Genetic Algorithms* (pp. 210-264).

Hush, D. R., & Horne, B. G. (1993, January). Progress in supervised neural networks. *IEEE Signal Processing Magazine*, 10, 8-39.

Ibrikci, T., Brandt, M. E., Wang, G., & Acikkar, M. (2002, October). Mahalanobis distance with radial basis function network on protein secondary structures. In *Proceedings of the 24ᵗʰ Annual Conference on Engineering in Medicine and Biology and the Annual Fall Meeting of the Biomedical Engineering Society, EMBS/BMES Conference, 2002* (Vol. 3, pp. 2184-2185).

Jamshidi, M., Vadiee, N., & Ross, T. J. (1993). *Fuzzy logic and control: Software and hardware applications.* Englewood Cliffs, NJ: PTR Prentice Hall.

Janson, D. J., & Frenzel, J. F. (1993, October). Training product unit neural networks with genetic algorithms. *IEEE Expert, 8*(5), 26-33.

Jiang, Y., McCorkell, C., & Zmood, R. B. (1995, December). Application of neural networks for real time control of a ball-beam system. In *Proceedings of the IEEE International Conference on Neural Networks, 5*(27), 2397-2402.

Karr, C. L. (1991). *Genetic algorithms for fuzzy controllers. AI Expert, 6*(2), 26-33.

Kassam, S. A., & Inhyok, C. (1993, November). Radial basis function networks in nonlinear signal processing applications. *Conference Record of The Twenty-Seventh Asilomar Conference on Signals, Systems and Computers* (Vol. 2, pp. 1021-1025).

Klopf, A. H. (1972). Brain funtion and adaptive sytems— a heterostatic theory. Teachnical report AFCRL-72-0164, *Air Force Cambridge Research Laboratories,* Bedford, MA, USA.

Leung, F. H. F., Lam, H. K., Ling, S. H., & Tam, P. K. S. (2003, August). Tuning of the structure and parameters of a neural network using an improved genetic algorithm. *IEEE Transactions On Neural networks, 14*(1), 793-799.

Ma, H (1995, March). Pattern recognition using Boltzmann machine. In *Proceedings of IEEE Southeastcon '95: 'Visualize the Future'*, (pp. 23-29).

Matsuda, S. (1993, October). The stability of the solution in Hopfield neural network. In *IJCNN '93-Nagoya, Proceedings of 1993 International Joint Conference on Neural Networks* (Vol. 2, pp. 1524-1527).

Meeden, L. A. (1996). An incremental approach to developing intelligent neural network controllers for robots. *IEEE Transactions on Systems, Man and Cybernetics, Part B, 26*(3), 474-485.

Mendel, J. M. (1995, March). Fuzzy logic systems for engineering: A tutorial. In *Proceedings of the IEEE, 83*(3), 345-377.

Mendil, B., & Benmahammed, K. (1998, May). Generalized adaptive defuzzifier. In *Fuzzy Systems Proceedings of the IEEE World Congress on Computational Intelligence* (Vol. 2, pp. 1680-1683).

Merzagora, A. C., Bracchil, F., Cerutti, S., Rossi, L., Bianchi, A. M., & Gaggiani, A. (2004). A radial basis function neural network for single sweep detection of somatosensory evoked potentials. In *Engineering in Medicine Conference Proceedings and EMBC 2004 and the 26th Annual International Conference of the Biology Society* (Vol. 1, pp. 427 - 430).

Minsky, M., & Papert, S. (1960). *Percepttons: An introduction to computational geometry.* MIT Press: Cambridge, MA.

Mitchell, M. (1996). *An introduction to genetic algorithms.* Cambridge, MA: MIT.

Mitchell, T. (1997). Decision tree learning. *Machine Learning.* McGraw-Hill.

Nan, G.-F., Li, M.-Q., & Kou, J.-S. (2004). Two novel encoding strategies based genetic algorithms for circuit partitioning. In *Proceedings of 2004 International Conference on Machine Learning and Cybernetics* (Vol. 4, pp. 2182-2188).

Pelayo, F. J., Rojas, I., Ortega, J., & Prieto, A. (1993, April). Current-mode analogue defuzzifier. *Electronics Letters, 29*(9), 743-744.

Phansalkar, V. V., & Sastry, P. S. (1994, May). Analysis of the back-propagation algorithm with momentum. *IEEE transactions on Neural Networks, 5*(3), 505-506.

Rosenblatt, F. (1958). The perceptron: A probabilistic model for information storage in the brain. *Psychological Review, 65,* 386-408.

Rumelhart, D. E., Hinton, G. E., & Williams, R. J. (1986). Learning internal representations by error propagation. In D. E. Rumelhart & J. L. McClelland (Eds.), *Parallel distributed processing: Explorations in the microstructures of cognition* (pp. 318 - 362). Cambridge, MA: MIT.

Sato, M. (1990). A real time learning algorithm for recurrent neural networks. *Biological Cybernetics, 62,* 237-241.

Scholkopf, B., Bartlett, P., Smola, A., & Williamson, R. (1999). Shrinking the tube: A new support vector regression algorithm. In M. S. Kearns, S. A. Solla, & D. A. Cohn (Eds.), *Advances in neural information processing systems 11.* Cambridge, MA: MIT.

Scholkopf, B., & Smola, A. J. (1998). *New support vector algorithms*. (NeuroCOLT Technical Report No. NC2-TR-98-031)

Smola, A., Murata, N., Scholkopf, B., & Muller, K. R. (1998). Asymptotically optimal choice of μ-loss for support vector machines. In L. Niklasson, M. Boden, & T. Ziemke (Eds.), *Proceedings of the 8th International Conference on Artificial Neural Networks - Perspectives in Neural Computing* (pp. 105-110). Berlin: Springer Verlag.

Smola, A., & Scholkopf, B. (1998, October). A tutorial on support vector regression. (NeuroCOLT Technical Report No. NC2-TR-98-030)

Suykens, J. A. K., Van Gestel, T., De Brabanter, J., De Moor, B., & Vandewalle, J. (2002, October). Weighted least squares support vector machines: Robustness and sparse approximation. *Neurocomputing, 48*(1-4), 85-105.

Suykens, J. A. K., Van Gestel, T., De Brabanter, J., De Moor, B., & Vandewalle, J. (2002). *Least squares support vector machines*. Singapore: World Scientific.

Tang, K. S., Man, K. F., & Chan, C. Y. (1994, December). Fuzzy control of water pressure using genetic algorithm. In *Proceedings of the IFAC Workshop on Safety, Reliability and Applications of Emerging Intelligent Control Technologies* (pp. 15-20).

Tang, K. S., Man, K. F., Kwong, S., & He, Q. (1996). Genetic algorithms and their applications. *IEEE Signal Processing Magazine, 13*(6), 22-37.

Trehern, J., Jack, M., & Laver, J. (1986). Speech processing with a Boltzmann machine. In *Proceedings of the IEEE International Conference on Acoustics, Speech, and Signal Processing* (Vol. 11, pp. 721-724).

Tumer, K., Ramanujam, N., Ghosh, J., & Richards-Kortum, R. (1998, August). Ensembles of radial basis function networks for spectroscopic detection of cervical precancer. *IEEE Transactions on Biomedical Engineering, 45*(8), 953-961.

Vapnik, V. (1995). *The nature of statistical learning theory*. New York: Springer-Verlag.

Vapnik, V., Golowich, S., & Smola, A. (1997). Support vector methods for function approximation, regression estimation, and signal processing. In M. Moser, M. Jordan & T. Petsche (Eds.), *Advances in neural information processing systems* (Vol. 9, pp. 281-287). Cambridge, MA: MIT.

Werbos, P. (1990, October). Backpropagation through time: What it does and how to do it. In *Proceedings of the IEEE, 78*(10), 1550-1560.

Werbos, P. J. (1974). *Beyond regression: New tools for prediction and analysis in the behavioral sciences*. Doctorial Dissertion, Appl. Math., Harvard University, MA.

Widrow, B., & Hoff, M. E. (1960). Adaptative switching circuits. In *1960 IRE WESCON Convention Record* (pp. 96-104). New York.

Winston, P. (1992). Learning by Building Identification Trees. In *Artificial Intelligence*, (pp. 423-442). Addison-Wesley.

Yan, L., Rad A. B., Wong Y. K., & Chan H. S. (1996, September). Model based control using artificial neural networks. In *Proceedings of the 1996 IEEE International Symposium on Intelligent Control* (pp. 283-288).

Zellner, A. (1984). *Basic issues in econometrics*. Chicago: University of Chicago.

Section II

Advances in Gait
Analysis and Modelling

Chapter V

Modelling of Some Aspects of Skilled Locomotor Behaviour Using Artificial Neural Networks

Stephen D. Prentice, University of Waterloo, Canada

Aftab E. Patla, University of Waterloo, Canada

ABSTRACT

Modelling the control of locomotor movements can take place at many different levels and represent gaits of different animal species. In many cases, these models attempt to capture the theoretical constructs for generating rhythmical motor patterns gained from neurophysiological studies. This chapter examines the use of artificial neural networks to gain insights into the control of walking movements. Two models discussed simplify the pathways and structures responsible for forming these fundamental cyclical movements, and capture the global transformations between intended goals and action. The use of computational models permits researchers to address certain questions that cannot be empirically tested using current experimental techniques.

INTRODUCTION

Legged locomotion affords great flexibility in travelling through complex environments as the use of isolated footholds permits a wider selection of travel paths. In contrast to vehicles or animals that utilize slithering, rolling or gliding movements and tend to maintain continuous ground contact, legged locomotion brings the ability to step over and around obstructions, as well as quickly alter the direction and location of force application. This flexibility of walking does come with the added cost of maintaining balance and support as the number of supporting limbs changes along with the configuration of the support base. The ability to coordinate the multitude of muscles acting at different joints to integrate the propulsive and postural objectives is impressive. Researchers have long been intrigued by how the nervous system controls this intricate task of legged locomotion.

Much of the knowledge regarding neural control of legged locomotion has been obtained through animal experiments. Isolation, lesion and stimulation studies have attempted to identify which neural structures and what information are necessary for the production of locomotor activity. These preparations have shown that the basic locomotor patterns occur in the absence of higher brain centres and that sensory information is not essential (see reviews by Delcomyn, 1980; Grillner, 1985; Patla, 1998). It is these findings which have formed the concept of central pattern generators (CPGs). It is proposed that the muscle activation patterns are produced by a group of neurons

Figure 1. A conceptual model for the control of locomotion

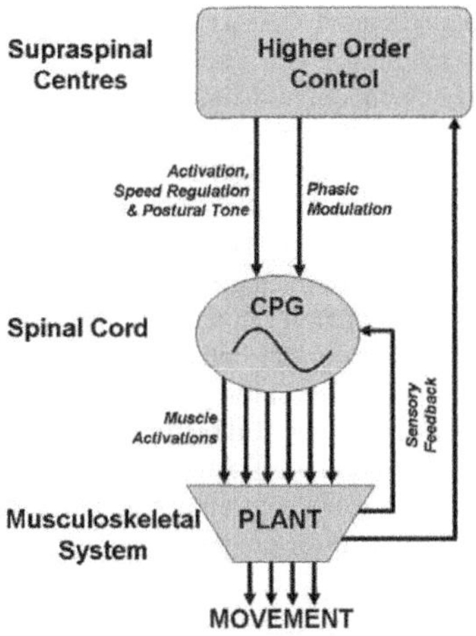

within the spinal cord and the release and maintenance of these patterns arises from tonic commands from higher brain centres. Recent work would suggest that sensory information has a major role in shaping the motor activation patterns during steady state walking and an even greater role when the basic locomotor patterns must be adapted to specific environmental events (Pearson, 2004; McCrae, 2001). The ability to perform voluntary modifications such as stepping over an obstacle, onto uneven surfaces or changing the length or direction of a stride relies greatly on a variety of descending pathways from cortical and brain stem structures (Drew et al., 2004 & 1996; Armstrong, 1988). The primary difficulty in understanding how legged locomotor control is achieved lies in the complexity of our nervous system and our inability to adequately monitor neural events. A conceptual model for the control of locomotion based on these studies is shown in Figure 1. The challenge has and continues to be decoding the neural circuitry involved and characterizing the various inputs and outputs, particularly in the feedback pathways. Modelling the locomotor control system offers an attractive alternative to study the principles of organization.

Introduction to Artificial Neural Networks

Model development has relied heavily upon techniques adopted from traditional engineering and control theory. However, these methods are not always adequate to represent certain systems. Traditional artificial intelligence (AI) approaches that utilize rule-based logic in conjunction with symbolic representation to endow machines with human faculties rarely perform certain tasks as well as humans and are unable to adapt to a variety of situations. Robotic control and other computational applications have been faced with similar challenges. Nervous systems prove superior in their ability to perform various tasks and adaptive behaviour is one of the more fundamental characteristics displayed by these biological systems. As a result, mathematical models inspired by biological systems have generated much attention in practical engineering and computational applications. System models concentrate on representing the operational and organizational principles of neural processing rather than striving for biological fidelity. Neural modelling is currently being approached at many different levels ranging from cellular models to circuit models to system models. Each type of model focuses on different aspects of neural processing. Compartmentalized models of specific activities of a single neuron require detailed information regarding a specific cell. These techniques are well suited for studying cellular dynamics, but the numbers of parameters required per neuron prohibit one from addressing the behaviour of a population of neurons. Circuit models utilize simpler models of the individual neuron which permit the study of small populations of neurons. However, they still require specific information concerning all connections and various cell parameters. Modelling complex behaviour such as locomotion requires the representation of an extremely large population of neurons and the acquisition of individual neural parameters and connection schemes would be an insurmountable task.

Artificial neural networks (ANNs) consist of a network of interconnected processing elements (PEs or units), which are inspired by computational capabilities of biological nervous systems. ANNs have many attractive features which are also evident in real neural circuits: the ability to process information in parallel, the ability to generalize to novel inputs, graceful degradation (fault tolerance), low sensitivity to noise and compactness (the ability of simple circuits to produce rather complex behaviour). The last feature was nicely illustrated by Braitenberg (1986), who developed simulated vehicles based on very simple connectionist control systems that exhibited rather complex behaviour. The message was clear: complex behaviours do not always need complex control architecture. From these observations, Braitnberg coined *"the law of uphill analysis and downhill invention"* referring to our tendency to overestimate the complexity of the control system based on analysis of its behaviour, and argued that understanding a system which we have constructed is a much easier process. The capabilities of the network are dictated by the dynamics of the individual processing units and their connection architecture. There are a variety of different types of ANNs. The simplest and most common connection schemes are feed-forward networks (Figure 2a) which typically consist of input, hidden and output layers of PEs and connections flow only in the direction from input to hidden to output. The strength of the connection between any two PEs is represented by a connection weight. The output of a PE is a function of the weighted inputs it receives (Figure 2b). Each input is a product of the connection weight and the output activation of the connected PE. The foundations for ANNs originated over 60 years ago (McCulloch & Pitts, 1943), but it was not until the development of various learning algorithms enabling the network to determine the connection weights (see review by Hertz et al., 1991) that their use became widespread in a variety of applications. Learning can be classified in two broad categories — supervised and unsupervised. Supervised learning often involves training sets of input/output pairs. The algorithm adjusts the connection weights until a satisfactory mapping is obtained, if such a mapping is possible. Unsupervised routines do not utilize a specific output measure to guide learning. Learning is accomplished by the input data itself or it may be

Figure 2. A typical feed-forward network showing input, hidden and output layers (A). The lower diagram (B) illustrates the operation of an individual processing element and includes an example of a sigmoidal transfer function between inputs and output.

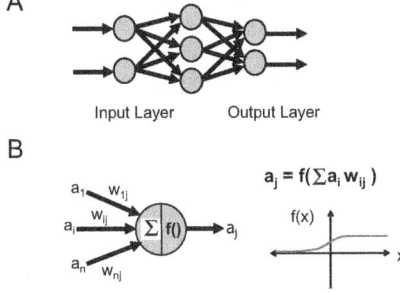

possible to use general goals of the network to aid in the learning process. The refinements of these learning algorithms have also made ANNs an attractive method for modelling neural processes.

Goals of the Chapter

The objective of this chapter is to investigate the usefulness of neurocomputational approaches in modelling some aspects of the human locomotor control system. ANNs capture some of the powerful features of biological input/output mapping that engineering methods have been unable to achieve. Models which embody the processing principles of biological nervous systems should lead to a better understanding of the control of locomotion. Two directions in modelling locomotor control are presented: a representational approach and a global transformation approach. It is intended that these two approaches would compliment each other and assist in answering some fundamental locomotor control issues.

REPRESENTATIONAL APPROACH FOR MODELLING THE CONTROL OF LEGGED LOCOMOTION

This approach is used to model the basic unit of locomotion, the central pattern generator (CPG) that produces the rhythmic movements of the limbs. Most of the evidence for the existence of locomotor CPGs comes from animal studies. It is a general organizational principle underlying locomotion in a wide range of vertebrate and invertebrate species (Delcomyn, 1980). Evidence for similar organizational principal in humans has come from developmental studies of infant stepping activity (Yang et al., 2004; Forssberg, 1985) and reflex activity and spontaneous rhythmic muscle activity in patients with complete spinal cord injuries (Bussel et al., 1989; 1988).

The neural substrate involved in this rhythm genesis lies within the spinal cord. Simple tonic commands from supraspinal structures are needed to release and modulate the speed and form of different gaits (Grillner, 1985; Rossignol, 1996; Rossignol & Debuc, 1994; Armstrong, 1988; Shik et al.,1966). The circuitry of the spinal cord alone can provide reasonably complex activation pattern to the muscles, controlling near-normal inter and intralimb movements. In decerebrate animals, a number of supraspinal centres have been identified which can initiate locomotor patterns of the spinal cord when stimulated. For example, the stimulation of the mesencephalic locomotor region (MLR) produces walking and modulation of the amplitude of this tonic input resulted in the scaling of speed or stride frequency of walking and ultimately to the production of additional modes of locomotion (e.g., walk, trot, gallop) (Shik et al.,1966). This presented a very elegant control design where the CPG circuits of the spinal cord would look after the timing and modulation of muscle activity and the supraspinal centres could adjust the speed of walking with a very simple command. The challenge in understanding the operations of these CPG networks for legged locomotion comes from our inability to detail the operations of this internal circuitry which has remained elusive.

Conceptual Framework for Representing Central Pattern Generation Functions for Legged Locomotion

Decoding the neuronal structures of these central pattern generators has been an important goal towards our understanding of how the nervous system controls locomotion. It is recognised that the CPGs for mammalian locomotion do not rely on pacemaker neurons for time keeping; rather the rhythmicity is an emergent property of neuronal networks. While such a network has been completely identified in the lamprey (Grillner, Wallen, Brodin & Lasner, 1991), similar success in the complex mammalian nervous system continues to be beyond the reach of current experimental techniques. Researchers have therefore used mathematical models to better understand the organisation of the CPGs (see Patla, Calvert & Stein, 1985). Oscillator models are a natural representation for a CPG. However, an appropriate oscillator must only be active in the presence of an external command and also have the ability to control the shape of its output pattern. Patla et al. (1985) proposed a model of the CPG using labile synthesised relaxation oscillator which could produce a specified complex periodic output when activated by a tonic input (see Bardakjian, El-Sharkaway & Diamant, 1983). Fundamental frequency signals (sine and cosine curve of the fundamental frequency) were modulated by two static nonlinear shaping functions to produce the complex periodic output. It was proposed that the CPG can be represented as two functional subsystems:

(a)   **Timing function** — the fundamental frequency oscillator producing both sine and cosine output at the fundamental locomotor frequency; and

(b)   **Shaping function** — nonlinear time independent functions providing the information storage about pattern shapes.

The objective of these models is not to capture the detailed biological fidelity (i.e., connectivity and membrane properties, etc.), but rather to ensure that these networks were based on the organisational and operational principles of CPGs gained from empirical research. Models with high neurological fidelity are primarily used to study small, specific, physiological circuits with known cellular properties. The CPGs for mammalian locomotion are at best ill defined and more likely a distributed phenomenon across a large population of neurons. Specific parameters required by for detailed cellular models are unavailable and make it difficult to both develop and justify their use to model the CPG circuits for mammalian locomotion. The processing elements of the networks while being crude representations of biological neurons capture the fundamental computational organisation within the nervous system (Selverston, 1993).

Using ANNs to Model the Timing and Shaping Functions of the Central Pattern Generator

We have developed neural network analogues (see Figure 3) for both the timing and shaping functions function of the CPG (Prentice, Patla, & Stacey, 1995, 1998). The timing network was developed to generate sinusoidal oscillations, bearing the fundamental frequency of locomotion (i.e., stride rate), based on a single tonic input. The role of the shaping network was to mould these simple oscillations into complex muscle activations required for steady state walking. Similar approaches have been proposed for developing

Figure 3. A neural network approach to modelling the timing and shaping functions of a CPG for human locomotion

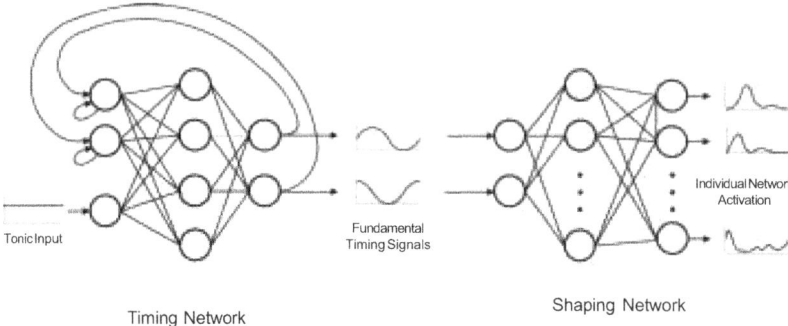

Tonic Input

Timing Network

Fundamental
Timing Signals

Shaping Network

Individual Network
Activation

The connection diagram illustrates the architecture of the proposed shaping network with the direction of processing flowing from the left to right. (Reproduced with permission from Prentice et al., 1998)

control systems for functional electrical stimulation of muscle in paralysed individuals (Abbas & Triolo, 1997; Srinivasan et al.,1992)

Specifically the requirements for the timing network were to produce oscillatory outputs in the form of sine and cosine functions whose frequency was proportional to the level of the tonic input consistent with MLR induced locomotion in the decerebrate cat (Shik et al.,1966). The release of these patterns should only occur in the presence of a non-zero input with the oscillatory output being an emergent property of the network architecture and not a result of specific pace-keeping properties of a neuron(s) in the network. The shaping network was proposed to be a static or time-independent transformation of the basic locomotor rhythm (represented by sine and cosine wave-forms) into the complex forms of the muscle activation time histories. This separation of timing and shaping functions permitted the independent examination of the temporal and amplitude modulation of the muscle activities and also required very different modelling approaches as they require distinct network dynamics.

ANN Architecture and Training for CPG Timing and Shaping Networks

Standard feed-forward ANNs (see Figure 2) are predominantly used to perform nonlinear input/output mappings and have become commonplace in many neural network applications. Typically, these networks do not contain any inherent temporal properties and as a result the mapping is static and is independent of the sequence in which the input patterns are presented. This model structure is appropriate for the shaping network as it is intended to be time independent. When time or any dependency on pattern presentation is crucial for the network's purpose, standard feed-forward

ANNs are inappropriate. This would be the case in representing the time keeping function of a CPG for locomotion.

Several methods have been developed to allow ANNs to capture temporal features (see Hertz, Krogh, & Palmer, 1991) and often involve alterations to input data, learning rules, and network architecture. Temporal information can be acquired by creating an input (context) buffer to allow the input of the current time step to be evaluated with previous time steps (Sejnowski & Rosenberg, 1987). Since the input for the present CPG model is tonic, the amplitude is constant for all time steps and, therefore, supplementing the input at a given time step with input from adjacent time steps would not provide any additional information. Another approach is to utilise recurrent backpropagation algorithms in which both the learning and PE activation equations are time sensitive (Pineda, 1989; Werbos, 1990). Williams and Zipser (1989) used a real time recurrent learning algorithm to develop a network that produces sine or square wave output. However, its use of time sensitive terms within individual cells or PEs makes them inappropriate for a representing timing of a CPG. Sequential ANNs retain the same equations as standard feedforward ANNs with timing being an emergent property of the network architecture. This is an attractive solution as much of the animal literature shows that locomotor pattern generation is the result of a network of stable neurons and not the result of "pacemaker" neurons. Sequential ANNs enhance the input of a feed-forward ANN via special feedback units which receive input from particular parts of the network. Jordan (1986) developed a model to learn sequences via feedback connections from the output units to units called state units and demonstrated that this class of ANN could be taught to produce limit cycles and act as oscillators. The state units provide feedback of the previous output state to act as additional input units that only receive internal input, while plan units in Jordan's model receive traditional external input.

The trainable connections of both the timing and shaping were established using a backpropagation algorithm (Rumelhart, Hinton, & Williams, 1986). We realised that backpropagation was not representative of how biological neural systems learn and it was employed here only as a means to establish the final connection weights and bias values. The fundamental basis of a CPG for mammalian locomotion appears to be largely an innate phenomenon and while learning would aid in fine tuning its operation over the lifespan, it is not clear how much learning influences the global organization of the CPG (Grillner & Wallen, 2004) Therefore, there is no intention to draw any conclusions with regards to the development of the CPG for mammalian locomotion.

Backpropagation is a supervised learning rule which utilizes gradient descent to solve the network connections and requires a training set containing a desired set of input-output patterns. Essentially, starting with random network connections the input vector is presented and based on current connections (see right side of Figure 3), the network will produce an output vector. The predicted output is compared to the actual or desired output and the error is then sent backward through the network to identify the connections most responsible for the errors. The input-output training data is presented repeatedly until the connections produce output signals with an error that is below a predetermined threshold. The real test to establish if the network has indeed captured the desired transformation is to reserve a separate *test* data set that was not used for training the network. The performance on test data set will indicate the ANN's ability to generalize to novel inputs. The specifics of the training and architecture of the timing and shaping models are discussed next.

Timing Network

The timing network employed a sequential ANN a single plan (input) unit, two state units, four hidden units, and two output units. The state units (S1 & S2) receive a feedback from output units (O1 & O2) and also have recurrent feedback onto themselves. These connections to the state units have fixed weights while the remainder of the network resembles a typical feed-forward network with trainable connection strengths. The connections from the output units to the state units were predetermined to provide an exact copy of the output while the recurrent connections of the state units were set to provide a weighted history. The input was represented as a tonic command to the plan unit with the two output waveforms forming sine and cosine signals, its frequency proportional to the level of input drive.

The training data set included a tonic level of plan activation (0.5, duration 1 sec.) followed by a step increase to double the magnitude (1.0, duration 1sec.) in order to generate sinusoidal outputs with frequencies of 1 and 2 Hz respectively. These frequencies were intended to roughly represent the range of stride frequencies of human walking and running. However, the specific frequencies used were not critical to the network's development and ability to predict a full range of desired output frequencies. During the training period, the value that was fed back to state units was the desired output instead of the calculated network output as suggested by Jordan (1986) and following training, the calculated output was always used to feed the state units.

The final network very closely replicated the oscillations of the training data showing appropriate phase and magnitude during steady state and transition periods evidenced by low % RMS difference (~30%) and high correlations (~0.95) when compared to the true sinusoids. This included a smooth transition when the tonic input was doubled to produce oscillations at twice the original frequency. Similar results were obtained when the order of the input activations were reversed forming a step decrease in input activation. The network was further tested across the entire range of tonic input activations (see Figure 4) to show that the network was able to scale the frequency of oscillations in a nearly linear fashion indicating that the network did indeed capture a generalisable transformation of input amplitude to output frequency. The network also failed to oscillate when the input activation was at or near zero, revealing that timing network was indeed labile and needs a requisite amount of activation to release its output patterns. The oscillation frequency also proved to be very stable when tested over a ten-second period (<0.027 Hz variation) and this stability remained even in the presence of 10% noise in the input activation level. The network's low sensitivity to noise is a desirable feature since tonic activations in physiological systems always contain some level of noise and if the system is responsive to these minor fluctuations the locomotor rhythm would become very unstable.

The proposed timing network (Prentice, Patla, & Stacey, 1995) reveals how a relatively simple arrangement of neurons could produce an oscillatory output whose frequency could be modulated by a single tonic input to the network. Indeed, this model met all of the requirements for modelling the time keeping function of a CPG for walking. First, the oscillations were a product of network configuration and not from temporal properties of individual processing elements. Second, it only oscillated in the presence of a non-zero input, and third, it could scale the output frequency in proportion to the magnitude of tonic input. It is remarkable that a generalized transformation with a

Figure 4. A raster of output waveforms, from one output unit O₁, are plotted for all input activations

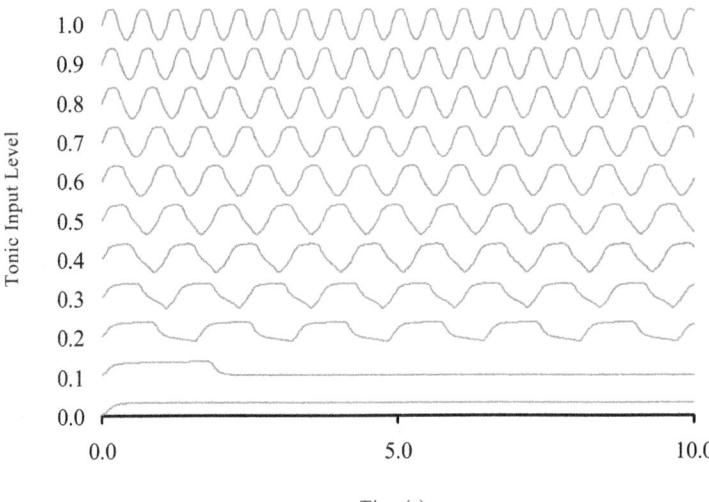

Time(s)

This figure demonstrates the effect of the level of input activation on output frequency and also shows the stability of the output oscillations over a ten second period (Adapted from Prentice et al., 1995)

capacity to adapt to a variety of input activations could be achieved from such a simple training set, containing two-fold step increase in input activation. The appropriate responses to novel inputs demonstrated that the model could indeed interpolate and extrapolate to unique input activation not used in training conditions. This general frequency modulation behaviour was anticipated by the nature of the training set including a sample of this scaling. However, the absence of oscillations with zero or near zero levels of input activation was not explicitly specified in the training data set. This was an important element, as locomotion is an episodic behaviour that requires the activation from supraspinal centres to release and maintain its rhythmic actions.

Shaping Network

We have proposed a standard feed-forward ANN to represent the shaping function of a CPG for human locomotion (Prentice, Patla, & Stacey, 1998). Specifically the inputs consisted of two sinusoidal inputs representing the basic walking frequency (stride rate) and the outputs of the model included eight representative muscle activations of one limb recorded during treadmill walking at different speeds. The time history data were represented through a series of input and output vectors where each vector pair depicted the data of a single time step. The initial model was formed by 2 input units, 16 hidden units and 8 output units and a reduced model with only four hidden units was used to probe the required complexity of ANN needed for this transformation. True sine and cosine waveforms were used as inputs in the formation of the shaping networks to permit a more accurate representation of the fundamental walking frequency and an independent evaluation of the shaping network performance.

The connections' weights and bias values for both shaping models were determined using a backpropagation algorithm (Rumelhart et al.,1986). Muscle activation patterns and stride information, linear Envelope EMG signals were recorded from the medial gastrocnemius, soleus, tibialis anterior, peroneus longus, biceps femoris, rectus femoris, gluteus medius and erector spinae from a single subject walking on a treadmill at five different walking speeds and separated into a training data set and a test data set. Strides were not normalized for time in order to preserve the temporal nature of successive strides. The average stride periods were used to determine the fundamental frequency or stride rate for each walking speed. The training set contained 12 strides from the following walking speeds: 1.2 & 1.8 m/s. Strides from the remaining speeds (1.4, 1.6 and 2.0 m/s), the test data set, include a range of speeds that were either within or outside the limits provided in the training set and were reserved to test the ability of the network to generalise its output when receiving inputs different than those used to train the network.

The length of network training was guided by monitoring the output error of the training set and the output error in the test set which was presented after every 10 passes of the training set during the training schedule. The training algorithm was suspended during each introduction of the test set. Therefore, the test set did not influence the network connections. Training continued until either the test or the training set no longer demonstrated a decreasing error. The output from the final network was then obtained by presenting both the training and test data.

Both the full and reduced networks successfully produced the muscle activation profiles from the simple timing oscillations across the range of walking speeds of both the test and training data sets (Figure 5). The ability to generalize to walking speeds both within and beyond those specified for training was an essential feature of the shaping function of the CPG. The accuracy of the predictions appeared to decrease in all muscles as walking speed increased, however the general ability to generate the phasic shape of activity were unaffected by speed. Thus, the models seem to be able to capture the general muscle activation pattern but have some difficulty in scaling the exact magnitude. A more functional evaluation of model performance was accomplished by comparing the predicted muscle activity to the normal variability (± 2 standard deviations about the mean) of muscle activity observed across the 12 strides at each of the various walking speeds. Predictions did not deviate very far from these limits. When it did fall outside these bounds it represented a loss of some of the finer phasing details.

Unexpectedly, a severe reduction in the complexity of the model to 4 hidden units did not substantially impair performance. In fact, the predictions were only slightly poorer than the full 16 hidden unit model. The operation of hidden units of feed-forward ANNs are somewhat comparable to principal component analysis and other factor analysis techniques where the basic features shared by a set of waveforms can be extracted and later used to reconstruct the individual waveforms. Thus, a simple representation of 4 feature signals, of the hidden units, was sufficient to capture the basic

Figure 5. Muscle activation patterns predicted by the sixteen hidden unit network (A) are shown for each of the three test walking conditions (1.4 m/s, 1.6 m/s & 2.0 m/s), while the leftmost column (B) shows the results for one of the test walking conditions (1.4 m/ s) using the four hidden unit network

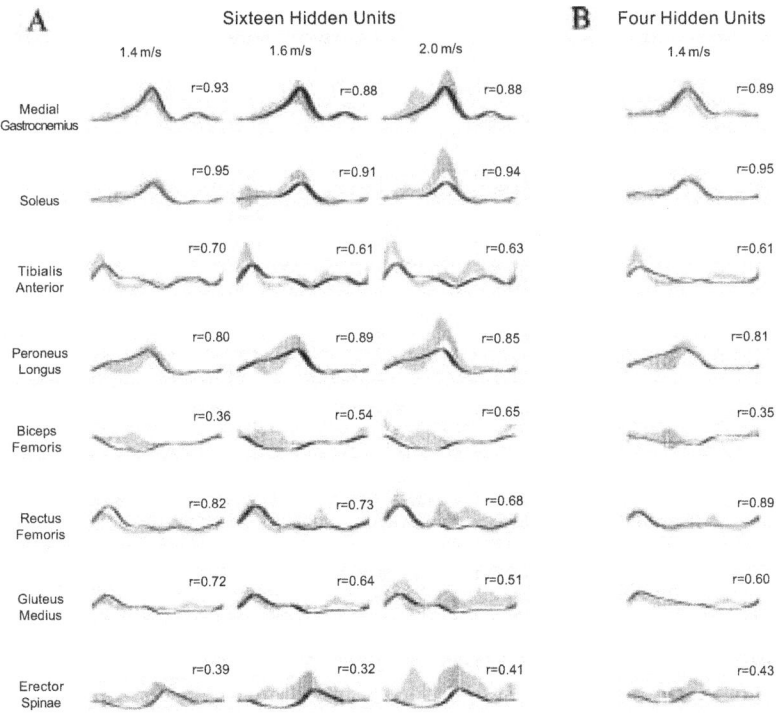

Each solid line represents the model's output from one stride, normalized for time between RFC and RFC while the shaded area is a two standard deviation band about the mean of the actual muscle activity recorded from the subject over 12 strides. The amplitude of activity for each muscle has been scaled to the peak activity found across all walking conditions. (Reproduced with permission from Prentice et al., 1998)

pattern of muscle activity since the added features of the full model only added subtle phasing information. Principal component analysis has been used by a number of researchers to identify core feature patterns shared by the ensemble of muscles in a single limb (Patla et al.,1985; Ivanenko et al.,2004). Their results indicated that only four to five feature signals were required to construct profiles accounting for the majority of muscle activations during locomotion. However, the differences in complexity between their feature signals and the activity of hidden units of these models make it difficult to make direct comparisons since there may not be a one to one correspondence. Optimization of number of hidden units to fully capture core activation patterns was not assessed directly and it would be beneficial to examine this aspect further.

If we examine the organization of the simpler four hidden unit network we can see how the activity of the four hidden units (HU_{1-4}) contribute to the eight muscle outputs. It is important to realize the network connections remain constant for all walking conditions. Figure 6 illustrates these connections along with the hidden unit and muscle activity patterns are shown for a single stride during the 1.4 m/s walking condition. Some hidden units are more active during specific periods within the stance phase (HU_1 & HU_4), while others exhibit a more phasic pattern (HU_2 and HU_3). The connection weights from these hidden units show excitation to those muscles that increase their activity during the same period and inhibitory actions to those muscles that show reduced activity in the same period. For example, the connections of HU_1 to the first four muscles: medial gastrocnemius, soleus and peroneus longus all peak in late stance and receive excitatory connections while the tibialis anterior shows an opposite relationship. Isolating the

Figure 6. Hidden unit activation patterns and their projections to the individual muscle outputs are shown for the four hidden unit network

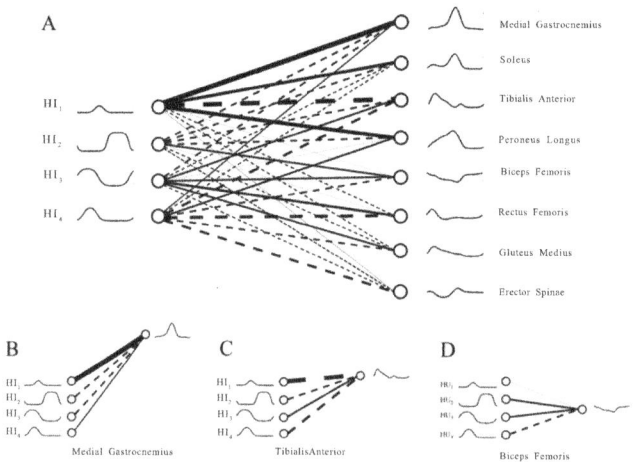

The top figure (A) illustrates all of the connection weights from the four hidden units (HU_{1-4}) to all eight muscles while the lower three figures (B-D) each isolate the connections to an individual muscle. The lines between units represent both excitatory (solid lines) and inhibitory (dashed lines) connections with the thickness of the line representing the relative strength of the connection. The hidden and muscle output activity profiles are taken from a single stride (RFC to RFC) of the 1.4 m/s walking condition (Reproduced with permission from Prentice et al. 1998)

network connections of specific individual muscles (Figure 6) further demonstrates the influence of hidden unit activation and network connectivity. Here the reciprocal nature of antagonist muscles, gastrocnemius and tibialis anterior, is reflected in the polarity of their connections from the hidden units (e.g., HU_2 & HU_4). The network connections also demonstrate how some muscles received more focussed projections (e.g., med. gastrocnemius) dominated by a particular hidden unit and others received more distributed connections (e.g., biceps femoris). Muscles that were poorly predicted tended to rely on more hidden unit features while better predictions were realized for muscles closely related to specific hidden unit activity. This may suggest that certain muscles may not be well represented in terms of hidden unit activity and the minimum number of features may not fully capture the organization needed for certain muscle groups.

The fact that the resulting muscle activation patterns do not exactly match that of the actual data is not unexpected. The proposed shaping models only received information regarding the fundamental locomotor rhythm. For modelling purposes, the driving oscillations were of a constant frequency equal to the subject's average stride frequency at a given walking speed. This representation resulted in the production of muscle activation patterns which did not vary from cycle to cycle. Even though the use of treadmill walking allows the collection of successive strides at a fixed belt velocity, the actual walking data do contain non-zero interstride variability. The actual walking trials will also be subject to variability in foot placement and postural demands during the 12 successive strides. These sources of variability inherent in actual overground walking would require additional activation to accommodate to the conditions of each individual stride. The shaping model did not receive any inputs conveying this information and their inability to represent certain aspects of the actual EMG profiles might reflect a CPGs dependence on inputs from other levels of the nervous system. The model utilizes only a simple timing input representing output of the time keeping function of a CPG energized by tonic supraspinal command. Certainly, it is well recognized that supraspinal and sensory information will play a substantive role even in the control of steady state locomotion (Pearson, 2004; McCrea, 2001). The EMG patterns used to train these models were obtained from a normal subject whose locomotor control system included this rich information. The use of fictive locomotion in curarized animal would eliminate this information and allow more rigorous development and assessment of the proposed models. Although the results of these models did not fully represent human locomotor activity for walking at different speeds, they have demonstrated how a simple arrangement of uncomplicated processing elements can produce complex muscle activations from a fundamental timing signal. This type of transformations should be well within the abilities of the neural circuitry at the spinal cord level.

If the models are to represent normal steady state locomotion, a better understanding of the role of supraspinal and sensory information and how they interact with the pattern generator is needed. Researchers have suggested that information arising from the motor cortex is important for the control of foot placement and the regulation of the timing of individual strides (see Drew, 1991, 1993) while input from other supraspinal centres have shown to be necessary to maintain and modify postural tone as well as modifying other aspects of locomotor activity on a step by step basis (see reviews by Drew et al., 2004, Orlovsky, 1991; Mori et al.,1991). Aside from characterizing these inputs, it must be established how this information influences the CPG network. It has been suggested that inputs to a CPG for rhythmic limb movements may affect either the timing or the patterning of muscle activity (Lennard & Hermanson, 1985; Koshland & Smith, 1989). A model that can incorporate these additional inputs and resolve effects on both the timing and shaping of muscle activity will be better suited to address the generation of normal locomotor behaviour.

These models may also have practical uses in the development of functional electrical stimulation systems for the restoration of gait in paraplegic patients. The current models could form the basis of a simple speed controller which would allow the patients to ambulate at different speeds based on a simple amplitude control. Any practical implementations will require further investigation into how these models will handle gait initiation and transient changes in speed.

A GLOBAL TRANSFORMATION APPROACH TO MODELLING THE GENERATION OF MUSCLE ACTIVATION PATTERNS FOR SKILLED LOCOMOTOR BEHAVIOUR

Skilled locomotor behaviour beyond steady state locomotion requires information from various levels within the nervous system (Figure 1). The difficulty in modelling this type of behaviour lies in the inability to adequately characterize the numerous inputs to the locomotor system. The global transformation approach circumvents this problem by using a kinematic movement plan to represent the complex inputs to the locomotor control system. It has been suggested that the nervous system may plan movements at a kinematic level and the necessary motor patterns are generated through various transformations within the nervous system (Georgopoulos, 1991; Scott, 2003). A neural network which emulates this transformation by generating muscle activation time histories from actual kinematic data recorded from a variety of different gait conditions is discussed next. Representations of this transformation, aside from adding to the fundamental understanding of locomotor control, also have implications for practical realizations of artificial control systems for use in rehabilitation medicine. This type of network has been proposed to assist in the control of functional electrical stimulation (FES) to restore gait in paralysed patients (Heller et al.,1993).

Background: Control of Limb Movements Using a Kinematic Movement Plan

The appropriate muscle activation patterns for locomotion are the result of a complex transformation of sensory information and central commands. Sensory signals provide information regarding both the external environment and the physical state of the body while central commands are expressions of the individual's intention. Understanding the relationship between these input signals and the muscle activity needed for skilled locomotor behaviour continues to be a challenge. The vast complexity of the mammalian nervous system has made it difficult to comprehend how such transformations are performed. The development of suitable models requires the characterization of both the inputs and outputs of the system of interest. For locomotion, the outputs of the CNS are readily available in the form of muscle activations accessible through electromyography (EMG). The information encompassing the various sensory inputs and central motor commands is more difficult to represent.

Animal studies have offered the ability to monitor and perturb various regions of the nervous system in intact specimens and identified various structures that may be responsible for initiating and maintaining the basic locomotor rhythm (Grillner, 1985; Rossignol, 1996; Rossignol & Debuc, 1994; Armstrong, 1988; Shik et al.,1966) and work continues to characterize the roles of various sensory and motor pathways and their interactions with the spinal locomotor circuitry. It is clearly evident that the rich array of both sensory signals and supraspinal inputs are needed for full expression of locomotor activities. Beyond the basic locomotor rhythm, input is needed to ensure stability, coordination and propulsion for walking across a level surface. The advantages of legged locomotion are manifested by how the CNS modifies the fundamental gait pattern to safely move through a changing environment. Inputs from supraspinal structures appear to play a much greater role in adjusting muscle activations in response to environmental conditions and are capable of either modulating or overriding the basic locomotor rhythm (Armstrong, 1988; Drew, 1991). This adaptive locomotor ability requires a more sophisticated level of control which utilizes sensory information regarding the physical environment in forming accommodation strategies. Much is still unknown as to how the various levels of the CNS contribute to planning these adjustments or how this information is transformed into the muscle activations required to produce the desired locomotor action. The inability to clearly identify and represent the specific inputs to the locomotor control system makes it extremely difficult to model this system.

Investigations of episodic goal-directed arm movements have suggested that cortical level activity is correlated with a movement plan. These findings have led to the proposal that movements are planned at a kinematic level and the nervous system transforms this information into the appropriate motor commands for individual muscles (Georgopoulos, 1991; Scott, 2003). Although these types of movements differ from rhythmic locomotor movements, there is some evidence that skilled locomotion requiring precise foot placement may be controlled in a similar fashion and that in fact it provided the neural substrate for the evolution of fine upper limb control (Georgopoulos & Grillner, 1989). Further evidence for kinematic organization during lower limb movements has been gained from recordings of the dorsal spinocerebellar tract (DSCT) in walking cats (Bosco & Poppele, 2000; Bosco et al.,2000); they showed that DSCT neurons carry information about global characteristics of limb movement. If movements are planned at a kinematic level, the complex inputs to the locomotor system may be approximated in terms of a movement plan. Although a true measure of an individual's intended action is not available, a kinematic description of the actual movement should be a good representation for modelling purposes. Using neural network models, Jordan and Rumelhart (1992) demonstrated that the transformation between a movement plan and the motor output may be a learned association that is obtained through the body's interaction with the environment.

This approach is intended to demonstrate the ability of ANNs to model the transformation of a kinematic movement plan into the individual muscle activations needed for a range of skilled locomotor activities. As ANNs capture the basic organizational and operational principles of biological nervous systems, a comparison between predicted and actual muscle activity will indicate the potential of the CNS to perform similar computations and the extent to which muscle activations are formed through feed-forward planning.

Model Architecture and Development

The proposed model was a feed-forward ANN where the input vectors consisted of kinematic profiles of the lower limb and the individual muscle activations formed the output vectors (Prentice et al.,2001). The time history data were represented through a series of input and output vectors where each vector pair depicted the data of a single time step. Most feed-forward ANNs utilize PEs which lack time dependant properties and thus the input vector at any given time step will be treated independently. The assumption here is the transformation from kinematic plan to muscle activation is largely time independent in that the desired/current kinematic state is sufficient to dictate the motor output. Some of the temporal dependencies will be embodied in the kinematic plan containing velocity and acceleration terms.

Specifically the model employed 21 input units, 14 hidden units and 8 output units (See Figure 7). The input vectors were based on a minimized postural representation of the lower limb segments and included the horizontal and vertical position of the hip joint, the horizontal and vertical toe position, the sagittal and frontal angles of the hip joint, and the sagittal angle of the knee joint. To provide additional information of the desired kinematic state of the lower limb, the input vectors also included the velocity and acceleration histories of each of the above inputs. The output vectors consisted of linear envelop EMG recordings of the medial gastrocnemius, soleus, tibialis anterior, peroneus longus, biceps femoris, rectus femoris, gluteus medius and erector spinae.

Kinematic and muscle activity data from a total of eleven different walking conditions including normal gait were used to develop and/or assess the model development. Each condition consisted of ten trials and was a variation of natural level walking to include changes in cadence, stride length, stance width or required foot clearance. Two sets of natural level walking data (2 x 10 strides) were obtained at a cadence of 96 steps/min, step length of 80 cm, step width of 30 cm and no obstruction during the swing phase. For the remaining 9 conditions only one stride parameter was changed per condition and provisions were made to maintain all others at natural level walking values. Three altered cadence conditions varied from their normal self paced cadence were collected (*slow*:-10 steps/min, *fast*: +10 steps/min. & *faster*: +20 steps/min). The remainder of the conditions required the subject to respond to a cue triggered at right heel contact (RHC) to make a prescribed adjustment for the subsequent RHC. A light cue would signal changes in foot placement (stride length: *short*: 50% & *long* 150% or stance width: *narrow*: 50% & *wide* 150%). The presentation of a physical obstacle triggered by the initial RHC-required modulation of limb elevation (obstacle height: *low*: 10cm, *medium*: 20, and *high*: 40cm).

The network connections for both models were determined using a backpropagation algorithm. The walking data were separated into a training data set and a test data set, with each containing data from each walking condition group: cadence, length, width and required foot clearance. The training set included the following walking data: normal,

Figure 7. Network architecture for the global transformation network

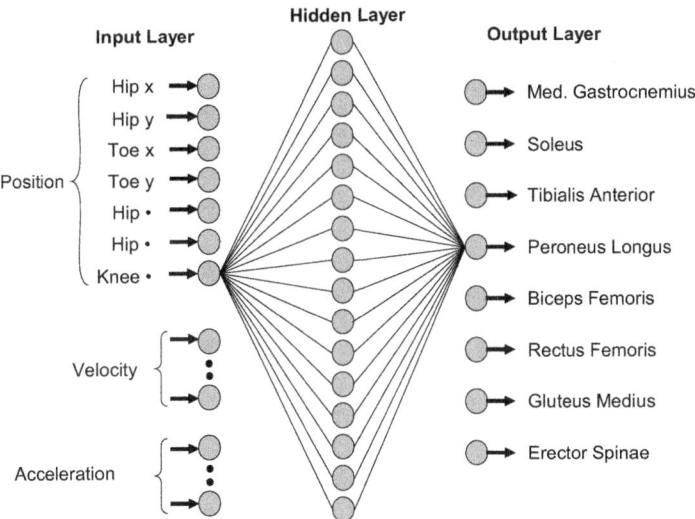

The input layer consisted of 21 units representing the position, velocity and acceleration of 7 key kinematic measures. Not all input units are shown for the velocity (n = 7) and acceleration (n = 7) inputs. All 14 hidden units are shown. The output layer generated the muscle time histories of 8 muscles of the lower limb and trunk. For simplicity, only the connections from one input unit to the hidden layer are shown and likewise projections from the hidden layer to one output unit are illustrated. The network was fully connected between each layer.

slow, faster, short, wide, low obstacle and high obstacle. This data set was repeatedly presented to the network to enable the learning rule to resolve the desired network parameters. The remaining conditions (normal, fast, long, narrow and low obstacle strides) formed the test data set and were reserved to assess the ability of the network to generalize to conditions other than those used to train the network. The selection of the test set data include a range of conditions that were either within or outside the limits provided in the training set. This would test the network's ability to both interpolate and extrapolate within each condition group.

The length of training was guided by monitoring the output error of the training set and the output error in the test set which was presented after every 10 passes of the training set during the training schedule. The training algorithm was suspended during each introduction of the test set. Therefore, the test set did not influence the network connections. Training continued until either the test or the training set no longer demonstrated a decreasing error. The output from the final network was then obtained by presenting both the training and test data set.

The muscle activation time histories predicted by this model closely matched those recorded experimentally for all conditions. Predicted muscle activation profiles for three conditions of the test set are shown in Figure 8. These conditions, normal, fast, and the medium height obstacle conditions represent the range of results from those conditions that were not used to train the model. The predicted muscle activations from individual trials are superimposed on a two standard deviation band of the actual muscle activations for the given condition. The predicted muscle activations captured the primary features

Figure 8. Muscle activation patterns predicted by the model for three different test conditions: normal walking, fast walking and steps over the medium obstacle

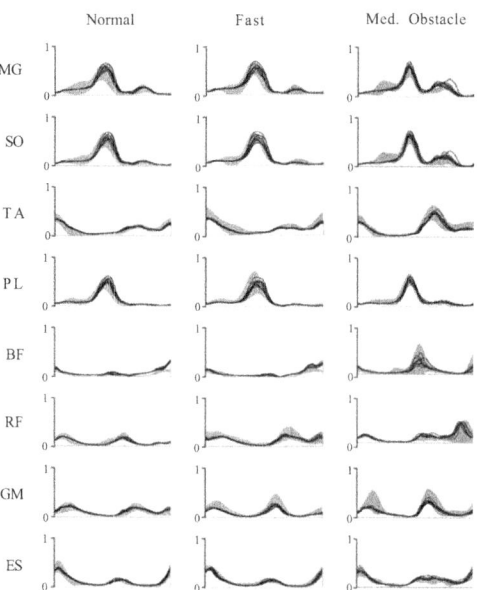

Each line (N=10) represents the predicted data from one stride, normalized for time between RHC and RHC. The shaded area is a two standard deviation band about the mean of the actual muscle activity recorded in each walking condition. The amplitude of each muscle has been scaled to the peak activity of the respective muscle across all walking conditions, including both test and training sets.

Note: The lower limit of the variability band has been truncated when values fell below zero representing a negative activation (reproduced from Prentice et al., 2001)

of every muscle over the different gait conditions and maintained the appropriate magnitude and temporal phasing required by each condition. There were very few conditions in which bursts of activity not seen in the actual recorded activity were seen in the model output and these cases were limited to specific muscles. The most noticeable was a small burst in the biceps femoris, which occurred in the middle of the gait cycle during some of the no obstacle conditions. An additional feature of the model's output was that a certain degree of the variability seen in the actual data had been removed. This was particularly evident in the early stance activity of the gastrocnemius and soleus muscles. However, it was also observed in other muscles at various points in the gait cycle.

There were only two cases out of 96 (gastrocnemius and soleus during the wide stance width condition) where the RMS difference between predicted and actual activation was greater or equal to 0.10. These low values indicated an extremely good match between predicted and actual activations of individual trials. Correlations between predicted and actual muscle activity evaluated the ability of the model to capture the temporal modulation of muscle activity. The correlations were very high with all but five muscle-gait combinations having values above 0.80 and most combinations exceeding 0.90. The weakest results were seen in the biceps femoris muscle which accounted for all five of the correlations below 0.80. These poor correlations which have been linked to the anomalous activity observed in the middle of the gait cycle.

The results from this model demonstrate the ability of ANNs to model the transformation between a kinematic movement plan and the necessary muscle activations. The model was not only successful for normal steady state locomotion but was also able to generate muscle activation patterns for conditions requiring changes in walking speed, foot placement and foot clearance. The model predicted well the actual muscle activations of the individual trials associated with the kinematic input. This was supported by the low RMS differences and high correlations calculated between the models' predictions and the empirically recorded activations.

Although these errors reflect the ability of the models to match their output to the actual muscle activity, they do not provide any insight into what effect these errors will have on the resulting movement. Errors in the level of activation of a particular muscle may have extremely different consequences depending on where in the gait cycle it occurs. For example, the level of tibialis anterior activity during early to mid-swing is critical in preventing tripping but changes in its activity during stance may have only a minimal effect on the resulting movement. Assessing errors in muscle activations would be simplified if limb movements could be calculated from muscle activation time histories. Many of the models which attempt to synthesize gait have only examined the mechanical system where joint torques are used to calculate limb motion and even these models have had only limited success. The inclusion of muscular dynamics adds further complications since the relationship between muscle activation and muscle force is not fully understood. Yamaguchi and Zajac (1990) have developed a model to predict locomotor movements from muscle activity. However, the muscle activations required to generate the expected limb motion were not very realistic. Until suitable models of musculoskeletal dynamics are developed, another method must be utilized to evaluate the behavioural outcome of the muscle activations predicted by the ANN models.

The predicted muscle activations of individual walking trials were compared to the normal variability associated with a particular walking condition to establish the functional outcome (Figure 8). It was assumed that activity predicted within the normal range observed experimentally would lead to more appropriate limb movements than those activations that were outside this region. A criterion of 25% of the gait cycle spent outside the normal variability was chosen to assist in the evaluation of these models. The majority of the predicted activations from the model remained within the range of normal variability for most of the stride. Only 25 of the 96 muscle-gait combinations spent more than 25% of the gait cycle outside of the two standard deviation band and ten were associated with a single muscle — biceps femoris. However, there was usually at least one muscle per condition which did not to meet this criterion. The only muscle to consistently produce activity outside this region for different conditions was the biceps femoris. The biceps femoris demonstrated a burst of hyperactivity during the middle of the gait cycle and was most notable in those conditions which did not include an obstacle. It appears that the biceps muscle was sensitized by the obstacle conditions where it exhibited the greatest range of activation. Although this erroneous activity may not facilitate normal locomotion, it may not jeopardize safe walking since an increase in activity during late stance and early swing should serve to increase limb clearance as in the obstacle conditions. Aside from the biceps femoris, these results indicate that the ANN models were able to produce the general pattern of muscle activity appropriate for the various locomotor conditions. The only exception may be the faster gait condition where six of the eight muscles did not meet the set criterion. However, upon examination of the profiles, RMS differences and correlation values the reason for the deviations for long durations from the normal variability seem to be due to a constant bias between the actual and predicted profiles.

The success of this ANN model to predict muscle activation time histories from a kinematic movement plan has implications towards both the fundamental understanding of the control of locomotion and practical realizations of artificial control systems to assist in the restoration of locomotion in paralysed individuals. It must be emphasized that the networks presented in this chapter are not intended to represent actual neural circuits. Instead, they are attempting to capture the general organizational and operational principles of the CNS. Therefore, the results of these models should reflect upon the capacity of the CNS to perform such transformations and they do not directly specify the actual structures involved.

GENERAL DISCUSSION AND CONCLUSION

The understanding of the control of human locomotion has traditionally relied on findings from animal experiments, where neural events may be monitored and/or manipulated during the performance of actual movements. These studies have provided and will continue to provide fundamental knowledge of the neural structures that contribute to the control of locomotion. Understanding of how the various levels of the nervous system interact to produce the movements necessary for locomotion has been limited by the vast complexity of the mammalian CNS. Mathematical modelling is one approach that has allowed researchers to study the dynamics of locomotor control. Models have often been established through the use of traditional engineering and control theory. However, these techniques have been limited in their ability to capture the essence of biological input/output transformations.

Two strategies to modelling the locomotor control system using ANNs were presented in this chapter: a representational approach and a global transformation approach. The main difference between these two approaches was based on the manner in which the inputs to the locomotor control could be characterized. The first approach was an attempt to embody some of the inputs and processes that have been proposed by the experimental literature regarding the formation of the basic locomotor pattern. The second approach modelled the system at a more abstract level to overcome the inability to adequately characterize many of the inputs required for skilled locomotor behaviour. It was intended that these two approaches would compliment each other in improving our understanding of the locomotor system and also facilitate practical implementation of artificial control systems for rehabilitation and robotic uses.

Both approaches, representational and global transformation, have also been proposed to assist the development of FES systems for gait restoration. The representational approach demonstrates that the generation of muscle activation time histories could be accomplished by a simple amplitude controller. The current state of the representational models provide only a limited repertoire of locomotor behaviour (changes in walking velocity), and are insufficient to meet the challenges of everyday mobility. The global transformation approach was able to generate muscle activation patterns for a wide range of gait conditions providing changes in walking speed, foot placement and foot clearance. It may be possible to develop a controller which could store movement plans for individual strides and permit patients to select from an array of strategies on a stride to stride basis. For example, such a controller may enable the patient to choose a particular walking speed and select different avoidance strategies as they are required. Although these approaches are not immediately suitable for FES control they do provide a framework for further research. The feasibility of these strategies will be dependant on a number of factors including: the availability of techniques to produce smooth stimulation profiles, the ability to customize the models for individual patients and the ability to resolve various temporal issues.

It is hoped that these models will serve to advance the understanding of the control of locomotion and at the same time provide suggestions for practical implementations of artificial control systems. Improvements to these models approaches will naturally follow as experimental work adds to the understanding of locomotor control and these models should in turn provide guidance for future research.

REFERENCES

Abbas, J. J., & Triolo, R. J. (1997). Experimental evaluation of an adaptive feedforward controller for use in functional neuromuscular stimulation systems. *IEEE Transactions on Rehabilitation Engineering, 5,* 12-22.

Armstrong, D. M. (1988). The supraspinal control of mammalian locomotion. *Journal of Physiology, 405,* 1-37.

Bardakjian, B. L., El-Sharkawy, T. Y., & Diamant, N. E. (1983). On a population of labile synthesized relaxation oscillators. *IEEE Transactions in Biomedical Engineering, 30,* 696-701

Bosco G., & Poppele, R. E. (2000). Reference frames for spinal proprioception: Kinematics based or kinetics based? *Journal of Neurophysiology, 83,* 2946-2955.

Bosco, G., Poppele, R. E., & Eian, J. (2000). Reference frames for spinal proprioception: Limb endpoint or joint-level based? *Journal of Neurophysiology, 83*, 2931-2945.

Braitenberg, V. (1984). *Vehicles: Experiments in synthetic psychology*. Cambridge, MA: Bradford.

Bussel, B., Roby-Brami, A., Azouvi, P. H., Biraben, A., Yakovleff, & Held, J. P. (1988). Myoclonus in a patient with spinal cord transection. *Brain, 111*, 1235-1245.

Bussel, B., Roby-Brami, A., Yakovleff, & Bennis, N. (1989). Late flexion reflex in paraplegic patients. Evidence for a spinal stepping generator. *Brain Research Bulletin, 22*, 53-56.

Delcomyn, F. (1980). Neural basis of rhythmic behaviour in animals. *Science, 210*, 492-498.

Drew, T. (1991). Visuomotor coordination in locomotion. *Current Opinion in Neurobiology, 1*, 652-657.

Drew, T. (1993). Motor cortical activity during voluntary gait modifications in the cat. I. Cells related to the forelimbs. *Journal of Neurophysiology, 70*, 179-199.

Drew, T., Jiang, W., Kably, B., & Lavoie, S. (1996). Role of the motor cortex in the control of visually triggered gait modifications. *Canadian Journal of Physiology and Pharmacology, 74*, 426-442.

Drew, T., Prentice, S., & Schepens, B. (2004). Cortical and brainstem control of locomotion. *Progress in Brain Research, 143*, 251-261.

Forssberg, H. (1985). Ontogeny of human locomotor control. I. Infant stepping, supported locomotion and transition to independent locomotion. *Experimental Brain Research, 57*, 480-493.

Koshland, G. F., & Smith, J. L. (1989). Mutamble and immutable features of paw-shake response after hindlimb deafferntation in the cat. *Journal of Neurophysiology, 62*, 162-173.

Lennard, P. R., & Hermanson, J. W. (1985). Central reflex modulation during locomotion. *Trends in Neurosciences, 8*, 483-486.

Georgopoulos, A. P. (1991). Higher order motor control. *Annual Review of Neuroscience 14*, 361-377.

Georgopoulos, A. P., & Grillner, S. (1989). Visuomotor coordination in reaching and locomotion. *Science, 245*, 1209-1210.

Grillner, S. (1981). Control of locomotion in bipeds, tetrapods and fish. In V. B. Brooks (Ed.), *Handbook of physiology, Sect. 1: The nervous system II, motor control* (pp. 1179-1236). American Physiological Society. Bethesda, MD: Waverly Press.

Grillner, S. (1985). Neurobiological bases of rhythmic motor acts in vertebrates. *Science, 228*, 143-149.

Grillner, S., & Dubuc, R. (1988). Control of locomotion in vertebrates: Spinal and supraspinal mechanisms. *Advances in Neurology, 47*, 425-453.

Grillner, S., & Wallen P. (2004). Innate versus learned movements — a false dichotomy? *Progress in Brain Research, 143*, 3-12.

Grillner, S., Wallen, P., Brodin, L., & Lasner, A. (1991). Neuronal network generating locomotor behavior in lamprey: Circuitry, transmitters, membrane properties, and simulation. *Annual Review of Neuroscience, 14*, 169-199.

Heller, B. W., Veltink, P. H., Rijkhoff, N. J. M., Rutten, W. L. C., & Andrews, B. J. (1993). Reconstructing muscle activation during normal walking: A comparison of symbolic and connectionist machine learning techniques. *Biological Cybernetics, 69,* 327-335.

Hertz, J., Krogh, A., & Palmer, R. G. (1991). *Introduction to the theory of neural computation.* Redwood, CA: Addison-Wesley.

Ivanenko, Y. P., Poppele, R. E., & Lacquaniti, F. (2004). Five basic muscle activation patterns account for muscle activity during human locomotion. *Journal of Physiology, 556,* 267-282.

Jordan, M. I. (1986). *Serial order: A parallel distributed processing approach* (Tech. Rep. No. 8604). University of California at San Diego, Institute for Cognitive Science.

Jordan, M. I., & Rumelhart, D. E. (1992). Forward models: Supervised learning with a distal teacher. *Cognitive Science, 16,* 307-354.

McCrea, D. A. (2001). Spinal circuitry of sensorimotor control of locomotion. *Journal of Physiology, 533,* 41-50.

McCulloch, W. S., & Pitts, W. H. (1943). A logical calculus of the ideas immanent in nervous activity. *Bulletin of Mathematical Biophysics, 5,* 115-133.

Mori, S., Sakamoto, T., & Takakusaki, K. (1991). Interaction of posture and locomotion in cats: Its automatic and volitional control aspects. In M. Shimamura, S. Grillner, & V. R. Edgerton (Eds.), *Neurobiological basis of human locomotion* (pp. 21-32). Tokyo: Japan Scientific Societies.

Orlovsky, G. N. (1991). Cerebellum and locomotion. In M. Shimamura, S. Grillner & V. R. Edgerton (Eds.), *Neurobiological basis of human locomotion* (pp. 187-199). Tokyo: Japan Scientific Societies.

Patla, A. E. (1988). Analytic approaches to the study of outputs from central patterns generators. In A. Cohen (Ed.), *Neural control of rhythmic movements in vertebrates* (pp. 455-486). New York: John Wiley & Sons.

Patla, A. E. (1996). Neuro-bio-mechanical bases for the control of human locomotion. In A. Bronstein, T. H. Brandt, M. Woollacott, & E. Arnold (Eds.), *Clinical aspects of balance and gait disorders* (pp. 19-40). UK.

Patla, A. E., Calvert, T. W., & Stein, R. B. (1985). Model of a pattern generator for locomotion in mammals. *American Journal of Physiology, 248* (Regulatory Integrative Comp. Physiol. 17), R484-R494.

Pearson, K. G. (2004). Generating the walking gait: Role of sensory feedback. *Progress in Brain Research, 143,* 123-129.

Pineda, F. J. (1989). Recurrent backpropagation and the dynamical approach to adaptive neural computation. *Neural Computation, 1,* 161-172.

Prentice, S. D., Patla, A. E., & Stacey, D. A. (1995). Modelling the time keeping function of the central pattern generator for locomotion using an artificial sequential neural network. *Medical & Biological Engineering & Computing, 33,* 317-322.

Prentice, S. D., Patla, A. E., & Stacey, D. A. (1998). Simple artificial neural network models can generate basic muscle activity patterns for human locomotion at different speeds. *Experimental Brain Research, 123,* 474-480.

Prentice, S. D., Patla, A. E., & Stacey, D. A. (2001). An artificial neural network model for the generation of muscle activation patterns for human locomotion. *Journal of Electromyography and Kinesiology, 11,* 19-30.

Rossignol, S. (1996). Neural control of stereotypic limb movements. In L. B. Rowell & J. T. Sheperd (Eds.), *Handbook of Physiology*, Section 12 (pp. 173-216). Exercise: Regulation and Integration of Multiple Systems. American Physiological Society.

Rossignol, S., & Debuc, R. (1994). Spinal pattern generation. *Current Opinion in Neurobiology, 4*, 894-902.

Rumelhart, D. E., Hinton, G. E., & Williams, R. J. (Eds.). (1986). Learning internal representations by Error Propagation. In *Parallel distributed processing: Explorations in the microstructure of cognition, Volume 1: Foundations*. Cambridge, MA: MIT.

Scott, S. H. (2003). The role of primary motor cortex in goal-directed movements: insights from neurophysiological studies on non-human primates. *Current Opinion in Neurobiology, 13*, 671-677.

Sejnowski, T. J., & Rosenberg, C. R. (1987). Parallel networks that learn to pronounce English text. *Complex Systems, 1*, 145-168.

Selverston, A. I. (1993). Modelling neural circuits: What have we learned? *Annual Review of Neuroscience, 16*, 531-546.

Shik, M. L., Severin, F. V., & Orlovskii, G. N. (1966). Control of walking and running by means of electrical stimulation of the mid-brain. *Biophysics, 11*, 756-765.

Srinivasan, S., Gander, R. E., & Wood, H. C. (1992). A movement pattern generator model using artificial neural networks. *IEEE Transactions on Biomedical Engineering, 39*, 716-722.

Werbos, P. J. (1990). Backpropagation through time: What it does and how to do it. In *Proceedings of the IEEE, 78*(10), 1550-1560.

Williams, R. J., & Zipser, D. (1989). Experimental analysis of the real-time-recurrent learning algorithm. *Connection Science, 1*, 87-111.

Yamaguchi, G. T., & Zajac, F. E. (1990). Restoring unassisted natural gait to paraplegics with functional neuromuscular stimulation: A computer simulation study. *IEEE Journal of Biomedical Engineering, 37*(9), 886-902.

Yang, J. F., Lam, T., Pang, M. Y. C., Lamont, E., Musselman, K., & Seinen, E. (2004). Infant stepping: A window to the behaviour of the human pattern generator for walking. *Canadian Journal of Physiology and Pharmacology, 82*, 662-674.

Chapter VI

Visualisation of Clinical Gait Data Using a Self-Organising Artificial Neural Network

Gabor J. Barton, Liverpool John Moores University, UK

ABSTRACT

The decision-making performance of gait experts varies depending on their background, training and experience. They have to analyse large quantities of complex gait data and this gives rise to an unbalanced use of the available information. These limitations inevitably lead to a biased interpretation. In this study, self-organising artificial neural networks were used to reduce the complexity of joint kinematic and kinetic data which form part of a typical instrumented gait assessment. Three dimensional joint angles, moments and powers during the gait cycle were projected from the multi-dimensional data space onto a topological neural map which thereby identified gait stem-patterns. Patients were positioned on the map in relation to each other and this enabled them to be compared on the basis of their gait patterns. The visualisation of large amounts of complex data in a two-dimensional map labelled with gait patterns is an enabling step towards more objective analysis protocols which will better inform decision making.

INTRODUCTION

The use of clinical gait analysis (CGA) in medical decision making is a controversial topic even today when it is widely recognised as a means to provide information about a patient's gait problems. On one hand, gait analysis is seen as a successful application of biomechanical concepts and methodologies applied to answer clinical questions related to the management of patients suffering from gait problems. The process of gait analysis is regarded as a collection of inter-linked methods that provide quantitative data about gait which are unavailable by any other means. Medical professionals can exercise evidence-based practices on the basis of objective information about their patients gained from performing gait analysis. The inter-disciplinary nature of CGA makes it possible to move beyond the pure mechanistic approach of the human body towards a movement disorder oriented view which leads to a pathology centred assessment of movement function thereby linking the biomechanical findings to medical interventions in a natural way.

On the other hand, it is well recognised that CGA has its limitations. As of today there are no widely accepted standards available in Europe that determine the technical, clinical and educational aspects of gait analysis, although there are some strong initiatives which move towards defining such essential measures. One of the most successful pioneering efforts is the standards laid down by the Clinical Movement Analysis Society of UK and Ireland (http://cmasuki.org). One would expect that certain gait abnormalities could be clearly defined through gait analysis, but even the sharing of a database on normal gait turned out to be problematic as witnessed by the participants of the Clinical Gait Analysis Web site in 1999 (http://www.univie.ac.at/cga/faq/norms.html). When it comes to purchasing the expensive services of a gait analysis laboratory then the customer looks for independent evaluations. One such report was published by the Technology Evaluation Centre on Gait Analysis for Pediatric Cerebral Palsy (Technology Evaluation Center, 2002) which concluded that such a service does not meet their criteria.

The benefits of CGA are shadowed by its limitations because the specific contribution of instrumented gait analysis to clinical decision making is not well known (Fuller et al.,2002). There is currently no clear evidence that medical intervention guided by CGA has better outcomes than intervention without CGA. Although there is insufficient evidence to support the claim that the use of gait analysis improves outcomes (Hailey and Tomie, 2000), there is evidence to suggest that gait analysis alters decision making in cerebral palsy (DeLuca et al.,1997; Kay et al.,2000; Cook et al.,2003) and it can promote higher agreement between surgeons in surgical planning (Fuller et al.,2002).

The outcome of the decision-making process is largely dependent on a number of factors. Perhaps the two main limitations of CGA are: (a) the subjectivity of interpretation due to varied levels of expertise, and (b) the inherent difficulty in comprehending large amounts of information. With regard to the first limitation, a gait analysis report is interpreted by different experts in various ways (Watts, 1994; Skaggs et al.,2000). Neurologists, specialists in physical and rehabilitation medicine, and physiotherapists adopted distinctly different strategies when analysing gait by focusing on the anatomical localisation of lesions, biomechanical findings and describing details of gait deviations, respectively (Watelain et al.,2003). Inconsistent interpretation of gait analysis data can also be the result of insufficient training. Toro et al. (2002) showed that 60.4% of a sample of physiotherapists working with patients suffering from gait problems had no formal training in the assessment of gait even though 70% of pediatric physical therapists had at least one patient who had undergone gait analysis. The teaching of gait analysis is constantly developing in response to research findings but the latest advances of gait analysis research are available to gait analysts only by participating in continuous training. A good example is the recent shift of focus from surgical treatment of bony deformities to the management of spasticity as the first line of treatment in spastic cerebral palsy (Gage & Novacheck, 2001). The level of expertise is another factor which determines decision making. Through experience, a gait analyst performs the basic mental tasks in the subconscious domain and spends more time understanding the delicate and unique details. Conversely, a newcomer is able to question the most basic phenomena and occasionally this leads to fundamental changes in our understanding of gait. The first reason for the variation of gait analysts' decision making, therefore, is probably subjectivity, which stems from their background, training and experience which all influence their interpretation of gait in an unknown way. A gait analysis team typically consists of individuals with a wide range of backgrounds, training and level of experience. The multi-faceted strategies of gait analysis are expected to enhance the overall performance of the team (Watelain et al.,2003), but it is of question whether the product of several subjective decisions can lead to an optimal decision.

With regard to the second limitation, the large quantity of information contained in a gait analysis report makes it difficult to balance attention among all pieces of information and this can lead to an inconsistent interpretation even within the specialist strategies as noted above. Miller (1956) showed that people can remember seven, plus or minus two, chunks in short term memory. As opposed to about seven pieces of information a gait analyst has to interpret curves of joint angles, moments, powers, electromyograms, together with details of patient history, results of joint ranges of motion, measures of muscle strength, levels of spasticity and energy consumption. An advanced student of gait analysis forms a hypothesis early on and knows exactly which pieces of information are necessary for confirmation (Simon et al.,1996). This speeds up decision making but at the same time carries the risk of over interpreting some possibly irrelevant details and neglecting some possibly relevant components.

There is a clear need for a method which overcomes these two limitations and enables large quantities of gait data to be analysed and interpreted objectively. Artificial neural networks possess these characteristics and have been found to be the most prevalent non-traditional methodology used for gait data analysis in the last 10 years (Chau, 2001). One form of artificial neural network is the self-organising map (SOM) (Kohonen, 2001) which employs unsupervised learning. The self-organising map clusters the input patterns based on the Euclidean distances among them. The result of the SOM's gradual adaptive learning process is that similar patterns are assigned to closely neighbouring parts of a rectangular grid. In the context of gait analysis, the SOM can be regarded as a software tool which reduces the amount of data with minimal loss of information content. The dimensionality reduction is done purely on the basis of the data presented to the SOM without any *a priori* definition of clusters and so the process is objective.

Koehle and Merkl (1996) classified patients into groups automatically with a SOM using the vertical component of the ground reaction force under both feet. The groups identified by the SOM agreed with the clinical classification of the patients. Lakany (2001, 2004) used SOMs successfully to group patients into clusters based on sagittal plane angles of the hip, knee and ankle joints. The pilot work of Barton (1999) visualised the multi-dimensional data space on a SOM producing single curves which represented four sagittal plane joint angles. Gait pattern labels were assigned to the SOM by calibrating the map with known data. In another study (Barton et al.,2000) the findings of physical examination (joint ranges of motion angles) were projected onto a self-organising topographical map. The arrangement of patients on the map suggested a grouping different from the patients' initial classification but the groups identified by the SOM matched the classification by gait experts based on full gait analyses.

Encapsulated decision-making algorithms, or "black boxes," are generally avoided in medical decision making because typically they do not provide any reasons for the outputs they generate in response to the inputs. The power of such automated methods should only be utilised in critical applications if the rules underlying the decisions can be formalised. Gait analysts use visual pattern recognition of graphically presented data in order to identify normal and abnormal gait patterns which are the building blocks of decision making in CGA and so the SOM has to maintain the graphical nature of the data and the rules extracted have to be related to gait patterns. The work of Barton (1999), Koehle and Merkl (1996), Lakany (2001; 2004) and Barton et al. (2005) have demonstrated that SOMs can meet these requirements and thus, it appears that the SOM has good potential in this field of analysis.

The targeted reader of this chapter is a gait analyst who is probably a physiotherapist, bioengineer or a clinical scientist involved in the daily care of patients suffering from gait problems. Gait analysts perceive the advanced analytical methods (including neural networks) as powerful tools which are unreachable to them due to their complexity and perhaps this explains why there is still a lack of using such methods in the practice of gait analysis. Through communication with such individuals it has become obvious that their focus is to apply a relatively simple and transparent method which gives a concise answer to their complex questions. An obvious reason why they avoid highly sophisticated methods is that their professional rules of conduct prescribes that they have to understand any new methods they introduce in their practice. Even though the original SOM has not been optimised specifically to analysing gait data, in this study the simplest form of the SOM was used which strikes a good balance between finding a powerful enough tool to tackle the problem and keeping the concepts and the methods simple enough to be accepted by the end-users. The program implementing Kohonen's concepts and written by his team (Vesanto et al.,2000) is a freely available software tool which is well documented both in a technical sense and as a tool used to solving practical problems. There are many advanced derivatives of the SOM described in the literature (e.g., Rubio & Gimenez, 2003; Yin, 2002; Marshland et al.,2002) which have a great potential enhancing the analysis of gait data but these methods were intentionally avoided as their mathematical complexity would make them unsuitable for the purposes of this study which was to provide a solution that can actually be used in the practical settings of clinical gait analysis without any substantial additions to the default SOM.

The aim of this study was to develop and explore a method which facilitates identification of gait disorders by visualising complex gait data in a simplified format, using self-organising maps.

METHODS

The database of the Derby Gait and Movement Laboratory was used (subsequent to ethical approval) which contained data of 612 patients and 10 normal subjects at that time and was generated with the GaitEliclinic software (BTS, Milan, Italy). Selected files of the database containing joint kinematic and kinetic data were used in this study.

In clinical gait analysis the interpretation of joint kinematics and kinetics is probably the most problematic because the 3D joint angles, moments and powers represent the complex dynamics of gait in the form of static line charts. In this study, therefore, three-dimensional joint angles, moments and powers (altogether 43 variables, Table 1) were presented to SOMs. Subsets of these files were processed and the generated results enabled an alternative analysis of 3D kinematics and kinetics. The stages of processing were as follows.

Normalisation of the Time Axis

The database contains up to four-seconds-long records of subjects walking on the walkway. Figure 1a shows bilateral sagittal plane knee joint angles as an example. Clearly the data are contaminated by artefacts as the subject enters and leaves the calibrated measurement area and so the full length data cannot be used.

Table 1. The 43 gait variables (3D joint angles, moments and powers) provided by the GaitEliclinic software (abbreviations: ab-add. = abduction-adduction; rot. = rotation; flx-ext. = flexion-extension; plflx. = plantarflexion)

Pelvic obliquity angle	Pelvic rot. angle	Pelvic tilt angle
Right hip ab-add. Angle	Right hip rot. angle	Right hip flx-ext. angle
Right knee ab-add. Angle	Right knee rot. angle	Right knee flx-ext. angle
Right ankle abd. Angle		Right ankle plflx. angle
Right hip ab-add. Moment	Right knee ab-add. moment	Right ankle ab-add. moment
Right hip rot. Moment	Right knee rot. moment	Right ankle rot. moment
Right hip flx-ext. moment	Right knee flx-ext. moment	Right ankle flx-ext. moment
Right hip power	Right knee power	Right ankle power
Left hip ab-add. Angle	Left hip rot. angle	Left hip flx-ext. angle
Left knee ab-add. Angle	Left knee rot. angle	Left knee flx-ext. angle
Left ankle abd. Angle		Left ankle plflx. angle
Left hip ab-add. Moment	Left knee ab-add. moment	Left ankle ab-add. moment
Left hip rot. Moment	Left knee rot. moment	Left ankle rot. moment
Left hip flx-ext. moment	Left knee flx-ext. moment	Left ankle flx-ext. moment
Left hip power	Left knee power	Left ankle power

Note: The three dimensional model used in GaitEliclinic does not provide ankle inversion angle

Data of one right gait cycle together with data from the left side for the same duration with a common absolute time axis could be selected (Figure 1b). This sounds especially attractive because the power of a SOM is the ability to handle the interactions across all data components, in this case data from the right and left sides of the body. Unfortunately, the database contained bilateral joint kinematics, but kinetics of only one side for each walk. This was due to having only one force platform installed, which is needed for the reconstruction of kinetics. To avoid using incomplete kinetics, this solution was also rejected.

Data of a gait cycle with right foot contact together with another gait cycle with a left foot contact would provide bilateral joint kinematics and kinetics with a common time line (Figure 1c). The disadvantage of this method is that the duration of gait cycles are not necessarily equal between the two sides.

Finally, one right and one left gait cycle can be normalised to the length of their respective gait cycles (Figure 1d). In this way kinematics and kinetics of both sides are available and normalisation makes the curves equal in length. The representation of curves in this format is widely used in most gait laboratories and allows a comparison between the performance of the SOM and that of gait analysts.

Figure 1. Four different ways of presenting bilateral gait data

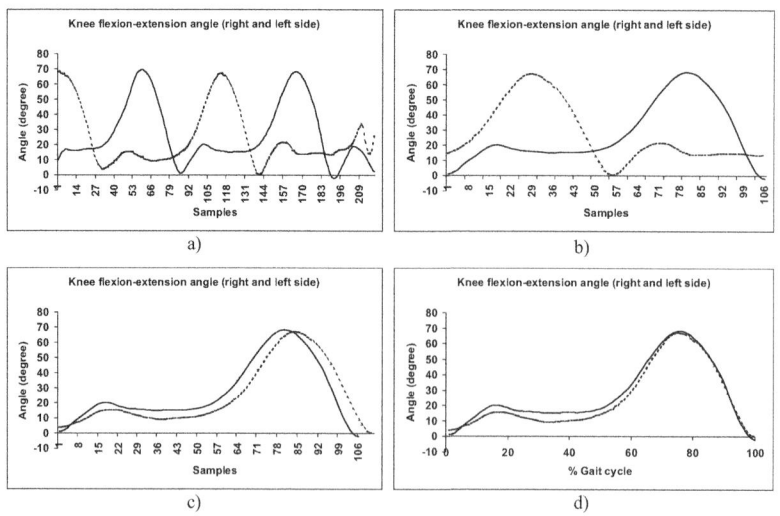

(a) The right and left side curves are shown on the same time axis, (b) one right gait cycle is selected together with data from the left side, (c) gait cycles of both sides shown with the same absolute time line, and (d) gait cycles of the two sides normalised to their respective gait cycle lengths

Normalisation of the Vertical Axis

In this study the range of each variable used by the Derby Gait and Movement Laboratory's clinical gait analysis printouts were used to determine the normalisation ranges of each curve (i.e., to re-scale each vertical axis to a range between 0-1). The generic recommendation by Kohonen (2000) was to normalise the SOMs' inputs so that the variance of each variable is 1 as this equalises the contribution of variables to the final decision of the SOM. For example, the dimensionless numerical range of the knee flexion-extension curve is about 60 as opposed to the range of pelvic tilt (5) or hip flexion-extension moment normalised to body mass (1.5) and so the knee flexion-extension curve would suppress the relatively smaller variations of the other curves unless the data are normalised.

When applied to gait data, the generic method of normalisation suggested by Kohonen would result in the unwanted effect of cancelling out curve offsets which in fact carry diagnostic information. Equinus gait, crouch gait, excessive femoral anteversion are some examples which result in offset curves of the ankle, knee and hip joints, respectively. The second argument against using conventional normalisation is that different sets of patient data can potentially have different minimum and maximum values which are used for normalisation. This means that different datasets would be normalised in different ways which would invalidate comparison between data sets.

Adjustment of Variables' Weights by means of Differential Normalisation

It is questionable if all 43 variables are needed to capture the uniqueness of a gait pattern. Most previous studies used sagittal plane data only (Lakany, 2001; Lakany, 2004) but many gait problems manifest themselves in the transverse and coronal planes (e.g., rotational bony deformities, pelvic retraction, leg length discrepancy). Also, many abnormal gait patterns are recognised by identifying deviations in moment and power curves (e.g., ankle clonus, knee flexion moment pattern with a flexed knee), which suggests that kinematic and kinetic data in all three planes carry discriminative information. In this study all 43 variables were used to present gait data to the SOM, but the relative weight of variables was controlled by means of differential normalisation.

The generic reason for normalising data is to control the relative contribution of variables to the total appearance of an input pattern. As the human expert's attention is influenced by the scaling of the curves, by using the same scales for the SOM its "attention" is made compatible with that of the human expert. Normalisation, however, can also be used to orient the SOM's focus onto certain parts of the input pattern. Multiplication of systematically selected components of the input pattern by a weight factor was used to change the emphasis on those components in relation to the rest of the input pattern.

Time Delay of Data

The temporal nature of gait data was maintained by grouping three time slices of data at t, t+5% and t+10% of the gait cycle (Figure 2). The method of dynamical embedding was used (Broomhead and King, 1986) in order to determine the optimal representation of gait dynamics. This approach consists of assembling the data into a matrix in which

Figure 2. A typical graph of all 43 curves of a patient normalised to the gait cycle (horizontal axis) and into the range 0-1 (vertical axis) using the clinical ranges of angles, moments and powers (the curves with values above 1 or below 0 had values off the scale on the conventional plots)

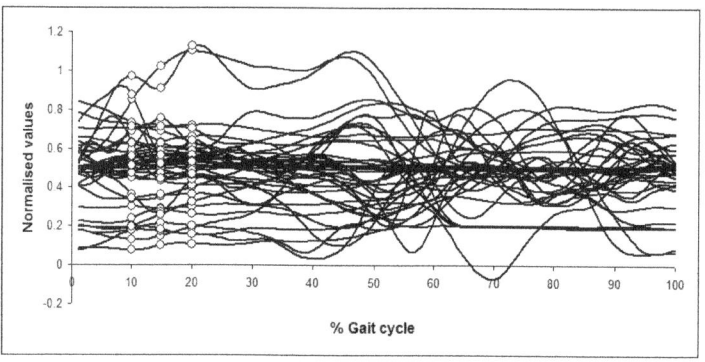

The three columns of circles illustrate the concept of using time delayed data at t, t+5% and t+10% of the gait cycle (10, 15 and 20% of the gait cycle in this example) in order to represent the dynamics of the curves. The last triplet of data can be taken at 90%.

consecutive rows comprise a windowed view of the time series data, for instance with samples 1 .. n in the first row, 5 .. 5+n in the second row, etc. It is then possible to calculate the eigenvalues of this matrix using singular value decomposition and, from this, to infer for how many windows, what window size and time shift between rows the eigenvalue spectrum of the matrix stabilises. This empirical analysis showed that the dynamics of individual gait signals, as contained in the eigenvalue decomposition of the embedding matrix, were well represented by three to five windows each containing one value with a spacing of five samples between windows, as there was little change to the eigenvalues by increasing those parameter values. With the exception of only a few curves with higher frequency content which were best represented by four or five samples, in most of the gait curves 3 samples at 5% intervals captured the curves' dynamics optimally.

Software Tool to Accomplish Pre-Processing of Data for Presentation

A program was written in Borland Delphi 6 which performed the above methodological tasks in a user-friendly environment (Figure 3). The program serves two purposes. Firstly, it is a data viewer which reads data files generated by the GaitEliclinic software and plots any subject's kinematics and kinetics on conventional line charts. If available, the program also displays SOM visualisation. Data with artefacts can be identified and list files can be saved which contain the codes of clean data files. Pre-defined subsets

Figure 3. The graphical user interface of the program displaying selected kinematics and kinetics curves of patient #569 who belongs to the list file stored in RL04_all.lst

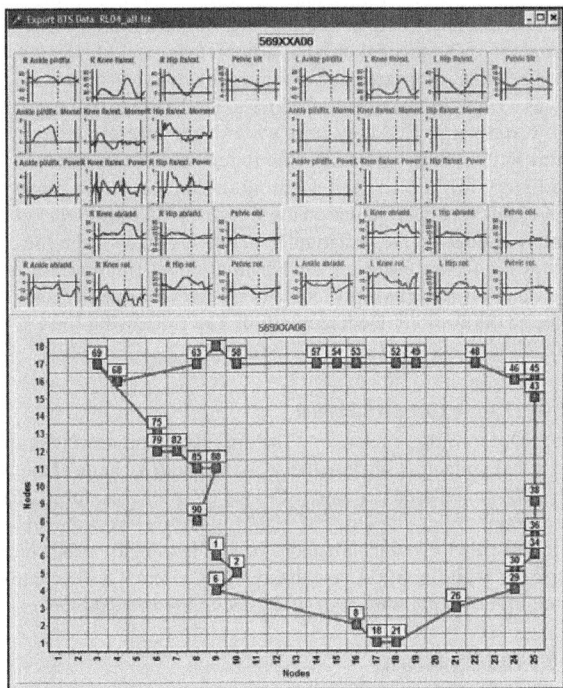

The single SOM curve corresponding to the same dataset is also displayed in the bottom chart. The numbers next to the series of activated SOM nodes indicate the time dimension.

of variables can be selected and the weights of variables can be adjusted. The second function of the program is to process the selected files and produce values normalised both to the gait cycle and to the ranges used in gait analysis printouts, modified by the weight factor. Ninety of the 100 time equidistant values measured during the gait cycle were taken in the ranges 1-90, 5-95 and 10-100 providing 129 columns (43 variables taken 3 times) and 90 rows of values. Any one row of data therefore contained values recorded at t, t+5% and t+10% of the gait cycle thereby representing the temporal nature of the data. This processing was performed for all gait cycles stored in the list file and then the blocks of data were concatenated into a single output file ready for loading into the SOM software.

SOM Training

The SOM takes the first line of the input file containing 129 (43*3) values which represent a data triplet at t, t+5% and t+10% of the gait cycle. The SOM algorithm finds the best matching node of the planar map and increases the weight of the connections between the input layer and the winning node, together with decreasing the weights away from the winner node in its hexagonal neighbourhood using the Mexican hat function (Kohonen, 2001). The same is repeated on the remaining rows until the SOM stabilises.

A program was written in Matlab which carries out the SOM training sequence by reading the data file, defining a SOM, training a SOM and generating the co-ordinates of the best matching units. The program calls the functions of SOM Toolbox 2.0 which is a free software library for Matlab implementing the self-organising map algorithm (Vesanto et al.,2000). In this implementation the size of the SOM (number of nodes and proportions of the two sides of the rectangular matrix of nodes), the initial weights connecting the nodes, and the criterion for termination of training are determined automatically as a function of the data presented to the SOM. Specific technical details are available in the SOM Toolbox Documentation. The processing time of a text file containing data of 71 subjects was less than two minutes.

SOM Trajectory Visualisation

As a result of SOM training, each row of the input file representing one time slice triplet of bilateral joint kinematics and kinetics was assigned to a node of the SOM. One series of 90 SOM nodes (an ordered sequence of 90 x;y co-ordinates) formed a trajectory which represents one gait cycle. The series of SOM trajectories were then reloaded into the pre-processing software tool which displays the conventionally used kinematic and kinetic curves together with the SOM visualisation. In essence the complexity of 43 curves is converted into a single trajectory of nodes on the trained SOM.

Codebook Visualisation

Following training, the internal weights of the SOM were stored in a codebook which represents the mapping of the 129 dimensional data space onto the two dimensional neural map. The codebook contains the weight vectors (129 values) associated with each node. These vectors are small sections of the gait cycle (t, t+5, t+10%) which developed as a result of the SOM's adaptive training process. The codebook vectors are the representations of the gait data which best describe the whole data set under the constraints of a two dimensional data space with a reduced size.

Another program was written in Borland Delphi 6 which visualises the codebook by reconstructing the small sections of the gait cycle stored in the weight vectors of the SOM nodes. The user can select ordered series of five nodes on the SOM and another chart plots the values stored in the weight vectors of the selected SOM nodes.

The codebook generated by the SOM Toolbox does not contain information about the temporal ordering of the nodes. To overcome this limitation additional Matlab code identified which parts of the gait cycle were assigned to the nodes of the map. One more column was added to the codebook which showed the average percentage of the gait cycle represented by each node. The grey-level of the nodes expresses the percentage values (0% – white; 90% – black). This additional piece of information provided the positioning of the reconstructed section within the gait cycle.

Several sets of data were created and processed following the above described sequence, in order to illustrate how the SOM's visualisation changes in response to different data presented to it and to demonstrate how the results can help to gain a better comprehension of gait data. Dataset 1 contained gait cycles of 10 healthy subjects. Dataset 2 contained all variables of the same 10 healthy subjects and the first 61 patients with various gait problems, and without measurement artefacts, who attended the Gait Laboratory. Dataset 3 is an extension of the second one containing data from the next 18 patients contained in the database (with artefact free gait data).

RESULTS

Figure 4a shows how the SOM arranged Dataset 1 as a result of training. Each SOM curve represents one gait cycle of one normal subject and the bold curve shows the mean of the 10 individual curves. Figure 4b shows the SOM representation of the same 10 normal subjects when used as a part of a larger set of data (Data set 2). In spite of the low inter-subject variation of normal gait, the SOM curves indicate large differences in Figure 4a but when the same data were processed together with largely different abnormal gait data (Figure 4b) then the SOM curves of the same normal subjects look more similar. The addition of abnormal gait data has also changed the dimensions of the SOM because the larger variations of gait patterns could only be handled by increasing the size of the SOM which occurred automatically on the basis of the data presented.

When examining the SOM projections of the 61 patients in Data set 2, it was found that six of them exhibited the same gait pattern (Figure 5a) as their SOM curves ran through a distinct area of the map which could be described as an oblique cluster running from bottom right to top left in the region bounded by nodes 11-17 horizontally and 9-15 vertically (Figure 5b). This area is placed in the middle-upper section of the map. There were some differences among the six curves at various parts of the gait cycle but they all ran through the same region during mid/late stance suggesting that they all exhibit the same gait pattern. The details of this gait pattern could be investigated in more detail by using the codebook visualisation.

Figure 6 shows a normal gait cycle reconstructed from the codebook vectors together with visualisation of the abnormal gait pattern identified above. The first series of five nodes were selected so that they evenly covered the path of the normal SOM curves (large circles on the top chart of Figure 6). The reconstructed sagittal plane angle, moment and power curves of the ankle, knee and hip joints show the typical appearance of normal gait. The second series of five nodes (large triangles on the top chart of Figure 6) were selected so that they covered the path determined by the SOM curves of the six patients who shared a common gait pattern (Figure 5a). The codebook vectors corresponding to the five nodes displayed the kinematics and kinetics of the abnormal gait pattern on the right side. This involved excessive hip and knee flexion in stance, knee extension moment pattern, slightly reduced ankle plantarflexion moment and power generation and increased knee power absorption in midstance. The pattern resembles quite closely the abnormal gait pattern known as crouch gait described by Sutherland and Davids (1993).

Figure 7a shows the SOM projections of the same six patients whose data were shown in Figure 5a but following training the SOM with a larger dataset containing the

Figure 4. (a) SOM trajectories of ten normal subjects; (b) The SOM projections of the same ten normal subjects when presented to the SOM together with gait data of 61 patients

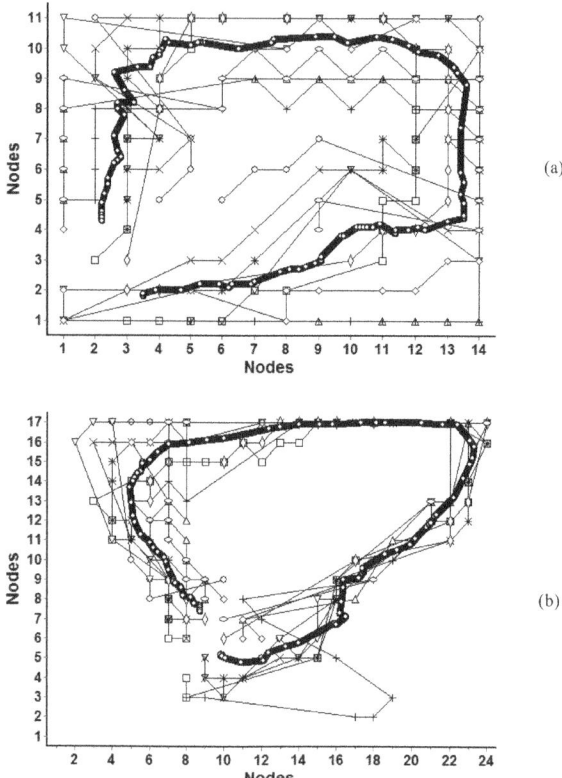

All curves run anti-clockwise and the bold curves show the means of individual curves

same 10 normal subjects and 79 patients (Data set 3). The curves looked slightly different as a result of including additional gait cycles but the location and orientation of the curves remained the same. All six curves ran through a region that corresponds to the previously identified abnormal gait cluster. The SOM curves of three of the additional 18 patients also travelled through the abnormal cluster (Figure 7b).

Figure 5. (a) Results of six patients whose SOM curves run through a cluster of nodes surrounded by the square; (b) The histogram shows the number of times each node was activated by the six curves

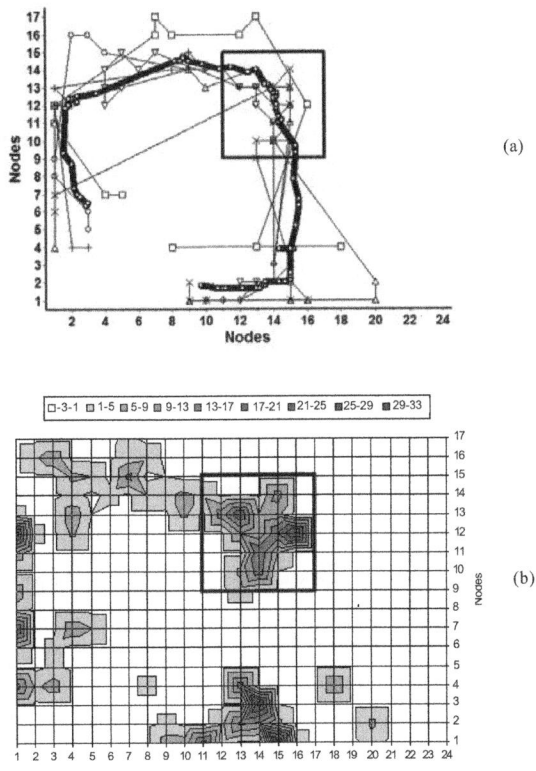

Crouch gait is mainly characterised by more than 30° knee flexion maintained throughout the stance phase of gait (Sutherland & Cooper, 1978) and so it is a gait pattern which is defined by joint kinematics (i.e., joint angles). However, walking with flexed knees has an effect on joint kinetics (i.e., moments and powers) as well. In order to test if the crouch gait pattern is a kinematic or kinematic/kinetic gait pattern, the relative contribution of kinematic and kinetic data to the overall gait pattern presented to the SOM

Figure 6. A representation of the normal gait cycle together with the abnormal gait pattern identified on Figure 5, as stored in the SOM weights (codebook vectors)

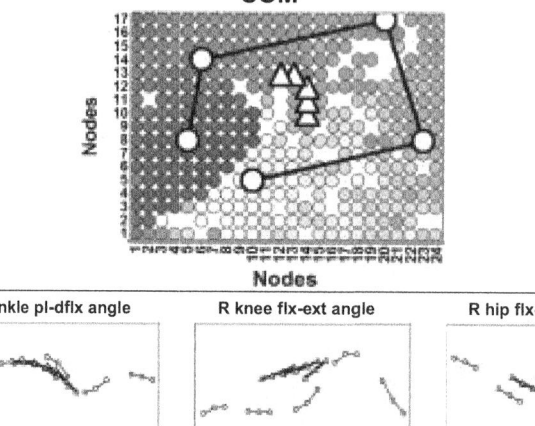

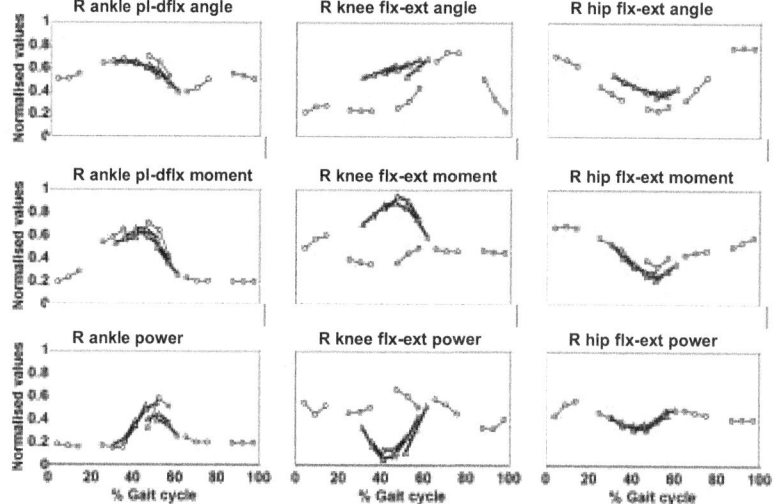

The user can select ordered sequences of five nodes on the upper chart. The temporal ordering of nodes is indicated by shading (light — early, dark — late). The lower charts show the corresponding five data triplets reconstructed from the codebook vectors representing a full normal gait cycle (circles) and the abnormal gait pattern (triangles). Angles, moments and powers of the right side are shown in the sagittal plane. All other variables are also available from a drop-down list.

was manipulated by differential normalisation. The weight of all moment and power curves was reduced to 0.8 thereby shrinking the kinetics curves to 80% of their amplitude around the mid-point (0.5) of the normalised range (0-1). When interpreting the SOM's visualisation of the same six subjects with crouch gait, it appears that a kinematics focused analysis resulted in a tighter cluster with a more uniform gait pattern (Figure 8).

Figure 7. (a) Data of those six patients whose curves run through a cluster of nodes shown in Figure 5 following training with a larger dataset containing more patients. Note the small changes when compared to Figure 5a; (b) Data of the same six patients together with three additional patients from the larger dataset whose curves cross the abnormal gait cluster

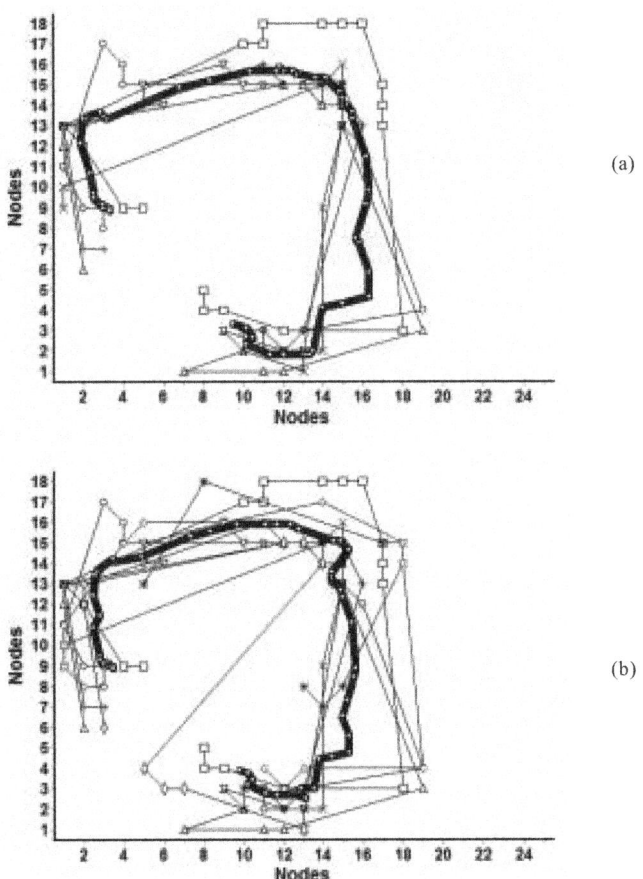

(a)

(b)

The bold curves show the means of the six and nine individual curves respectively

Figure 8. Histogram of the crouch gait pattern as visualised by the SOM following training with kinematics focused input patterns

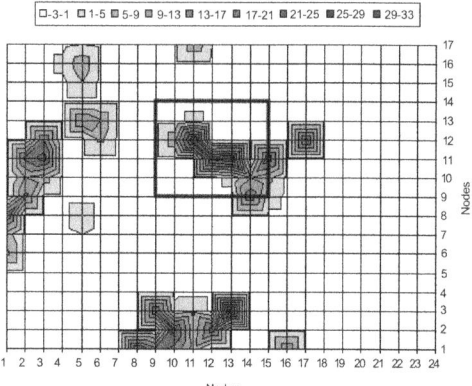

The moment and power curves were multiplied by 0.8 in order to reduce their contribution to the overall pattern. Note the differences from the histogram on Figure 5b which shows a larger crouch cluster with more detail reflecting the sub-groups of the gait pattern

DISCUSSION

The complexity of gait data can be reduced by projecting a large number of joint kinematic and kinetic curves into a single curve on the SOM. The shape of this curve follows the best matching gait patterns stored in the SOM's codebook vectors and so in spite of its simplicity the single curve is tightly linked to the underlying complex patterns hidden inside the SOM. The dimensionality reduction of the SOM allows it to handle the patient's gait as a whole and to focus on gait patterns. The fine detail is also available by visualising the codebook vectors of the SOM which provides the reasoning why the single curve takes a certain shape.

It is an essential requirement that the visualisation of the SOM must be consistent, otherwise two different SOM projections could be attributed both to changes in the gait pattern and variations of the SOM's performance. The classic algorithm of the SOM involves randomisation of all the weights before training (Kohonen, 2000). The consequence of randomisation is that the projection of the multidimensional space onto the two-dimensional surface can be different every time a new training occurs. A disadvantage of the early software package SOM_PAK (Kohonen et al.,1996) was that the results were not consistent for this reason. The software used in this study initialises the SOM weights using the two first eigenvectors calculated from the data to be presented to the

SOM (Vesanto et al.,2000) and this makes training repeatable as long as the same gait data are presented to the SOM.

The trained SOM can be regarded as a simplified representation of relative distances among gait patterns residing in a multidimensional data space as the SOM algorithm is based on the Euclidean distances among input patterns. The deviation of a patient's SOM curve from that of a normal subject is a reflection of the overall difference in their gait patterns. The expression of multidimensional distance between gait patterns on a two-dimensional grid enables ranking of subjects. The SOM weights represent the continuous spectrum ranging from normal to abnormal and the SOM curves position the subject on this scale to the appropriate location.

The SOM consists of discrete nodes and so converts the continuous spectrum of gait patterns into a discrete scale. The resolution of the SOM grid (i.e., the number of nodes) is determined by the amount of data presented to it. Even so, the level of detail captured by the SOM can be controlled by presenting different amounts of the multidimensional data space. This is illustrated by the effect of using only normal data as opposed to using normal and abnormal data together (Figure 4a and 4b). Any data presented to the SOM will cover the available SOM surface. If normal gait cycles close to each other in a multidimensional space are presented then the level of detail visualised by the SOM will increase compared to the same SOM presented with a mix of normal and abnormal data. The responsiveness of the SOM's resolution was also illustrated by the altered SOM curves in response to some more gait cycles of additional patients being added to the input file (Figures 5a and 7a).

The adaptability of the SOM to the input data can be further utilised to zoom in on small details. Gait data of a group of patients with similar clinical problems can be packaged into a separate data file which can be presented to a new SOM. Following training the resultant curves will look different but will show the ranking of subjects within the group in more detail.

The ranking can be applied longitudinally as well, as a patient's gait changes over time. The movement of the SOM curve away from or towards the normal reference can be an indicator of how well the patient performs during the recovery and the rehabilitation phase. The effect of treatment and different rehabilitation techniques could be measured using the SOM projections as overall measures of how well the patient walks.

Differential normalisation enables a sensitivity analysis of gait data presented to a SOM. Systematic modification of the input variables' weighting changes the SOM's visualisation and this can highlight delicate details of gait patterns. By reducing the weight of moment and power curves, the sub-grouping of the crouch cluster was eliminated suggesting that the primary features of the crouch gait pattern can be captured by joint kinematics but the fine detail of how individuals cope with the condition is contained in joint kinetics.

All SOM curves (for normal and abnormal gait) are more or less closed curves. The joint angles, moments and powers at the beginning and the end of a gait cycle are nearly identical and so are mapped to the same location of the SOM. The gap between the first and last nodes of the SOM curve is due to only 90% of the gait cycle being presented to the SOM.

Querying the codebook vectors of a trained SOM reveals how the normal and abnormal gait data are mapped onto a simple grid of nodes as a result of the SOM's iterative training process. The data fragments belonging to each node represent abstract gait patterns. These patterns which exist in the codebook only cannot be found in any of the gait cycles presented to the SOM as they are the condensed forms of all those patterns which are close to each other. The convergence of individual gait patterns towards stem-patterns is due to the large ratio between the number of input patterns and the number of nodes. For example the 6,390 sets of 129 values (71 gait cycles *90 samples) were arranged on a SOM with 408 nodes (24*17) which represents a 94% compression. Every node of the SOM stores one such elementary stem-pattern which contains three time slices of 43 curves. The actual patterns of real gait cycles gravitate to those stem-patterns which match them best. An examination of the stem-patterns can provide an insight into the distribution of gait patterns in relation to each other.

The cluster indicated on Figure 5 could be called the crouch gait cluster as the reconstructed stem-pattern matches the gait pattern which is commonly known as crouch gait. The SOM contains several such clusters all representing normal and abnormal gait patterns. Some of the clusters can be associated with conventional gait patterns known in gait analysis but certainly there are clusters which represent previously unidentified complex gait patterns which involve 3D joint angles, moments and powers during certain sections of the gait cycle. These unnamed clusters are the ones which are difficult to identify unless a multivariate data processing tool can visualise them. The naming of these clusters should be related to their topology on the SOM as it is unlikely that a short name can express all the features which make up the gait pattern. A SOM labelled with multidimensional gait patterns would give a deeper insight into the grouping of subjects.

CONCLUSION AND FUTURE DIRECTIONS

In this study, the power of self-organising artificial neural networks was used to visualise complex gait patterns in the form of single curves. The SOM operates by converging gait data to stem-patterns which are arranged on a relational map in the context of the total data space presented to the SOM during training. The method enables identification of existing gait patterns and opens up the possibility of defining new gait patterns which are otherwise difficult to find in the multidimensional data space. The method gives repeatable dimensionality reduction with a resolution which can be controlled by careful selection of the input data. The multidimensional ranking of subjects is possible both cross-sectionally and longitudinally. Sensitivity analysis of the SOM's response can be achieved by differential normalisation.

The method used in this study may provide an alternative representation of gait analysis results which can cope with the complexity of the data and can help to make decision making more repeatable and so more objective. Perhaps the true potential of clinical gait analysis can only be realised if the complexity of the data describing movement is coupled with matching analytical tools, as suggested by Simon (2004), who defined the benefits and limitations of clinical gait analysis and described several attempts to overcome the limitations. If the methods are powerful enough to meet the demands and can fit into the thinking of practicing gait analysts at the same time, then gait analysis may become a more widely accepted tool.

REFERENCES

Barton, J. G. (1999). Interpretation of gait data using Kohonen neural networks. *Gait & Posture, 10*, 85-6.

Barton, J. G., Lees, A., Lisboa, P., & Attfield, S. (2005). Visualisation of gait data with Kohonen self-organising neural maps. Under Review in *Gait and Posture*.

Barton, J. G., Lisboa, P., & Lees, A. (2000). Topological clustering of patients using a self organising neural map. *Gait & Posture, 12*, 57.

Broomhead, D. S., & King, G. P. (1986). Extracting qualitative dynamics from experimental data. *Physica, 20D*, 217-36.

Chau, T. (2001). A review of analytical techniques for gait data. Part 2: Neural network and wavelet methods. *Gait & Posture, 13*, 102-20.

Cook, R. E., Schneider, I., Hazlewood, M. E., Hillman, S. J., & Robb, J. E. (2003). Gait analysis alters decision-making in cerebral palsy. *Journal of Pediatric Orthopedics, 23*, 292-5.

DeLuca, P. A., Davis, R. B., 3rd, Ounpuu, S., Rose, S., & Sirkin, R. (1997). Alterations in surgical decision making in patients with cerebral palsy based on three-dimensional gait analysis. *Journal of Pediatric Orthopedics, 17*, 608-14.

Fuller, D. A., Keenan, M. A., Esquenazi, A., Whyte, J., Mayer, N. H., & Fidler-Sheppard, R. (2002). The impact of instrumented gait analysis on surgical planning: treatment of spastic equinovarus deformity of the foot and ankle. *Foot & Ankle International, 23*, 738-43.

Gage, J. R., & Novacheck, T. F. (2001). An update on the treatment of gait problems in cerebral palsy. *Journal of Pediatric Orthopedics, Part B, 10*(4), 265-74.

Hailey, D., & Tomie, J. A. (2000). An assessment of gait analysis in the rehabilitation of children with walking difficulties. *Disability and Rehabilitation, 22*, 275-80.

Kay, R. M., Dennis, S., Rethlefsen, S., Reynolds, R. A., Skaggs, D. L., & Tolo, V. T. (2000). The effect of preoperative gait analysis on orthopaedic decision making. *Clinical Orthopaedics and Related Research, 372*, 217-22.

Koehle, M., & Merkl, D. (1996, April 24-26). Identification of gait patterns with self-organizing maps based on ground reaction force. In *Proceedings of European Symposium on Artificial neural networks ESANN'96*. Bruges, Belgium, April 24-26 (pp. 73-78).

Kohonen, T. (2001). *Self-organizing maps*. Berlin: Springer.

Kohonen, T., Hynninen, J., Kangas, J., & Laaksonen, J. (1996). SOM_PAK: The self-organizing map program package. Espoo: Helsinki University of Technology.

Lakany, H. M. (2001). Human gait analysis using SOM. In N. Allinson (Ed.), *Workshop on self-organising maps: Advances in self-organising maps* (pp. 29-38). London: Lincoln.

Lakany, H. M. (2004). A generic kinematic pattern for human walking. *Neurocomputing, 35*, 27-54.

Marsland, S., Shapiro, J., & Nehmzow, U. (2002). A self-organising network that grows when required. *Neural Networks, 15*(8-9), 1041-1058.

Miller, G. A. (1956). The magical number seven, plus or minus two: Some limits on our capacity for processing information. *Psychological Review, 63*(2), 81-97.

Rubio, M., & Gimenez, V. (2003). New methods for self-organising map visual analysis. *Neural Computing & Applications, 12*(3-4), 142-152.

Simon, S. R. (2004). Quantification of human motion: Gait analysis - benefits and limitations to its application to clinical problems. *Journal of Biomechanics, 37*(12), 1869-1880.

Simon, S. R., Smith, P. J., Nippa, J. C., Johnson, K. A., Stern, L. S., & Sriram, M. G. (1996). Applications of intelligent multimedia technology in human motion analysis. In G. F. Harris & P. A. Smith (Eds.), *Human motion analysis: Current applications and future directions* (pp. 407-438). New York: IEEE.

Skaggs, D. L., Rethlefsen, S. A., Kay, R. M., Dennis, S. W., Reynolds, R. A., & Tolo, V. T. (2000). Variability in gait analysis interpretation. *Journal of Pediatric Orthopedics, 20*, 759-764.

Sutherland, D. H., & Cooper, L. (1978). The pathomechanics of progressive crouch gait in spastic diplegia. *The Orthopedic Clinics of North America, 9*(1), 143-154.

Sutherland, D. H., & Davids, J. R. (1993). Common gait abnormalities of the knee in cerebral palsy. *Clinical Orthopaedics and Related Research, 288*, 139-147.

Technology Evaluation Centre. (2002). Gait analysis for pediatric cerebral palsy. *TEC Assessment Programme, 16*(19).

Toro, B., Nester, C. J., & Farren, P. (2002). A survey of the use of gait assessment amongst UK physitherapists. *Gait & Posture, 16*, 87-88.

Vesanto, J., Himberg, J., Alhoniemi, E., & Parhankangas, J. (2000). SOM Toolbox for Matlab 5. Espoo: Helsinki University of Technology.

Watelain, E., Froger, J., Barbier, F., Lensel, G., Rousseaux, M., Lepoutre, F. X., & Thevenon, A. (2003). Comparison of clinical gait analysis strategies by French neurologists, physiatrists and physiotherapists. *Journal of Rehabilitation Medicine, 35*, 8-14.

Watts, H. G. (1994). Gait laboratory analysis for preoperative decision making in spastic cerebral palsy: Is it all it's cracked up to be? *Journal of Pediatric Orthopedics, 14*, 703-704.

Yin, H. J. (2002). Data visualisation and manifold mapping using the ViSOM. *Neural Networks, 15*(8-9), 1005-1016.

Chapter VII

Neural Network Models for Estimation of Balance Control, Detection of Imbalance, and Estimation of Falls Risk

Michael E. Hahn, Montana State University, USA

Arthur M. Farley, University of Oregon, USA

Li-Shan Chou, University of Oregon, USA

ABSTRACT

Gait patterns of the elderly are often adjusted to accommodate for reduced function in the balance control system. Recent work has demonstrated the effectiveness of artificial neural network (ANN) modeling in mapping gait measurements onto descriptions of whole body motion during locomotion. Accurate risk assessment is necessary for reducing incidence of falls. Further development of the balance estimation model has been used to test the feasibility of detecting balance impairment using tasks of sample categorization and falls risk estimation. Model design included an ANN and a statistical discrimination method. Sample categorization results reached accuracy of

0.89. Relative risk was frequently assessed at high or very high risk for experiencing falls in a sample of balance impaired older adults. The current model shows potential for detecting balance impairment and estimating falls risk, thereby indicating the need for referral for falls prevention intervention.

INTRODUCTION

In the last 20 years, artificial neural network (ANN) modeling has been used in various applications of disease classification and diagnosis. Recent research has used ANN theory to classify gait patterns (Barton & Lees, 1997; Gioftsos & Grieve, 1995; Holzreiter & Kohle, 1993), estimate dynamic balance control (Hahn, Farley, Lin, & Chou, 2005), and provide diagnostic classification of falls risk (Hahn & Chou, 2005). These most recent efforts have relied on simple three-layer, back-propagation networks and statistical discrimination models. The objective of this chapter is to provide examples of recently successful application of ANN theory in the classification of gait patterns, estimation of dynamic balance control, and diagnostic classification of relative falls risk in the elderly.

BACKGROUND

Traumatic falls in the elderly are prevalent, debilitating and costly, with over 35% of the elderly population experiencing falls (Coogler, 1992) and approximately $20.2 billion spent in treatment each year (American Academy of Orthopaedic Surgeons [AAOS], 1998). Accurate assessment of the risk of falls is critical to reducing the incidence of falls. As humans age, gait patterns are known to adjust, accommodating for reduced function in the balance control system and a general reduction in skeletal muscle strength (Fiatarone and Evans, 1993; Grimby and Saltin, 1983). The temporal-distance (T-D) measures of gait (gait velocity, stride length, stride time, step width) have been used in evaluation of overall function and determination of gait dysfunction in the elderly (Elble, Thomas, Higgins, & Colliver, 1991; Ferrandez, Pailhous, & Durup, 1990; Heitmann, Gossman, Shaddeau, & Jackson, 1989; Judge, Davis, & Ounpuu, 1996; Leiper & Craik, 1991; Maki, 1997; Menz, Lord, & Fitzpatrick, 2003). These studies showed that while T-D measures of gait do provide an overall impression of walking performance, there is substantial inter-subject variability in the measures. Such variability may contribute to a lack of power in accurately predicting the risk of falling in the elderly. The effect of aging on muscle activation and strength in the elderly has been shown to result in higher electromyographic (EMG) signal amplitudes during gait (Finley, Cody, & Finizie, 1969; Shiavi, 1985). However, the resulting force production in aged subjects is highly variable (Galganski, Fuglevand & Enoka, 1993; Grabiner & Enoka, 1995). No previous studies have examined the effect of T-D parameters and EMG activity on control of whole body stability.

Many studies have attempted to predict falls prospectively, with varying results. Previous studies (Maki, Holliday, & Topper, 1994; Topper, Maki, & Holliday, 1993) used measures of static posturography to indicate risk of falls. Results from their work showed that control of medio-lateral sway may be a strong predictor of falls in the elderly. Another study (Graafmans et al., 1996) reported that general mobility impairments are strongly associated with recurrent falls. No measures of diagnostic accuracy were reported in these studies.

More predictive models were reported using logistic regressions which combined Berg Balance scores with a self-reported history of imbalance to predict risk of falls (Shumway-Cook, Baldwin, Polissar, & Gruber, 1997). Results from this retrospective study produced a sensitivity of 91% and specificity of 82%. Further results (Shumway-Cook, Brauer, & Woollacott, 2000) indicated that the timed-up-and-go test was also accurate in assessing risk of falls (sensitivity = 87%; specificity = 87%). These two studies showed that relatively simple clinical measures could predict risk of falls with reasonable accuracy.

Recently, a few research groups have reported fall-risk screening in community-dwelling elderly using logistic regression models based primarily upon a previous history of falls. One study (Tromp et al., 2001) relied on reports of visual impairment and urinary incontinence in addition to falls history, to estimate risk using the receiver operating characteristic (ROC) measure (Swets, 1988), producing overall diagnostic accuracy of 0.71 for predicting recurrent falls. Another study (Stalenhoef, Diedriks, Knottnerus, Kester, & Crebholder, 2002) relied on history of falls, measures of postural sway, hand grip strength, and "a depressive state of mind" to estimate risk of recurring falls (ROC = 0.79). These studies indicated improvement in accuracy when balance control and strength were considered.

Other groups investigated risk estimation using measures familiar to the in-patient hospital or nursing home setting. One study used a Poisson regression model, resulting in five variables which predicted risk: age, gender, morbidity predisposition, surgical procedure and length of stay (Halfon, Eggli, Van Melle, & Vagnair, 2001). No values of accuracy were reported. Another study reported that in institutionalized settings (long-term care) the primary predictor of falls was that of nurses' opinion/prediction (Izumi, Makimoto, Kato, & Hiramatsu, 2002). Interestingly, this study also reported that the primary indicator of falls risk in a general hospital setting was mobility (including independent ambulation). Results from their study were quite low in prediction accuracy; ranging from 34.8% to 45.2%.

ISSUES AND PROBLEMS

Problems in Assessing Dynamic Balance Control in Older Adults

Control of whole-body stability has been studied over the years by analysing motion of the whole-body center of mass (COM), the patterns of which were reported to be quite consistent during locomotion (Jian, Winter, Ishac, & Gilchrist, 1993; MacKinnon & Winter, 1993; Prince, Winter, Stergiou, & Walt, 1994; Winter, 1995). More recent studies have demonstrated an ability to distinguish elderly individuals with balance impairment from their age-matched healthy peers, using measures of medio-lateral (M-L) COM motion during obstacle crossing (Chou, Kaufman, Hahn, & Brey, 2003; Hahn &

Chou, 2003). Accurate estimation of the whole-body COM requires three-dimensional reconstruction of a multiple segment biomechanical model. This technical requirement alone may restrict broad application of assessing dynamic instability.

In many clinical settings, gait analysis can be performed with accuracy in measures of gait velocity, stride length, stride time and step width. Additionally, with many brands of inexpensive hardware/software currently available, relative magnitude of muscle activations may be measured with surface EMG during locomotion and other activities of daily living. Using T-D and EMG data to predict dynamic stability would be advantageous by reducing the necessity for a multiple camera motion analysis system (more costly), and reducing the time commitment of data-processing and analysis. A model is, therefore, needed which would allow accurate description of whole-body balance control, given simple measures of gait such as EMG and T-D parameters.

One approach for mapping interactions between gait measurements and balance control is to construct a nonlinear model using an artificial neural network (ANN). Biological nervous systems are capable of learning by adjustment of the synaptic connections between individual neurons. The ANN is modelled in a similar fashion, allowing the network to be trained by exposure to a set of input data where the output values are known. Weights of the ANN interconnections are iteratively adjusted to attempt correction of the final processed output to match that of known values. Once a network has been trained to a satisfactory level, the knowledge gained by this learning process is stored in the connection weights (synapses), allowing a trained network to solve new problems similar to the task it was trained on. The primary advantages of ANNs in solving real-world classification problems are: (1) their resilience in the face of noise and variability within a dataset, and (2) the ability to map relationships between variables that would not otherwise be noticeable. They have been used with high success in problems that are either too complex for conventional methods or are of an exploratory nature (Chau, 2001).

Application of ANN models in musculoskeletal biomechanics has dealt with joint angles and joint moment estimations in gait simulation (Sepulveda, Wells, & Vaughan, 1993) and estimation of muscle recruitment in static conditions (Nussbaum, Chaffin, & Martin, 1995). Sepulveda and colleagues used traditional back-propagation algorithms to successfully map the relationship between EMG and joint angles, and between EMG and joint moments during gait. Nussbaum et al. also reported success using a back-propagation algorithm to map lumbar muscle recruitment during moderate static exertions. Koike and Kawato (1995) used an architecturally complex ANN model to estimate isometric joint torques and trajectory from surface EMG in upper limb motions. More recent efforts showed promising results with use of a simple, three-layer ANN, using an adaptive learning rate back-propagation algorithm in the determination of elbow joint torque from EMG activity (Luh, Chang, Cheng, Lai, & Kuo, 1999).

Use of ANN modelling has expanded over the years to address a wide variety of questions related to human motion. Previously successful efforts have shown that relatively simple network architectures with back-propagated error correction are capable of mapping linear and non-linear relationships between biomechanical variables, allowing greater understanding of the neuromuscular and musculoskeletal relationships involved.

Problems Involved in Faller Classification and Risk Estimation

Many of the previous studies which have addressed faller classification and risk estimation restricted predictive variable selection to static measures of posture, or clinical estimates of overall health, rather than dynamic musculoskeletal measures which may better represent balance control during activities when falls are likely to occur. One limitation of those studies, which did include dynamic or musculoskeletal strength measures, is that the models did not allow for prospectively testing the efficacy of interventions. This limitation may be significant, in that application of risk-estimation results can not be effectively used in falls-prevention, unless the model allows interventions to be examined prospectively. Exercise and strength training have become more common in recent years as an intervention for maintaining functional mobility in the elderly. Although improvements in strength and walking function were observed, studies only revealed that general exercise and strength training had a beneficial effect on fall incidence and could not determine which type of exercise was most effective (Province, 1995; Wolfson et al., 1996). There is, therefore, an ultimate goal to develop a model that can not only accurately assess the risk of falling of an older individual but also have the ability to predict possible outcomes of the prescribed muscle strengthening intervention.

The first step is to establish a model that can provide a more accurate estimation of falls risk in the elderly by taking into account the level of muscular challenge (electromyography), dynamic gait performance (temporal-distance parameters) and measures of whole-body dynamic stability. A few approaches could be used to develop such a model, linear regression or principal component analysis, for example. However, both of these techniques restrict the input predictors to only those which explain a high amount of variability in the system. Furthermore, both of these models rely on linear relationships between variables. One approach which allows inclusion of more variables and is tolerant of non-linear relationships is to use ANN theory to map dynamic measures of gait and whole-body stability onto estimates of individual risk. Recent years have seen an increase in the use of ANN analyses applied to human locomotion (Chau, 2001). Most of the studies have used back-propagation learning algorithms (Rumelhart, Hinton, & Williams, 1986) and relatively simple ANN designs (Haykin, 1994; Lafuente, Belda, Sanchez-Lacuesta, Soler, & Prat, 1998; Prentice, Patla, & Stacey, 1998; 2001; Savelberg & de Lange, 1999; Sepulveda et al., 1993).

EXAMPLES OF ANN THEORY APPLICATION

Collection Protocol

Inclusion criteria for the young (20-32 years of age) and healthy elderly (64-85 years of age) samples required no histories of significant head trauma, neurological disease (e.g., Parkinson's, post-polio syndrome, diabetic neuropathy), visual impairment not correctable with lenses, musculoskeletal impairments (e.g., amputation, joint replacement, joint fusions, joint deformity due to rheumatoid arthritis), or persistent symptoms of vertigo, light-headedness, unsteadiness. Healthy elderly subjects were noted to be active community members, with many of them currently involved in recreational sporting activities. Ten elderly subjects with complaints of imbalance during walking or a history of falls were recruited from the local community. Imbalance for these individuals was self-reported. A history of falls was defined as an occurrence of two or more falls. A fall was defined as any event in which the individual lost their balance and made contact with the floor (i.e., did not simply fall back into a chair after trying to stand up). Subject responses to inclusion criteria were confirmed by a consulting physiatrist. Three subjects with imbalance were diagnosed with either unilateral or bilateral vestibular weakness. All subjects with balance disorders were community-dwelling and able to walk more than 100 m without the use of gait aides at the time of testing. The experimental protocol was approved by the Institutional Review Board and experimental procedures were explained to all subjects prior to testing, with verbal and written consent obtained.

Input data consisted of normalized electromyography (EMG) data of the lower extremities (gluteus medius, vastus lateralis, medial gastrocnemius), temporal-distance (T-D) measures of gait (gait velocity, stride length, stride time, step width), and medio-lateral (M-L) motion (displacement and peak velocity) of the whole body center of mass (COM). Input data were averaged within each subject, over three individual gait trials. Input data were collected using the following gait protocol and analysis procedures (Hahn & Chou, 2003; Hahn, Lee, & Chou, 2005).

The gait protocol included walking on level ground with no obstructions and stepping over an obstacle corresponding to 2.5% of each subject's height. All subjects performed the trials at a self-selected pace, while barefoot. Whole body motion data were collected using a six-camera ExpertVision system (Motion Analysis Corp., Santa Rosa, CA). The COM position was estimated throughout the obstacle crossing stride, using a weighted-sum approach with a 13-link anthropometric model (Chou, Kaufman, Brey, & Draganich, 2001; Jian et al., 1993; Meglan, 1991). Velocities were calculated using the generalized, cross-validated spline algorithm (Woltring, 1986). Ranges of M-L displacement and peak M-L velocity values were compiled due to previous findings indicating their ability to distinguish between individuals with and without balance impairment (Chou et al., 2003).

Dynamic EMG measures were taken from pre-amplified surface electrodes (Motion Lab Systems, Inc., Baton Rouge, LA) placed bilaterally over the bellies of the gluteus medius (GM), vastus lateralis (VL) and medial gastrocnemius (GA). Activation magnitude of each muscle during gait was normalized to values taken during maximal effort manual muscle testing (MMT). Maximal GM activation was tested in 30° of hip abduction, while side lying. For VL maximum, subjects were seated with the knee in 45° of flexion. Maximal GA activation was tested in neutral ankle position, with the subject fixed to a table in prone position. MMT procedures were performed by one examiner for each muscle group, bilaterally. Subjects were verbally encouraged to ensure maximal recruitment.

During gait trials, only the peak phases of activity were used as EMG inputs into the model. Peak phases of activity corresponded to periods of double support for the GM and VL and single support for the GA. Gait velocity, stride length (normalized to body height), stride time and step width were measured during the obstacle-crossing stride and used as inputs into the ANN model. From the various measurements taken during the gait analysis protocol, a total of twelve variables were available for input into the ANN model (2 COM, 6 EMG and 4 T-D measures).

Table 1. Subject information: age, height and mass, Mean (SD)

Group	Female	Male	Age (yrs)	Height (cm)	Mass (kg)
Young (n = 11)	6	5	24.5 (3.3)	171.5 (7.2)	68.9 (6.8)
Healthy Elderly (n = 19)	8	11	71.5 (5.6)	167.9 (9.8)	73.3 (14.1)
Elderly/Imbalance (n = 10)	7	3	78.5 (4.7)	162.0 (9.8)	72.2 (14.4)

Estimation of Dynamic Balance Control with ANN

Although ANN modelling has been used in studies of human locomotion (Holzreiter & Kohle, 1993; Lafuente et al., 1998; Prentice et al., 1998; 2001; Savelberg & de Lange, 1999; Sepulveda et al., 1993; Su & Wu, 2000; Wu, Su, Cheng, & Chou, 2001), no previously published work has addressed the ability of such models to map the interaction between basic gait measurements and descriptions of dynamic balance control. The purpose of this study was to demonstrate the effectiveness of an ANN model in mapping gait measurements (normalized lower extremity EMG signals and basic T-D parameters) onto whole body measures of dynamic stability (motion of the COM). It was hypothesized that a relatively simple ANN architecture would be capable of accurately mapping interactions between these variables.

Input/output data of the ANN model were obtained from a database of previously collected subjects (Chou et al., 2003; Hahn & Chou, 2003; Hahn & Chou, 2005; Koshida, 2002). The subject pool (n=40) consisted of 11 healthy young adults, 19 healthy elderly adults, and 10 elderly adults with complaints of imbalance (Table 1). Inclusion criteria and empirical data were as described previously in the common collection protocol.

An ANN model was designed and implemented using Matlab's Neural Network Toolbox (v. 6, The MathWorks, Inc., Natick, MA) to estimate M-L COM motion using input from EMG magnitudes of the lower extremity and T-D values during a low-level, obstacle-crossing task. The ANN system consisted of a supervised 3-layer feed-forward neural network (Figure 1) with two cells in the output layer to represent measures of sideways dynamic stability (M-L COM range of displacement and peak velocity), up to 20 processing units in the hidden layer, and an input layer of EMG amplitudes and T-D parameters, normalized to zero mean and unity standard deviation.

The hidden and output units sum incoming connections and derive an outgoing activation signal based upon a sigmoidal transfer function in the hidden units and a pure linear transfer function in the output units. For the output units, the result of the transfer function is the normalized value of the predicted variable. Activation in the hidden units was calculated as:

$$a_m = sig(n) \tag{1}$$

where *sig* is a sigmoidal transfer function:

$$n = 2/(1 + e^{(-2*n)}) - 1 \tag{2}$$

and *n* is the result of the weight summation:

$$n = \sum_{p=1}^{numP} (i_p w) + b \tag{3}$$

In Equation 3, i_p is the input signal from each p parameter, w is the synaptic weight on each input signal, and b represents the bias of the hidden unit. The same calculation steps were carried out in the output units, except with a pure linear transfer function (see Figure 1).

Error correction was performed on the synaptic weights and biases using the Levenberg-Marquardt learning algorithm (Levenberg, 1944; Marquardt, 1963), known to provide rapid convergence and robust generality in small networks being trained on small-to-moderate subsets. Error correction in this algorithm can be described with the following general equation:

$$x_{k+1} = x_k - \alpha_k g_k \tag{4}$$

where x_k represents the vector of current weights and biases, g_k the current gradient value, α_k the current learning rate, and k indicates the iteration index. The Levenberg-Marquardt algorithm allows for iterative adjustment of the learning rate in order to control the rate of convergence. Ideally, the algorithm will adjust the learning pattern from an initial "gradient-descent" style of convergence, with large adjustments made to the weights and biases, to a more controlled convergence allowing smaller adjustments to be made with greater accuracy in the final, converged solution.

The full data set consisted of all subject groups collected, with 3 trials representing each subject (40 subjects * 3 = 120 cases). The training set was selected randomly from the entire data set at a given proportion (i.e., 0.70), with the remainder of subjects entering the testing set. Individual subjects were selected rather than individual trials, so that it was not possible for a particular subject's data to be used in training and testing. Moderate bootstrap re-sampling was utilized, with 20 re-sampling instances used for each of the parameter settings listed below. Bootstrap re-sampling was chosen as a method for improving model generalizability due to the small size of the total data set. Bootstrapping is a valid way to thoroughly test the available data set while still allowing random selection for network training. It does not, however, increase the variety of data in the initial data set.

The model then ran iteratively on the training set with error correction proceeding until the mean squared error (MSE) of the predictions became less than a preset value. The training set proportion (Pr), training error goal (E), and number of hidden units (H) were manipulated to assess which settings provide the best estimation accuracy. Using three values for each setting [Pr={0.6, 0.7, 0.8}; E={0.01, 0.001, 0.0001}; H={5, 10, 20}], a total of 27 setting conditions were tested.

After successful training of the ANN, the model's output was transformed back to real-world units (cm, cm/s) and overall model performance was assessed by correlation analysis to detect how well the model estimations fit the target values (reported as the correlation coefficient (R): R_1, for M-L displacement; R_2, for peak M-L velocity) of the entire subject pool. Separate analyses were performed for EMG and T-D inputs, as well as when EMG and T-D were combined as complimenting inputs. Mapping accuracy results were statistically compared using a two sample t-test.

Figure 1. General diagram describing the neural network architecture. A tangential sigmoid transfer function was used in the hidden layer, and a linear transfer function in the output layer.

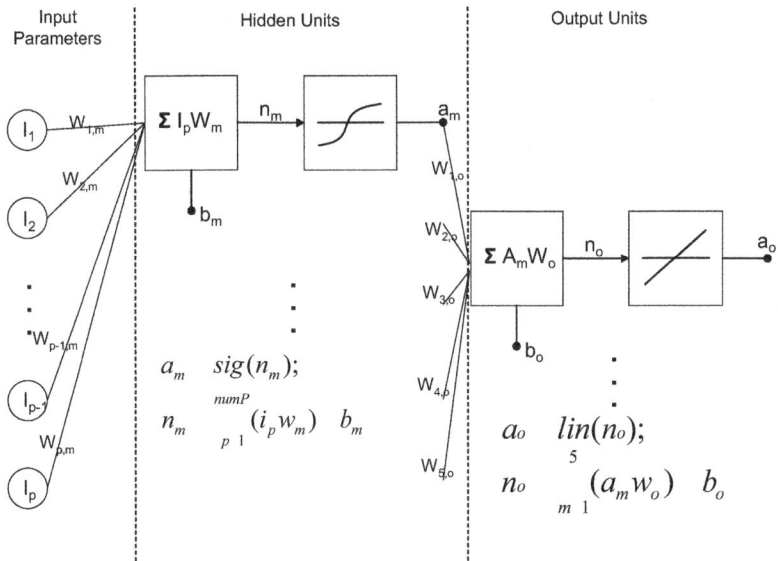

Comparisons between young and healthy elderly adults showed no significant differences in the measures of M-L COM displacement or peak M-L velocity (Figure 2). Elderly adults with balance impairment allowed significantly greater M-L displacement and peak velocity, compared to both the young ($p < 0.001$ and $p = 0.014$, respectively), and the healthy elderly ($p = 0.002$ and $p = 0.013$, respectively). Empirical results of the COM variables were then compiled as target values for the ANN model output.

The model architecture was shown to be efficient by requiring minimal processing time. Generally, training sessions required less than 5 minutes to reach the target training goals. Based on correlation analysis, the ANN performed reasonably well in terms of prediction accuracy; mean R_1 values ranged from 0.64 to 0.89 in the estimation of M-L COM motion, and R_2 values from 0.57 to 0.82 for the estimation of peak M-L COM velocity (Table 2). When EMG data were used as the sole input to the model, R-values ranged from 0.57 to 0.82. With T-D parameters as the sole model input, R-values ranged from 0.64 to 0.80. When the two types of data were entered together both R-values improved, ranging from 0.73 to 0.89 and from 0.65 to 0.82 for R_1 and R_2, respectively. When mapping results from the highest training proportion setting (Pr = 0.8) were compared, significant differences existed in the goodness of fit (R-values) when comparing between the EMG-only and EMG/T-D combined input (p=0.006) and between T-D input and combined input (p=0.001). Other mapping result differences were not significant.

Table 2. Mapping results showing the effect of training proportion and input data type. The goal and hidden unit values reflect the settings producing highest accuracy. R-values are reported in Mean (SD) of 20 attempts.

Input	Proportion	Goal	# hidden units		R_1	R_2
EMG	0.6	0.001	10		0.65 (0.12)	0.57 (0.16)
	0.7	0.01	20		0.72 (0.12)	0.67 (0.16)
	0.8	0.01	20	a	0.82 (0.09)	0.77 (0.10)
T-D	0.6	0.001	20		0.64 (0.11)	0.64 (0.13)
	0.7	0.001	20		0.72 (0.12)	0.75 (0.11)
	0.8	0.01	20	b	0.80 (0.10)	0.80 (0.10)
EMG & T-D	0.6	0.0001	20		0.73 (0.09)	0.65 (0.13)
	0.7	0.01	20		0.79 (0.09)	0.75 (0.09)
	0.8	0.0001	20	a,b	0.89 (0.06)	0.82 (0.09)

[a] *Significant difference between EMG and combined input types (p=0.006)*
[b] *Significant difference between T-D and combined input types (p=0.001)*

As training proportion increased from 0.6 to 0.8, the goodness of fit for the ANN mapping accuracy improved for both the M-L COM displacement and peak M-L velocity, regardless of input type. No noticeable trends developed regarding which training error goal (E) performed best for the mapping networks. Of note, however, is that when EMG and T-D parameters were combined as input, the model performed better with a more strict training goal (E = 0.0001).

For each type of input data (EMG, T-D alone, and combined), it was determined that more hidden units provided quicker convergence within the model, improving the ability to generalize. With only 5 hidden units the EMG input mapping to COM motion took an average of 460.5 epochs (SD, 77.7) to converge, whereas when the same mapping was performed with 20 hidden units, the system converged in 7.0 epochs on average (SD, 1.2). Similarly, when T-D input was mapped with only 5 hidden units an average of 483.8 epochs was required (SD, 10.1), compared to 12.1 epochs (SD, 2.7) with 20 hidden units. When EMG and T-D measures were combined as input, mapping solutions were reached in 226.6 epochs on average (SD, 129.3) with 5 hidden units, compared to an average of 4.4 epochs (SD, 0.5) with 20 hidden units.

This study sought to demonstrate the ability of an ANN model to accurately map muscular activation levels and temporal-distance parameters onto whole body measures of dynamic stability during gait. Results supported the hypothesis that a relatively simple ANN model architecture is adequate for estimating dynamic stability from the basic measures of normalized EMG activation and T-D parameters.

Given the relatively small size of the sample data set, the model performed reasonably well, with average correlation coefficient (R_1) values from 0.64 to 0.89 in estimation of M-L COM motion, and correlation coefficient (R_2) values from 0.57 to 0.82 for the estimation of peak M-L COM velocity. Further investigation is needed to confirm this stability estimation in a larger, more diverse sample set. By the very nature of the empirical data used in this model (EMG and T-D), variability within subject group was high enough that more distinctive categorization *a priori* would be beneficial in the initial validation of this model. However, the strength of using neural network theory to model the relationships between these variables and balance control lies in its ability to map non-linear functions in systems that are not clearly defined.

As ANN models are allowed to learn, they have a tendency to solidify the weightings between input parameters, the processing units, and the output. This tendency can sometimes lead to interpretations regarding the predictive strengths of the input parameters. Further examination of the network connections between the input layer and the layer of hidden units revealed no definitive weighting patterns. When EMG data were the sole inputs, the muscles which received strong weighting (values of 2 or

Figure 2. Empirical results of M-L COM displacement (a) and peak M-L COM velocity (b) during the obstacle crossing stride at an obstacle height of 2.5% body height

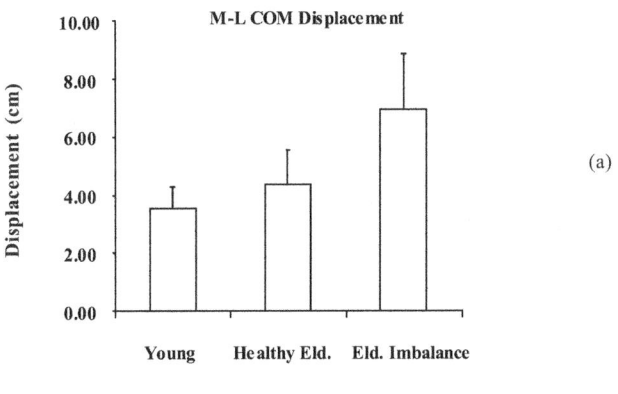

(a)

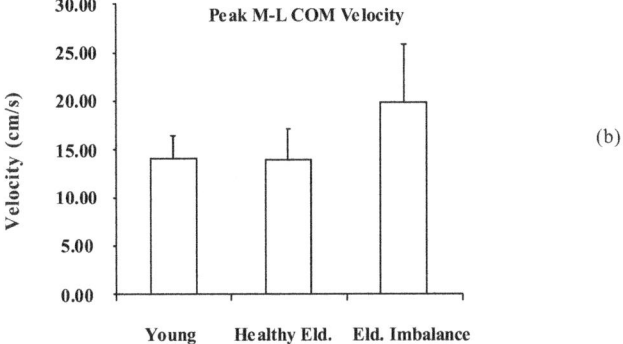

(b)

Elderly subjects with imbalance exhibited greater displacement and peak M-L velocity than either healthy young or healthy elderly subjects (p < 0.014)

greater) would vary between training attempts. The leading limb gluteus medius was the only muscle to continuously receive strong weighting with each training attempt. With T-D input, stride length consistently received stronger weightings, while step width rarely received strong weightings. The lack of strong weightings for step width may appear counterintuitive, as step width may be assumed to have some effect on M-L COM motion. However, recent findings have demonstrated that change in step width does not necessarily indicate greater sway or dynamic instability (Chou et al., 2003; Krebs, Goldvasser, Lockert, Portney, & Gill-Body, 2002) as its increase, if any, is variable within groups.

Network weighting for gait velocity and stride time varied between training attempts. As gait velocity is a function of stride length and stride time, it is not surprising that its contributive weighting was sporadic in nature. Inclusion of gait velocity as an input may not be warranted due to its association with the other variables. When EMG and T-D were combined as inputs, no distinct patterns could be detected. Overall, these findings indicate that the present ANN model did not rely on solidifying weight values to reach accurate output estimations. This could be due in part to the rapid convergence of the network. With rapid convergence, the tendency of a network to solidify weightings would be inhibited. The present model's success in prediction accuracy was likely dependent on the generalized weightings of its network connections. However, this reduced the ability to examine network weightings for patterns of learning in the network. Future investigations into this and similar models could include the use of class estimators as utilized in heteroscedastic probabilistic neural network models.

Additionally, it is interesting to note the effect of increasing the number of hidden units on the model's performance. As the number of hidden units increased, the number of epochs necessary for the network to converge on a solution decreased. Rapid convergence of a system to an accurate solution indicates an enhanced ability to generalize the function in a broader application. Indeed, findings from this study indicate that accuracy was improved with more hidden units, indicating enhanced generality overall (see Table 2). The ability to generalize may be similar to the concept of plasticity in natural neural pathways. Early in the learning of a task, the neural system has a number of pathways/solutions to choose from (i.e., plastic), but the more corrections and repetitions that it takes to converge on a solution, the more rigid the final pathway becomes. In this way, an ANN model that can converge quickly (e.g., 5 epochs, with 20 hidden units) will have a greater number of possible pathways to the solution, compared to an ANN model which converges slowly (e.g., 500 epochs, with 5 hidden units).

Faller vs. Non-Faller Categorization (via ANN)

One of the strengths of multi-layer neural network models is their robust ability to map inputs non-linearly onto resulting outputs. For this reason, neural network modeling theory was chosen in this study as the method for assessing the level of functional mobility through gait measures, leading to a categorization of faller/non-faller based on individual cases.

Various combinations of input data were tested for their ability to accurately categorize fallers: EMG, T-D and COM parameters alone, then EMG and T-D together, and finally EMG, T-D, COM parameters were all entered as input predictors. The reason for solely entering the EMG and T-D data as "paired" input was that these data were

Figure 3. Flow chart describing the development of the two-system model

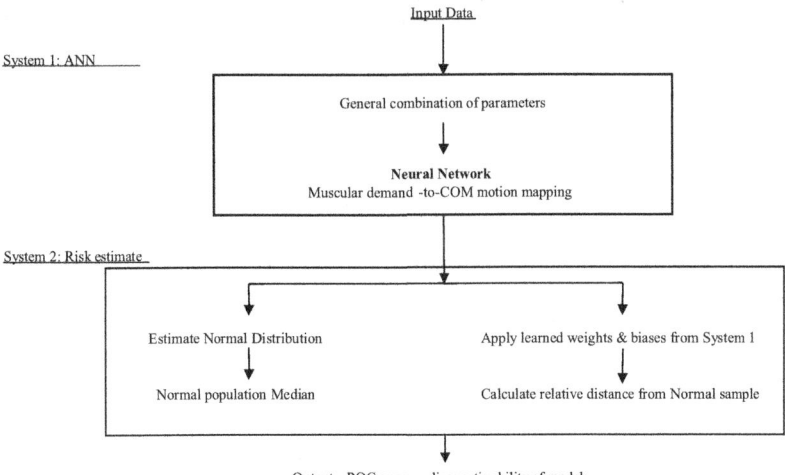

Output: ROC curve - diagnostic ability of model;
Clinical interpretation of relative risk value

shown in previous work to accurately predict COM motion variables when entered into a simple ANN structure (Hahn et al., 2005). Paired entry of EMG and T-D data therefore serves as a substitution method in the event that COM data cannot be compiled for analysis.

The model used in this study was designed to assess interactions between muscular demands, gait performance, and balance control during walking. It was comprised of two systems (Figure 3). The first system, an ANN model, estimated the category of non-faller/faller (0 or 1, respectively), based on the empirical inputs listed above. The second system discriminately classified the relative risk that an individual will experience falls, using output from the first system. Used in combination, these two systems can demonstrate the relative distance from a healthy, normative distribution of age-matched subjects, thereby giving a scaled estimate of risk.

The ANN structure was a feed-forward, three-layer network with scaled error correction conducted via the Levenberg-Marquardt backpropagation algorithm (Levenberg, 1944; Marquardt, 1963). The first layer consisted of a variable number of input units (specific to the data type being tested). The second layer held a variable number of "hidden" units ($H=\{5, 10, 20\}$). These hidden units are generally considered to be the information-processing level of artificial networks, allowing parallel distributed processing to take place (Rumelhart et al., 1986). The final layer of the ANN consisted of a sole output unit. This output unit reflected the network's estimate of the faller/non-faller category. The network was trained using a set of training error goals ($E=\{0.01, 0.001, 0.0001\}$). The training proportion was set at 0.7 for all tests. Twenty tests (with randomly selected training samples) were made for each of the nine combinations of ANN settings

(H and E) that were possible (resulting in 180 training attempts made for each combination of input parameters). The hidden and output units summed incoming weighted connections, and processed an outgoing activation signal with a sigmoidal transfer function in the hidden units, and a pure linear transfer function in the output units (Haykin, 1994). Figure 1 shows a detailed diagram of the ANN structure used.

The value of the ANN's output unit provided an estimate of the likelihood that an individual be categorized as healthy, or as a faller. For the output units, the result of the transfer function was the predicted category value for either 0, or 1. An output value of 0 indicated healthy, and a value of 1 indicated the faller category. Further details on the ANN model design have been reported previously (Hahn et al., 2005).

After training was completed in each attempt, cases from the testing set (30% of the initial subject group) were input into the network with fixed ("learned") weights/biases. Each case then received an output value estimating their likelihood of being either a healthy subject, or a faller (0 or 1, respectively). A distribution function was generated from the ANN output for the healthy subject group. Categorization required selection of a decision line at some value that was thought to best delineate between healthy elderly and fallers' output data. In this study, the median of the healthy subjects' output distribution was conservatively selected as the decision line (X_0). A new value for X_0 was calculated with each training session. A distribution function was also generated for the ANN output of the group of elderly fallers, for initial validation of the categorizations made by the decision line. Figure 4 depicts how these distributions are shaped and how a decision line is used.

The effectiveness of categorization to either side of the decision line was measured by the receiver operating characteristic (ROC) value calculated for the system. The ROC is a method derived from signal detection theory that has been described as providing "a precise and valid measure of diagnostic accuracy" (Swets, 1988). Specifically, the ROC value is calculated as the area under a curve which is generated from the specificity and

Figure 4. Two ANN output distribution curves representing healthy elderly subjects on the left, and those with reduced balance control on the right

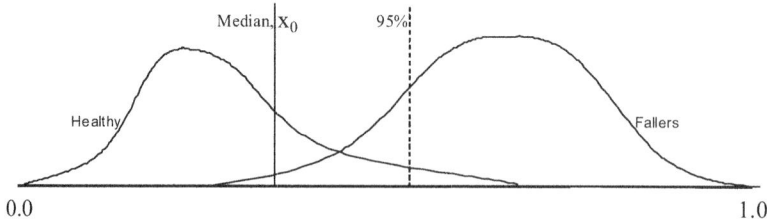

Two viable decision lines are shown, at the median of the healthy subjects' distribution and at the upper 95% mark of the healthy distribution. Categorizations made using the 95% decision line would result in a lower proportion of false positives, at the cost of reducing the proportion of true-positives.

Figure 5. Representative ROC curve showing 89% overall classification accuracy

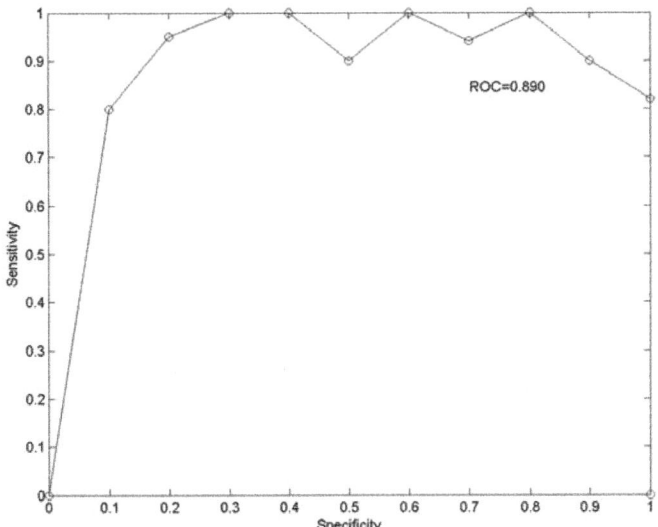

Sensitivity represents the proportion of true-positives; Specificity represents the proportion of false positives. This curve was generated from ANN settings of H = 5, E = 0.001, and the combined input data of EMG and T-D.

sensitivity results of multiple diagnostic testing sessions. Specificity, as defined by Swets (1988), was calculated as the proportion of false-positives (individuals categorized as being a "faller," when they are in fact a "non-faller"). Sensitivity is defined as the proportion of true-positives (individuals appropriately categorized as "fallers"). For each test, discrete values for specificity and sensitivity were calculated. A curve was thus shaped from the specificity (x-axis) and sensitivity (y-axis) values calculated from each of the multiple tests (n=20) conducted for each type of input. The area under the resulting curve is interpreted as the overall diagnostic accuracy of the classification system (ROC value). This method of assessing diagnostic accuracy has been used widely in medical diagnostics (Swets, 1988), and more recently in studies of fall-risk estimation (Stalenhoef et al., 2002; Tromp et al., 2001). A representative ROC curve is presented in Figure 5.

Accuracy of the diagnostic network was moderately high, with top ROC values ranging from 0.689 to 0.890. Accuracy was lowest when COM or EMG data were used as the sole input. When T-D was the sole input, diagnostic accuracy increased substantially. Accuracy was further increased when the data types were combined, with overall

Table 3. Accuracy results of the sample categorization (ROC value)

Input	Goal	# hidden units	ROC
EMG	0.01	5	0.689
T-D	0.001	5	0.807
COM	0.0001	5	0.702
EMG, T-D	0.001	5	0.890
EMG, T-D, COM	0.001	5	0.884

For each input data type, training goal and number of hidden units resulting in highest accuracy are reported

Table 4. Effect of the training error level (E) on diagnostic accuracy (ROC)

Input	E = 0.01	E = 0.001	E = 0.0001
EMG	0.532	0.532	0.503
T-D	0.634	0.612	0.637
COM	0.570	0.552	0.557
EMG, T-D	0.708	0.687	0.652
EMG, T-D, COM	0.653	0.661	0.714

Values are averaged from the settings of hidden unit number

best accuracy being achieved when the combined input of EMG and T-D data were used. Table 3 provides details of the categorization accuracy results. For each input data type, the training goal and number of hidden units are reported as the settings that resulted in the highest accuracy.

There was no discernible trend in the effect of training error on resulting ROC values. For most input data combinations, the response of the ANN to more specific training (lower E value) resulted in no change, or very slight change to the positive or negative. The only combination of input data which resulted in improved diagnostic performance was the combination of EMG, T-D and COM. As the training error was reduced, the overall diagnostic accuracy increased for this combination of inputs. Table 4 provides descriptive data of these findings.

The effect of increasing the number of hidden units produced a very discernible trend (Figure 6). For every combination of input data five hidden units produced the highest ROC value. As H increased to 10, there was a sharp decrease in diagnostic accuracy for all input combinations. Finally, as H increased to 20 a marked improvement in ROC values occurred, though still less than the original setting of H=5. The combined inputs of EMG and T-D, and EMG, T-D and COM consistently produced the highest ROC values.

This study sought to test the ability of an ANN model, recently developed to map muscular inputs and gait measurements onto balance control, in classifying faller categories in the elderly. The task chosen in this study was to group a random sample

Figure 6. Overall effect of increasing the number of hidden units (H)

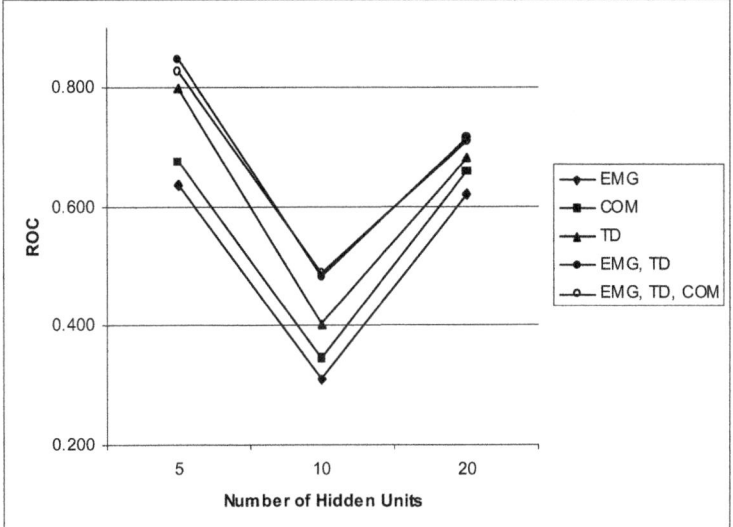

The ROC value was highest with all data input types, when H=5; decreased sharply when increased to H=10; and then improved substantially when increased to H=20. Values are averaged from the three different settings of training error (E)

of healthy elderly adults and elderly with balance impairments into the categories of "healthy," or "faller."

In group categorization the model performed respectably, with overall diagnostic accuracy reaching 0.89. Categorization in the present study indicates that 9 out of 10 elderly subjects with imbalance could be detected as being at risk of falls. Other studies that have reported faller categorization accuracy using the ROC measure indicated diagnostic accuracies of 0.71 (Tromp et al., 2001) and 0.79 (Stalenhoef et al., 2002). These two studies relied on reports of visual impairment and urinary incontinence in addition to falls history (Tromp et al., 2001), as well as the history of falls, measures of postural sway, hand grip strength, and "a depressive state of mind" (Stalenhoef et al., 2002) to categorize older individuals as fallers or non-fallers.

Examination of model accuracy arising from each type of input and the ANN settings that produced higher accuracy yields some interesting discoveries. In a recent study (Chou et al., 2003), it was determined that variability of individual T-D parameters was enough to cause insignificance in the differences between healthy older adults and those with balance disorders. Interestingly, the present model showed improvement in group categorization accuracy when it included multiple T-D parameters as input. When COM parameters were the sole input, categorization accuracy was at 0.702, lower than EMG or

T-D parameter on their own. This finding indicates that while M-L COM motion may show less variability within groups (allowing significant differences to be detected) (Chou et al., 2003). It is perhaps not variable enough to allow robust generality in group categorization through ANN modeling. To an extent, as dataset variability increases, the accuracy of ANN modeling also increases.

Results also showed that accuracy of the model's group categorization improved when different types of data were combined as inputs. Combination of EMG and T-D data produced the highest ROC value (0.890), while combination of all data types (EMG, T-D, and COM motion) yielded a slightly lower value (0.884). One explanation of decreased ROC when COM data were included may be that inclusion of COM data in this level of the model creates redundancy among the inputs, making the input-output relationship less clear within the model.

Improvement in group categorization with a more diverse dataset is not surprising in ANN models. It is thought that the more variety a neural network sees in data patterns, the more robust it will be in identifying similar patterns when new data is encountered. In the second task of estimating individuals' relative risk of falls, it appears that the relative distance index (D_r) provided adequately spaced values for delineation of risk. In fact, the D_r values were spread across a wider range than expected. Instead of being confined within the arbitrary boundaries of 0.0 and 1.0, index values reached beyond -1.25 and 1.25. As the ANN model used in this study converged to solution rapidly, it is not surprising that successful discrimination between two categories necessitated a wide range of index values. If the network converged more slowly, it may be that the index values would fit a 0.0 – 1.0 range. However, this would likely result in "overtraining" by the network. Overtraining often results in a reduced ability for the network to generalize. The wide range of index values simply requires wider category delineations for levels of risk (i.e., low, moderate, high, and very high).

Risk Estimation Due to Strength of ANN Categorization

After assessing functional mobility and faller categorization, it is desirable to estimate the individual level of risk for falls in each case. Once an individual's risk of falls is estimated, it is possible to prescribe interventions best suited to that individual. The architecture of a neural network model has the potential to allow for simulation of improvements in balance control as a result of changing muscular strength inputs.

Network output data from the testing cases (which the network had not been trained on) were compared against X_0. If the testing case' predicted value (X_1) was below X_0 it was inferred that the test case was similar to the normal, healthy population. If X_1 was greater than X_0, the inference was that the case was away from normal. The relative distance (D_r) of the predicted value from X_0 was used as an index for risk estimation, defined as:

$$D_r = \frac{1 - X_1}{1 - X_0} \tag{5}$$

Using this distance metric, it is possible for the difference between each case and the normal healthy group to be quantified, (i.e., relative risk of falling). For example, if one case receives ANN estimation identical to X_0, it will result in a D_r of 1.0 (identical to healthy

Figure 7. Example of risk estimation for an individual case, where the ANN output is ~0.8

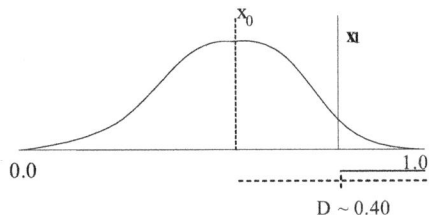

$D_r \sim 0.40$

In that case, the D_r value would be ~0.40, indicating a relatively moderate risk of falling. X_0 indicates the median value (decision line) of the healthy subjects' distribution; X_1 indicates the predicted value of an example test case.

group median). However, a case resulting in ANN estimation of X_1 greater than X_0, would result in a lower value of D_r, indicating a substantial relative distance from the healthy comparison group. In this way, as D_r approaches or exceeds 0.0, the risk of falling increases, and as D_r approaches or exceeds 1.0, the relative risk of falling decreases. Therefore, the calculated D_r value is relatively inverse to the output of the ANN. A similar calculation of a distance metric has been used previously to categorize ankle function in patients with arthrodesis (Su & Wu, 2000). Figure 7 provides an example of how an individual case can be interpreted using this distance measure.

For delineation of the relative levels of risk, general category boundaries were set as follows:

Very low risk	$D_r \geq 1.00$,
Low risk	$1.00 > D_r > 0.50$,
Moderate risk	$0.50 > D_r > 0.25$,
High risk	$0.25 > D_r > -1.00$,
Very high risk	$D_r < -1.00$

Risk estimation using the relative distance value (D_r) produced varied results across the subjects with balance disorders, and across the different types of input data. When EMG data were used alone as input, a majority of the subjects were estimated to be at moderate ($D_r < 0.50$) to high risk ($D_r < 0.25$), however two of the subjects received D_r values of greater than 1.10 (indicating no greater risk of falls, compared with healthy peers). With T-D data as input, all ten subjects were estimated to be at high risk of falls ($D_r < 0.25$), with one subject estimated at very high risk ($D_r = -1.86$). Entry of COM data resulted in an estimate that seven subjects were at high risk ($D_r < 0.001$), two at moderate risk ($D_r < 0.50$), and one subject at relatively low risk ($D_r = 0.70$). When EMG and T-D data were combined as input, nine subjects were estimated to have high risk of falls ($D_r < 0.20$) and one subject was estimated to be at very high risk ($D_r = -1.44$). With the combination of all three data types as input (EMG, T-D and COM), nine of the subjects were estimated to have high

Table 5. Relative risk values (D_r), for the 10 elderly subjects with balance impairment

Balance Impaired Subject	EMG		T-D		COM		EMG, T-D		EMG, T-D, COM	
1	-0.42	(h)	-0.68	(h)	0.70	(l)	-0.60	(h)	0.09	(h)
2	0.44	(m)	-1.86	(vh)	-0.42	(h)	-0.39	(h)	-0.73	(h)
3	-0.95	(h)	-0.32	(h)	-0.48	(h)	-1.44	(vh)	-0.74	(h)
4	-0.20	(h)	-0.66	(h)	0.00	(h)	0.10	(h)	0.68	(l)
5	-0.02	(h)	0.03	(h)	0.46	(m)	0.18	(h)	-0.34	(h)
6	1.47	(vl)	-0.85	(h)	-0.48	(h)	-0.48	(h)	-0.54	(h)
7	1.12	(vl)	-0.86	(h)	-0.42	(h)	-0.16	(h)	-0.16	(h)
8	0.00	(h)	-0.97	(h)	-0.42	(h)	-0.52	(h)	-0.77	(h)
9	-0.65	(h)	0.13	(h)	-0.80	(h)	-0.13	(h)	-0.50	(h)
10	-0.23	(h)	-0.28	(h)	0.27	(m)	-0.02	(h)	-0.44	(h)

Lower values indicate greater relative risk of falls. Risk category distinctions are indicated in parentheses; vl = very low, l = low, m = moderate, h = high, vh = very high.

risk of falls ($D_r < 0.10$), while one subject was estimated to have relatively low risk ($D_r = 0.68$). Details of the risk estimation results are provided in Table 5.

This study sought to test the ability of an ANN model, recently developed to map muscular inputs and gait measurements onto balance control, in stratifying the relative risk of falls in the elderly. An applied feasibility task was constructed in which the level of relative risk was estimated for each individual with balance impairment (n=10).

Results from the relative risk estimation task revealed a unique estimation pattern for subject #4 (Table 5). When the single data types were used as predictive inputs, the subject was estimated to have "high" risk, but when all three data types were entered, the subject's risk was then estimated to be "low." This pattern is unique compared to the other subjects' estimations. One explanation of the low risk estimate when data were grouped is that the individual demonstrated conflicting trends with each type of input data, thus serving to cancel the predictive strengths of any one type of data. Having a subject display conflicting input data reveals the difficulty involved in applying a diagnostic technique universally. There is a small likelihood that some individuals will always fall outside of the detection pattern established by a diagnostic algorithm.

Ideally, a risk prediction model should not only yield a yes/no answer, but also provide an estimate of the severity of risk. One limitation of this study is that the relative severity of each subject with balance impairment was not quantified previously. If previous measures of risk severity had been available, the risk estimation from the present model could be sufficiently validated. However, it may be said that there are presently no "gold-standard" methods for assessing relative risk of falls, and so direct validation of our model predictions to another technique is not currently possible. Given that many of our subjects reported previous falls and/or had demonstrated significant impairment in balance control, it is more likely that they would be categorized as moderate to high-risk. It would be interesting to compare risk estimation levels of individuals who have not fallen previously, but are beginning to show signs of imbalance.

In this study, ANN modelling theory was chosen due to its ability to include a broad range of predictive variables and its strength in mapping non-linear relationships. The primary limitation in using an ANN model lies in the relative "black-box" nature of its input/output relationships. This limitation is reduced to some extent by the *a priori* selection of input variables that have been reported in previous work to be influential on the overall function of balance control.

Linear regression models and principal component analysis could have been selected, but were passed over because they restrict variable inputs and rely on the assumption of linearity between input and output variables. Due to the ANN architecture used in this model, it may be possible to simulate the potential improvements that can arise in balance control from increasing the muscular strength inputs. Thus, targeted improvements at the muscular level could be investigated prospectively by varying its corresponding magnitude at the input layer.

Furthermore, the ANN structure is adaptable, so that measures of sensory motor function could be entered into the prospective intervention set, allowing highly targeted predictions in balance improvement to be made. Further studies will be conducted to explore the use of this ANN model in predicting efficacies of different interventions of muscle strengthening in balance improvement. The ability to test the efficacy of various interventions before they are implemented is a primary strength of this model.

SOLUTIONS AND RECOMMENDATIONS

Overall, the simple three-layer model used in these studies appears to be effective as a mapping tool for the estimation of balance control during locomotion. Effectiveness was revealed by the robust performance of the model with different network parameters. When EMG and T-D data were combined as input, R-values ranged from 0.73 to 0.89 in estimating M-L COM displacement. Previous studies have used simple ANN models in human locomotion, reporting general acceptance of training proportion set to 0.70, and model accuracy up to 0.897 (Su & Wu, 2000; Wu et al., 2001). The accuracy achieved through this model is therefore deemed acceptable, considering the range of training proportions used (from 0.60 to 0.80). Moreover, with simple 3-layer architecture, the ANN model is computationally efficient. Further validation of this model is needed with inclusion of a larger, more diverse sample set.

These studies also demonstrate the effectiveness of an ANN model in classifying individual cases of balance control as being within healthy ranges, or having high enough risk to be categorized as a "faller." It was further indicated that relative risk can be estimated for individuals with balance impairment. Findings of this study were based on a small sample of impaired subjects, compared to a relatively small sample of healthy peers. Therefore, further validation of both the group categorization task ("healthy" vs. "faller") and the relative risk estimation task is suggested in a larger, more diverse sample of healthy and balance-impaired elderly adults.

FUTURE TRENDS

In this section, future developments in applications of ANN and other machine-learning techniques to problems of fall prediction and intervention selection are discussed. Three important directions could be immediately expanded from our current models. First is the expansion of input variables to the learning systems. Since falls in the elderly have a multifactorial etiology, multidimensional assessments in reducing falls and fall-related injury will likely be more effective. Therefore, inclusion of quantifiable intrinsic and behavioral measures of risk factors, such as neurological, sensory, cognitive, and musculoskeletal functions as well as general health and functioning, gait and balance performance will expand the range of subsystems evaluated as possible contributing factors to fall likelihood.

A second future direction for the research reported here is to develop models of the impact of aging and therapy intervention upon the most relevant input variables, as determined by the research effort above. The neural network models we propose above would at best be effective at indicating whether someone is a likely faller today. What we would like to do is to predict whether someone is a likely faller in the future and if so to propose interventions that could lessen that likelihood. What is needed is an analysis of existing longitudinal data on input parameter values to develop models of aging impacts upon these parameters. Artificial neural networks can be applied to this problem as well, with inputs being input parameter values at one subject age and outputs being values of these same parameters at a later point (with this age is another input variable). Artificial neural networks, including recurrent networks that feed back current outputs as inputs, have been shown to be effective time series predictors (Liao, Moody, & Wu, 2001).

Third, one of the chief criticisms of neural networks has been that they tend to be black-box models relating values of input to output variables by a complex system of weights and activation functions. Simple, selective scanning of edge weights in such networks often give misleading indications as to the relative role of inputs in determining outputs, due to the complex interaction of weights in the computation determining output values. Recently, statistical sensitivity analysis techniques have been proposed for analyzing the contents of such networks to determine the relative importance of input variables in determining outputs (Montano & Palmer, 2003). Input variable values are varied in a controlled manner allowing extraction of individual variable contributions. Another approach to semantic analysis is based upon search through subsets of input variables to determine subsets producing locally optimal neural networks. Methods proceed either by starting with all input variables and eliminating these one at a time as long as this improves the resultant networks or by starting with single input variables and adding variables one at a time as long as this improves the resultant networks performance (Witten & Frank, 2000). These methods tend to give upper and lower bounds on subsets of effective, most relevant input variables.

CONCLUSION

A series of ANN models have been developed which demonstrate promise for accurate estimation of balance control using simple measures of gait, accurate detection of balance impairment in older adults, and accurate estimation of relative risk of falls at the individual level. The models developed from this work appear to be effective as: (1) a mapping tool for the estimation of balance control during locomotion, (2) a balance impairment detection tool, and 3) a tool for estimating individual cases of falls risk. Model effectiveness was supported by the robust performance of the models with different network parameters and varying types of input data. Application of these models has great potential for impacting the detection of balance impairment and further estimating the risk of falls, thereby reducing incidence of falls and enhancing quality of life in the elderly population. If level of impairment is well estimated, many elderly individuals who are at greater risk of falls might be identified prior to traumatic fall events. These individuals could then begin targeted balance-improving interventions to reduce their relative risk of falls.

With expansion of the data set available for input, the effectiveness of these models is expected to further improve. Future tests of the ability to use current estimates of balance impairment and predict the effectiveness of balance-improving interventions will enhance the application of these models. Lastly, unpacking the "black box" of the ANN will allow more precise estimates of balance control and subsequent falls risk, by a focusing of input variables to those providing the most relevant predictions.

REFERENCES

American Academy of Orthopaedic Surgeons. (1998). *Don't let a FALL be your last TRIP*.

Barton, J. G., & Lees, A. (1997). An application of neural networks for distinguishing gait patterns on the basis of hip-knee joint angle diagrams. *Gait & Posture, 5*, 28-33.

Chau, T. (2001). A review of analytical techniques for gait data. Part 2: Neural network and wavelet methods. *Gait & Posture, 13*, 102-120.

Chou, L. S., Kaufman, K. R., Brey, R. H., & Draganich, L. F. (2001). Motion of the whole body's center of mass when stepping over obstacles of different heights. *Gait & Posture, 13*, 17-26.

Chou, L. S., Kaufman, K. R., Hahn, M. E., & Brey, R. H. (2003). Medio-lateral motion of the center of mass during obstacle crossing distinguishes elderly individuals with imbalance. *Gait & Posture, 18*, 125-133.

Coogler, C. E. (1992, April/May). Falls and imbalance. *Rehabilitation Management, 53*.

Elble, R. J., Thomas, S. S., Higgins, C., & Colliver, J. (1991). Stride-dependent chages in gait of older people. *Journal of Neurology, 238*, 1-5.

Ferrandez, A. M., Pailhous, J., & Durup, M. (1990). Slowness in elderly gait. *Experimental Aging Research, 16*, 79-89.

Fiatarone, M. A., & Evans, W. J. (1993). The etiology and reversibility of muscle dysfunction in the aged. *Journal of Gerontology, 48S*, 77-83.

Finley, F. R., Cody, K. A., & Finizie, R. V. (1969). Locomotion patterns in elderly women. *Archives of Physical Medicine and Rehabilitation, 50*, 140-146.

Galganski, M. E., Fuglevand, A. J., & Enoka, R. M. (1993). Reduced control of motor output in a human hand muscle of elderly subjects during submaximal contractions. *Journal of Neurophysiology, 69*, 2108-2115.

Gioftsos, G., & Grieve, D.W. (1995). The use of neural networks to recognize patterns of human movement: Gait patterns. *Clinical Biomechanics, 10*, 179-183.

Graafmans, W. C., Ooms, M. E., Hofstee, H. M., Bezemer, P. D., Bouter, L. M., & Lips, P. (1996). Falls in the elderly: A prospective study of risk factors and risk profiles. *American Journal of Epidemiology, 143*, 1129-1136.

Grabiner, M. D., & Enoka, R. M. (1995). Changes in movement capabilities with aging. In J.O. Holloszy (Ed.), *Exercise and sports sciences reviews* (pp. 65-104). Baltimore: Williams and Wilkins.

Grimby, G., & Saltin, B. (1983). Mini-review: The aging muscle. *Clinics in Physiology, 3*, 209-218.

Hahn, M. E., & Chou, L. S. (2003). Can motion of individual body segments identify dynamic instability in the elderly? *Clinical Biomechanics, 18*, 737-744.

Hahn, M. E., & Chou, L.S. (2004). Age-related reduction in sagittal plane center of mass motion during obstacle crossing. *Journal of Biomechanics, 37*, 837-844.

Hahn, M. E., Farley, A. M., Lin, V., & Chou, L. S. (2005). Neural network estimation of balance control during locomotion. *Journal of Biomechanics*, In press.

Halfon, P., Eggli, Y., Van Melle, G., & Vagnair, A. (2001). Risk of falls for hospitalized patients: A predictive model based on routinely available data. *Journal of Clinical Epidemiology, 54*, 1258-1266.

Haykin, S. (1994). *Neural networks: A comprehensive foundation*. New York: MacMillan College.

Heitmann, D. K., Gossman, M. R., Shaddeau, S. A., & Jackson, J. R. (1989). Balance performance and step width in non-institutionalized elderly female fallers and nonfallers. *Physical Therapy, 69*, 923-931.

Holzreiter, S. H., & Kohle, M. E. (1993). Assessment of gait patterns using neural networks. *Journal of Biomechanics, 26*, 645-651.

Izumi, K., Makimoto, K., Kato, M., & Hiramatsu, T. (2002). Prospective study of fall risk assessment among institutionalized elderly in Japan. *Nursing and Health Sciences, 4*, 141-147.

Jian, Y., Winter, D. A., Ishac, M. G., & Gilchrist, L. (1993). Trajectory of the body COG and COP during initiation and termination of gait. *Gait & Posture, 1*, 9-22.

Judge, J. O., Davis, R. B. III, Ounpuu, S. (1996). Step length reductions in advanced age: The role of ankle and hip kinetics. *Journal of Gerontology, 51*, M303-312.

Koike, Y., & Kawato, M. (1995). Estimation of dynamic joint torques and trajectory formation from surface electromyography signals using a neural network model. *Biological Cybernetics, 73*, 291-300.

Koshida, S. (2002). *Identifying biomechanical challenge during locomotion in the elderly*. MS Thesis, University of Oregon, Eugene, OR.

Krebs, D. E., Goldvasser, D., Lockert, J. D., Portney, L. G., & Gill-Body, K. M. (2002). Is base of support greater in unsteady gait? *Physical Therapy, 82*, 138-147.

Lafuente, R., Belda, J. M., Sanchez-Lacuesta, J., Soler, C., & Prat, J. (1998). Design and test of neural networks and statistical classifiers in computer-aided movement analysis: A case study on gait analysis. *Clinical Biomechanics, 13*, 216-229.

Leiper, C. I., & Craik, R. L. (1991). Relationships between physical activity and temporal-distance characteristics of walking in elderly women. *Physical Therapy, 71*, 791-803.

Levenberg, K. (1944). A method for the solution of certain non-linear problems in least squares. *Quarterly of Applied Mathematics, 2*, 164-168.

Liao, Y., Moody, J., & Wu, L. (2001) Applications of artificial neural networks to time series prediction. In Y. H. Hu & J. Hwang (Eds.), *Handbook of neural network signal processing* (Chapter 9). New York: CRC.

Luh, J. J., Chang, G. C., Cheng, C. K., Lai, J. S., & Kuo, T. S. (1999). Isokinetic elbow joint torque estimation from surface EMG and joint kinematic data: Using an artificial neural network model. *Journal of Electromyography and Kinesiology, 9*, 173-183.

MacKinnon, C. D., & Winter, D. A. (1993). Control of whole body balance in the frontal plane during human walking. *Journal of Biomechanics, 26*, 633-644.

Maki, B. E. (1997). Gait changes in older adults: Predictors of falls or indicators of fear. *Journal of the American Geriatric Society, 45*, 313-320.

Marquardt, D. W. (1963). An algorithm for least squares estimation of nonlinear parameters. *Journal of the Society of Industrial and Applied Mathematics, 11*, 431-441.

Meglan, D. A. (1991). *Enhanced analysis of human locomotion.* PhD Dissertation, The Ohio State University, Columbus.

Menz, H. B., Lord, S. R., & Fitzpatrick, R. C. (2003). Age-related differences in walking stability. *Age and Ageing, 32*, 137-142.

Montano, J. J., & Palmer, A. (2003). Numeric sensitivity analysis applied to feedforward neural networks. *Neural Computation and Applications, 12*, 119-125.

Nussbaum, M. A., Chaffin, D. B., & Martin, B. J. (1995). A back-propagation neural network model of lumbar muscle recruitment during moderate static exertions. *Journal of Biomechanics, 26*, 1015-1024.

Prentice, S. D., Patla, A. E., & Stacey, D. A. (1998). Simple artificial neural network models can generate basic muscle activity patterns for human locomotion at different speeds. *Experimental Brain Research, 123*, 474-480.

Prentice, S. D., Patla, A. E., & Stacey, D. A. (2001). Artificial neural network model for the generation of muscle activation patterns for human locomotion. *Journal of Electromyography and Kinesiology, 11*, 19-30.

Prince, F., Winter, D. A., Stergiou, P., & Walt, S. E. (1994). Anticipatory control of upper body balance during human locomotion. *Gait & Posture, 2*, 19-25.

Province, M. A. (1995). The effects of exercise on falls in elderly patients: a preplanned meta-analysis of the FICSIT trials. *Journal of the American Medical Association, 273*, 1341-1347.

Rumelhart, D. E., Hinton, G. E., & Williams, R.J. (1986). Learning representations by back-propagation errors. *Nature, 323*, 533-536.

Savelberg, H. H., & de Lange, A. L. (1999). Assessment of the horizontal, fore-aft component of the ground reaction force from insole pressure patterns by using artificial neural networks. *Clinical Biomechanics, 14*, 585-92.

Sepulveda, F., Wells, D. M., & Vaughan, C. L. (1993). A neural network representation of electromyography and joint dynamics in human gait. *Journal of Biomechanics, 26*, 101-109.

Shiavi, R. (1985). Electromyographic patterns in adult locomotion: A comprehensive review. *Journal of Rehabilitation and Research Development, 22*, 85-98.

Shumway-Cook, A., Baldwin, M., Polissar, N. L., & Gruber, W. (1997). Predicting the probability of falls in community-dwelling older adults. *Physical Therapy, 77*, 812-819.

Shumway-Cook, A., Brauer, S., & Woollacott, M. (2000). Predicting the probability of falls in community-dwelling older adults using the timed up & go test. *Physical Therapy, 80*, 896-903.

Stalenhoef, P. A., Diedriks, J. P. M., Knottnerus, J. A., Kester, A. D. M., & Crebholder, H. F. J. M. (2002). A risk model for the prediction of recurrent falls in community-dwelling elderly: A prospective cohort study. *Journal of Clinical Epidemiology, 55*, 1088-1094.

Su, F-C., & Wu, W-L. (2000). Design and testing of a genetic algorithm neural network in the assessment of gait patterns. *Medical Engineering and Physics, 22*, 67-74.

Swets, J. A. (1988). Measuring the accuracy of diagnostic systems. *Science, 240*, 1285-1293.

Topper, A. K., Maki, B. E., & Holliday, P. J. (1993). Are activity-based assessments of balance and gait in the elderly predictive of risk of falling and/or type of fall? *Journal of the American Geriatric Society, 41*, 479-487.

Tromp, A. M., Pluijm, S. M. F., Smit, J. H., Deeg, D. J. H., Bouter, L. M., & Lips, P. (2001). Fall-risk screening test: A prospective study on predictors for falls in community-dwelling elderly. *Journal of Clinical Epidemiology, 54*, 837-844.

Winter, D. A. (1995). Human balance and posture control during standing and walking. *Gait & Posture, 3*, 193-124.

Witten, I. H., & Frank, E. (2000). *Data mining: Practical machine learning tools*. San Diego, CA: Academic.

Wolfson et al. (1996). Balance and strength training in older adults: Intervention gains and Tai Chi maintenance. *Journal of the American Geriatric Society, 44*, 498-506.

Woltring, H. J. (1986). A FORTRAN package for generalized, cross-validatory spline smoothing and differentiation. *Advances in Engineering Software, 8*, 104-113.

Wu, W-L., Su, F-C., Cheng, Y-M., & Chou, Y-L. (2001). Potential of the genetic algorithm neural network in the assessment of gait patterns in ankle arthrodesis. *Annals of Biomedical Engineering, 29*, 83-91.

Chapter VIII

Recognition of Gait Patterns Using Support Vector Machines

Rezaul Begg, Victoria University, Australia

Marimuthu Palaniswami, The University of Melbourne, Australia

ABSTRACT

Automated gait pattern recognition capability has many advantages. For example, it can be used for the detection of at-risk or faulty gait, or for monitoring the progress of treatment effects. In this chapter, we first provide an overview of the major automated techniques for detecting gait patterns. This is followed by a description of a gait pattern recognition technique based on a relatively new machine-learning tool, support vector machines (SVM). Finally, we show how SVM technique can be applied to detect changes in the gait characteristics as a result of the ageing process and discuss their suitability as an automated gait classifier.

INTRODUCTION

Human walking is a complex process involving numerous interactions of the various elements of the musculoskeletal and neural systems. Gait analysis is a recognized procedure that is conducted to diagnose disability and to assess the effectiveness of a particular rehabilitation procedure. There are mainly two approaches adopted while analysing gait-related data: statistical techniques and machine-learning approaches (e.g., neural networks, fuzzy logic, evolutionary technique (e.g., genetic algorithm) or support vector machine). There have been also some attempts to combine more than one of the machine-learning techniques (to work in a hybrid model) so that these techniques can complement each other to deliver better outcomes. The vast majority of the works in gait analysis have been based on traditional statistical techniques because of their clear advantage of providing much greater insight into the models and the ease of studying the effects of various variables. However, machine-learning approaches are increasingly becoming widespread in recent times, especially because of their non-linear modelling capabilities. Another advantage of a machine-learning approach is that the predictive models can be built without much *prior* knowledge about the distribution of the data, which makes it particularly suitable for modelling complex motions such as those involved in the dynamics of human gait.

In this chapter, a relatively new machine-learning technique based on statistical learning theory (Vapnik, 1995), support vector machine (SVM), is described and then we apply it to the task of automated recognition of gait pattern changes due to ageing. The SVM technology has been shown to have a much better ability in building superior predictive models. This is due to the fact that in SVM techniques, an optimal separating hyperplane is determined that is expected to result in superior generalization performance on unseen data. It is also very effective in situations when the dimension of the input data is relatively high and the number of observations available for developing and training the models is limited (c.f., Zavaljevski et al., 2002).

BACKGROUND

Swing Phase and the MFC During Gait

The swing phase is an important phase of the gait and its analysis can offer much useful information, especially when studying the gait characteristics related to trips and related falls. Tripping during walking has been identified to account for >50% of all falls (Owings et al., 1999). Minimum foot clearance (MFC) during walking (see Figure 1), which occurs during the mid-swing phase of the gait cycle is defined as the minimum vertical distance between the lowest point on the shoe and the ground. The MFC has been identified as an important gait parameter in the successful negotiation of the walking environment, including obstacles and uneven surfaces. There are two main reasons for this. First, it has been estimated that the mean MFC value is quite low in magnitude (1.11 - 1.29 cm) in both young and older adults. Second, during this MFC event the foot travels at its maximum velocity (Winter, 1991). This small mean MFC value combined with the variability in MFC data (.5-.62cm) has the potential to cause tripping during walking, especially for unseen obstacles or obstructions. The literature relating to the MFC event

Figure 1. Toe clearance during one gait cycle beginning at heel contact (0%)

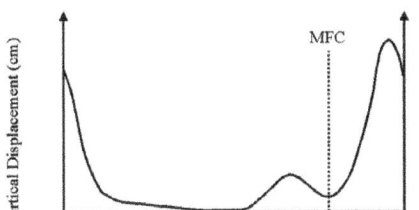

The minimum foot clearance (MFC) event occurs during the mid swing phase (Adapted from Taylor, 2003)

has revealed a strong rationale for being associated with tripping during walking and has, therefore, implications for trip-related falls in the older population. While there are many gait variables (e.g., basic stride-phase variables, joint angles, ground reaction forces, electromyographic data, etc.) that show age-related effects (e.g., Princea et al., 1997), it has been suggested (Karst et al., 1999) that more sensitive gait variables such as foot clearance data (e.g., MFC) during walking should be used to describe age-related declines in gait for predictors of falls risks.

Machine Learning Approaches for Gait Recognition

Early identification of at-risk gait is important and provides the opportunity to undertake preventative measures (e.g., to prevent falls in older individuals). At present, research in the area of automatic identification of gait types from their gait features is less common. Majority of the research in the area of automated gait recognition has been undertaken using neural networks technology to classify various gait types. One of the earlier works in the area was undertaken by Holzreiter & Kohle (1993), who applied neural networks for the classification of normal and pathological gait using force platform recordings of foot-ground reaction forces taken during normal gait. In this study, altogether 8,173 pairs of foot strikes from 94 healthy subjects and 131 pathological populations with various diseases were used to train and test the neural network models. The input features included vertical foot-ground reaction forces that were converted to their equivalent fast Fourier transform (FFT) coefficients. A 3-layer feed-forward neural network topology was used with 30 input nodes and a single classification output node. The network was trained with a back propagation algorithm with random weight initialisation. The network was trained 80 times for different initialisation of weights and using a different proportion of the data set (either 20/40/60 or 80%), the remaining data being used to test the "predictability" of the network. Accuracy of gait recognition was found to be about 95%, which demonstrates that neural networks are able to learn from examples of diagnosed gait patterns to be used for diagnosis of future unseen gait patterns.

Barton and Lees (1997) applied neural networks to differentiate simulated gait (e.g., leg length discrepancy) using features extracted from lower–limb joint-angle measures. Wu et al. (1998) applied a back propagation neural network to classify gait patterns of patients with ankle arthrodesis and those of normal subjects, and reported superior classification outcomes by the neural networks compared to statistical techniques (e.g., linear discriminant analysis) (98.7% vs. 91.5%). There have been a number of other attempts to classify various gait patterns using neural networks, some of which may be found in the review articles (Chau, 2001; Scholnhorn, 2004).

Automatic classification of gait patterns has also been undertaken with the help of fuzzy clustering techniques. Su et al. (2001) used fuzzy cluster paradigm to separate walking patterns of 10 healthy subjects and 10 patients with ankle arthrodesis. Features extracted relate to Euler angles from foot-segment motion kinematics and the various clusters generated represented distinct walking patterns of the normal and pathological subjects. O'Malley et al. (1997) were able to demonstrate the usefulness of the fuzzy clustering technique to show improvement in the surgical intervention in the cerebral palsy (CP) children using basic temporal-distance gait data: stride length and cadence. Using gait data from 88 spastic diplegia CP children and from 68 normal children, their technique assigned each child's gait data a membership function related to one of the five different clusters. Improvement as a result of treatment was identified by a shift in the membership function and the cluster belonging of the subject.

Genetic algorithm (GA) is based on the "survival of the fittest" principle in natural systems (Su et al., 2000) and has also been applied to facilitate the task of gait pattern recognition. In particular, GA technique can be used in conjunction with neural networks for the determination of optimal number of hidden layer neurons, as well as for the selection of good input features. Su et al. (2000) have used GA for searches in appropriate space to determine optimal parameters of neural networks for gait recognition in healthy and pathological populations. In their application to differentiate a group of normal and pathological (ankle arthrodesis) gait patterns from measures of force platform recordings, GA-based neural network was found to perform superiorly compared to a traditional neural network (98.7% vs. 89.7%). Such findings suggest the clear advantage that can be obtained by optimising the parameters in classifiers and also selecting the relevant and important features for classification tasks, and in those respects the suitability of GA in gait pattern recognition tasks.

Support vector machine (SVM) is a machine-learning tool and has emerged as a powerful technique for learning from data and for solving various classification and regression problems (Vapnik, 1995). This has been particularly effective for binary classification applications, as evidenced by their demonstrated high success rates in other biomedical areas (c.f. Zavaljevski, 2002). SVM has been applied for the classification of young and elderly gait patterns (Begg et al., 2005) and also for gender classification task from video gait sequence data (Lee & Grimshaw, 2002).

SUPPORT VECTOR MACHINES

Overview of SVM

Support vector machines are a relatively new machine learning tool and have emerged as a powerful technique for learning from data and for solving classification and regression problems. In particular, it is shown to be very effective for solving binary classification problems. SVMs originate from Vapnik's statistical learning theory (Vapnik, 1995), and their major advantage over neural networks is that they formulate the learning problem as a quadratic optimisation problem whose error surface is free of local minima and has global optimum (Kecman, 2002). An important concept in SVM is data transformation to a higher dimensional space, known as the "feature space" for the construction of optimal separating hyperplane (OSH). As illustrated in Figure 2, the two classes are non-separable using a linear hyperplane in input space, but can be linearly separated in a non-linearly mapped feature space. SVMs perform this non-linear mapping into a higher dimensional feature space by means of a *kernel* function and then construct a linear OSH between the two classes in the feature space. Figure 3 shows an example of such SVM solution in a 2-class problem using optimal separating hyperplane. Those data vectors nearest to the hyperplane margins in the transformed space are called the support vectors. SVMs thus concentrate valuable information relating to the classification task from the training data into the support vectors. SVMs minimize the structural risk (i.e., the chance of misclassification of a previously unseen data) rather than minimizing the empirical risk (i.e., the error in the misclassification on training data as adopted in neural networks) (Kecman, 2002). The principle is based on the fact that generalization error rate on the test data is bounded by the sum of training error rate and a term that depends on Vapnik-Chervonenkis (VC) dimension. VC dimension is a measure of complexity of the dimension space.

SVM Theory

The detailed theory behind SVM theory has appeared in many textbooks and journal articles (e.g., Kecman, 2002; Vapnik, 1995). In brief, the problem of pattern recognition may be stated as follows.

Figure 2. Example of linearly non-separable two class data in input space, but can be linearly separable in a higher dimension feature space by transforming data using ϕ

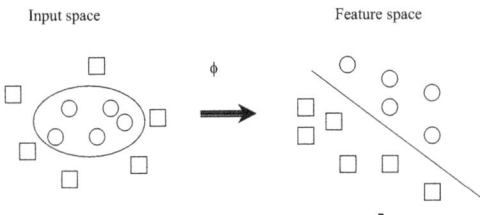

Figure 3. An example of two-class (+ and -) problem with optimal separating hyperplane and the maximum margin

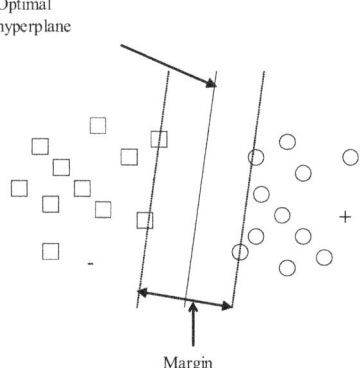

The circles and squares represent samples of class +1 and –1, respectively

Let us assume a training set of the form (Begg et al., 2005):

$$\Theta = \{(\mathbf{x}_1, d_1), (\mathbf{x}_2, d_2), \ldots, (\mathbf{x}_N, d_N)\}$$

where:

$$\mathbf{x}_i \in \mathfrak{R}^{d_L} \tag{1}$$
$$d_i \in \{+1, -1\}$$

Also, we assume that g was generated by some unknown but well defined function:

$$g : \mathfrak{R}^{d_L} \to \{+1, -1\} \tag{2}$$

And possibly corrupted as a result of noise, the learning machine can be thought of a function as follows:

$$f(\beta) : \mathfrak{R}^{d_L} \to \{+1, -1\} \tag{3}$$

And adjusts the parameter set b in order to minimize the misclassification risk, namely:

$$R(\beta) = \int_{\mathbf{x}} L\big(g(\mathbf{x}), f(\beta)(\mathbf{x})\big) dF(\mathbf{x})$$

where:

L = the loss function
F = the (unknown) probability distribution of $\mathbf{x}$ (4)

Support vector machines are a recently developed technique to solve this type of classification problems. Much of the credit goes to Vapnik (1995) for undertaking early work on this. The two fundamental concepts behind the SVM method are optimal margin classification and the kernel trick.

To construct SVMs, users must select a kernel function. So far, no analytical or empirical study has conclusively established the superiority of one kernel over another. Some of the commonly used kernels are shown in Table 1. The kernel function implicitly maps the data from input space into a (usually much higher dimensional) feature space via the non-linear function, thus, avoiding the curse of dimensionality. The kernel function K (x,y) is related to the non-linear feature mapping function $\varphi(\mathbf{x})$ by:

$$K(\mathbf{x}, \mathbf{y}) = \varphi(\mathbf{x})^T \varphi(\mathbf{y})$$

where:

$$\mathbf{x} \in \Re^{d_L}$$ (5)
$$\varphi(\mathbf{x}) \in \Re^{d_H}$$

Optimal margin classification involves finding a linear separating hyperplane in feature space. The fitness of a hyperplane is usually measured by the distance between the hyperplane and those training points lying closest to it (the support vectors). This will allow us to completely specify our decision surface in terms of these support vectors. The distance between the support vectors of the two different classes, as measured perpendicularly to the decision surface (optimal separating hyperplane) is referred to as the margin of separation (see Figure 3).

Table1. Some of the commonly used kernel functions and their mathematical formulae

Kernel Function	Mathematical Formula
Linear	$K(\mathbf{x}_i, \mathbf{x}_j) = \mathbf{x}_i \cdot \mathbf{x}_j$
Polynomial	$K(\mathbf{x}_i, \mathbf{x}_j) = (\mathbf{x}_i \cdot \mathbf{x}_j + 1)^d$, d is the degree of polynomial
Radial basis function (RBF)	$K(\mathbf{x}_i, \mathbf{x}_j) = \exp\left(-\dfrac{\|\mathbf{x}_i - \mathbf{x}_j\|^2}{2\sigma^2}\right)$, σ is the width of RBF function
Spline	$K(\mathbf{x}_i, \mathbf{x}_j) = 1 + \mathbf{x}_i \cdot \mathbf{x}_j + \dfrac{1}{2}(\mathbf{x}_i \cdot \mathbf{x}_j) \, min(\mathbf{x}_i, \mathbf{x}_j) - \dfrac{1}{6} min(\mathbf{x}_i, \mathbf{x}_j)^3$

The major problem that affected the use of this approach in common applications was how to deal with non-separable data in a convincing way. Following on from the work of Guyon et al. (1993), Cortes and Vapnik (1995) showed how a soft margin approach based on slack variables could be used to solve this problem in a simple but effective way. Thus, the primal form of the problem can be written as:

$$\min_{\mathbf{w},\xi,b} \Phi(\mathbf{w},\xi) = \frac{1}{2}\mathbf{w}^T\mathbf{w} + C\mathbf{1}^T\xi$$

such that:

$$d_i\left(\mathbf{w}^T\boldsymbol{\varphi}(\mathbf{x}_i) + b\right) \geq 1 - \xi_i$$
$$\xi \geq 0$$

where:

$$f(\mathbf{x}) = \mathrm{sgn}\left(\mathbf{w}^T\boldsymbol{\varphi}(\mathbf{x}_i) + b\right) \tag{6}$$
$$\mathbf{w}^T\boldsymbol{\varphi}(\mathbf{x}_i) + b = 0 \ \ \text{for the decision surface}$$

Let us consider the Wolfe dual, that is:

$$\min_{\alpha} \tfrac{1}{2}\boldsymbol{\alpha}^T\mathbf{G}\boldsymbol{\alpha} - \mathbf{1}$$

such that:

$$\mathbf{d}^T\boldsymbol{\alpha} = 0 \tag{7}$$
$$0 \leq \boldsymbol{\alpha} \leq C\mathbf{1}$$

where:

$$\mathbf{G}_{ij} = d_i d_j K\left(\mathbf{x}_i, \mathbf{x}_j\right)$$
$$f(\mathbf{x}) = \mathrm{sgn}\left[\sum_i \alpha_i d_i K\left(\mathbf{x}_i, \mathbf{x}\right) + b\right]$$
$$b = d_i\left(1 - (\mathbf{G}\boldsymbol{\alpha})_i\right), \ 0 < \alpha_i < C$$

Where the variable C is used to tune the trade-off between minimizing empirical risk (i.e., training errors) and the complexity of the machine. Schölkopf et al. (1995) introduced the concept of the Vapnik-Chervonenkis (VC) dimension of a SVM, and proved that the cost function is a trade-off between empirical risk minimization and minimization of the

VC dimension. For a given SVM, there is a probability of $1 - \eta$ that the following bound will hold:

$$R < R_{emp} + \sqrt{\frac{h\left(\log\dfrac{2N}{h}+1\right)-\log\left(\dfrac{\eta}{4}\right)}{N}}$$

where:

$h = $ VC dimension of SVM

$$h \leq \min\left\{\left[\frac{D^2}{\rho^2}\right], d_H\right\}+1 \tag{8}$$

$$R_{emp} = \frac{1}{2N}\sum_{i=1}^{N}\left|d_i - f(\alpha)(\mathbf{x}_i)\right|$$

$\rho = $ margin of separation
$D = $ diameter of the smallest ball containing all training vectors

Some of the important characteristics of SVMs are: SVMs are able to condense information from a large training set into a very small number of points (i.e., the support vectors). These support vectors contain all the necessary information that is required to define the decision surface. Also, Schölkopf et al. (1995) demonstrated via experiments conducted on support vector machines that SVMs would tend to extract much of the same set of support vectors regardless of what kernel function is used. This was illustrated by giving an example where one SVM is trained using only the support vectors found by another, with only a minimal loss of accuracy. Further details of SVM pattern recognition techniques can be found in Kecman (2002) and Burges (1998).

APPLICATION OF SVM FOR YOUNG/OLD GAIT CLASSIFICATION

In this section, we apply SVM technique for the case of separating young and elderly gait (i.e., automated recognition of young and elderly gait types from their respective features). Gait features selected for this application were minimum foot clearance (MFC) data that were collected during steady state walking on a treadmill. In the following, data collection procedure is first described very briefly followed by application of the SVM technique to classify them and evaluation of the classification performance.

MFC Gait Data Collection

The details of the MFC data collection procedure can be found in Begg et al. (2005), however, a brief overview is provided below. Briefly, MFC data of 58 healthy adults (30 young with a mean age of 28.4 years and 28 elderly with a mean age of 69.2 years) were taken from the gait database of the Biomechanics Unit of Victoria University to develop the SVM models. The young adults were from the academic community of Victoria University and the elderly participants were volunteers from various local senior citizen clubs. All subjects undertook informed-consent procedures as approved by the Victoria University Human Research Ethics Committee. The subjects had no known injuries or abnormalities that would affect their gait.

Foot clearance (FC) data were collected during steady state self-selected walking on a treadmill using the PEAK MOTUS 2D motion analysis system (Peak Technologies Inc, USA). A 50Hz Panasonic F15 video camera, with a shutter speed of 1/1000s, was positioned 9m from the treadmill, perpendicular to the plane of foot motion to record unobstructed treadmill walking. Two reflective markers were attached to each subject's left shoe at the 5th metatarsal head (MH) and the great toe (TM). Each subject completed about 20 minutes of normal walking at a self-selected comfortable walking speed. The MH and TM markers were automatically digitised for the entire walking task and raw data were digitally filtered using optimal cut-off frequency, which used a Butterworth filter with cut-off frequencies ranging from 4 – 8Hz. The marker positions and shoe dimensions were used to predict the position of the shoe/foot end-point, that is, the position on the shoe travelling closest to the ground at the time when minimum foot clearance (MFC) occurs using a 2D geometric model of the foot (Best et al., 1999). MFC magnitude was calculated by subtracting ground reference from the minimum vertical coordinate during the mid-swing phase of the gait cycle (see Figure 1).

Gait Feature Extraction Using MFC Histograms and Poincaré Plots

Each subject's MFC data was plotted as histograms showing individual MFC data and their respective frequencies. Features describing major statistical characteristics of these distributions were extracted as illustrated in Figure 4.

Gait features were also extracted using Poincaré Plots of the MFC data which is a plot of MFC data between successive gait cycles (i.e., between MFC_n and MFC_{n+1}) (Figure 5). These plots show variability of MFC data and measures of short- and long-term variability in MFC data. Any lack of MFC control due to ageing is expected to be reflected in these plots. Poincaré plots with high correlation coefficient is attributed to high level of control between strides, whereas a low correlation shows less control since one stride is not affected by the previous stride. These plots have been widely used in heart rate variability research indicating relationships of R-R intervals to identify plots representative of sound and abnormal heart function (Brennan et al., 2001). In gait, these plots show unique relationships of MFC events between successive gait cycles and would describe performance of the locomotor system in controlling the foot clearance at this critical event.

Figure 4. MFC distribution of one elderly subject showing MFC data and their relative frequencies

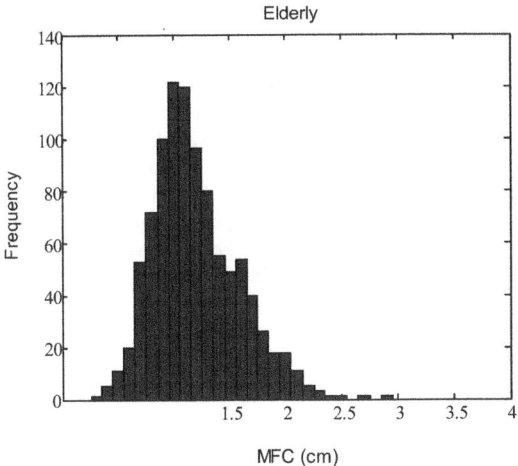

Features such as mean, median, standard deviation, minimum, maximum, etc., were extracted from these plots.

Figure 5. MFC Poincaré plot for the same subject whose MFC distribution is displayed in Figure 4

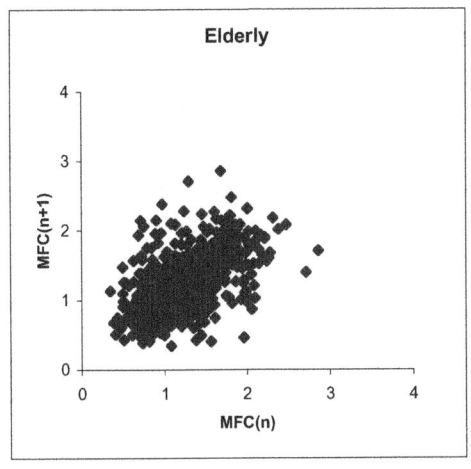

Key features extracted from the plot along the major (long) and minor (short) axes of these plots

Development of the SVM Model and Testing Protocol

Features extracted from these two types of plots can be used to train SVMs, and later on for testing their ability to discriminate between the two age groups. Cross-validation (3-fold) technique was used to test the ability of the classification system using various combinations of the testing and training data sets (Begg et al., 2005). Software routines were developed in Matlab 6.0 (The MathWorks, Natick, MA). Three kernel functions were tested: *linear, polynomial* and *radial basis function (RBF)* (see Table 1). Receiver operating characteristic (ROC) plots were used to display relationship between true positive rates or sensitivity (i.e., positively labeled test data classified as positive) versus false positive rates (i.e., negatively labeled test data classified as positive) as the threshold level of classification is varied from 0 (false positive rates = 1 – specificity). ROC plots offer a better insight into the classification performance and the area under the ROC curve provides a measure of the overall performance of the classifier (i.e., larger the ROC area the better is the classification accuracy).

CLASSIFICATION OUTCOMES

Accuracy of Gait Class Prediction

Table 2 presents mean success rates of the SVM classifier for the cross-validation tests. The average success rate was found to be >81% with a maximum accuracy of 85% in discriminating the young/old gait patterns obtained with linear and RBF kernels. Tests with *Polynomial* kernel provided lower accuracy, whereas when compared with a standard back propagation neural network's results, these results proved to be superior. The results of both SVM and neural network classification tools suggest that the SVM-linear and SVM-RBF performed superiorly when applied to separate young/old gait patterns. In a gender classification task using video sequence images, Lee and Grimson (2002) also reported superior SVM performance with linear kernel compared to a polynomial kernel. SVMs eliminate many of the problems experienced by neural networks, such as local minima and over-fitting, thereby, leading to a better classification performance by an SVM over a neural network. In addition, its ability to produce stable and reproducible results makes it a good candidate for solving many classification problems as evidenced by the recent surge in the use of this technique in many areas.

Table 2. Accuracy of prediction (%) of prediction of cross-validation tests using linear, polynomial (d=3) and radial basis function (RBF) kernels (Adapted from Begg et al., 2005, ©IEEE 2005)

Classifier	Linear	Polynomial	RBF
Mean	83.3	61.4	81.7

Figure 6. ROC (receiver operating characteristics) curves showing true positive (sensitivity) and false positive rates (1 – specificity) for various thresholds using linear, polynomial (d=3) and RBF (g=1) kernels (Adapted from Begg et al., 2005, ©IEEE 2005)

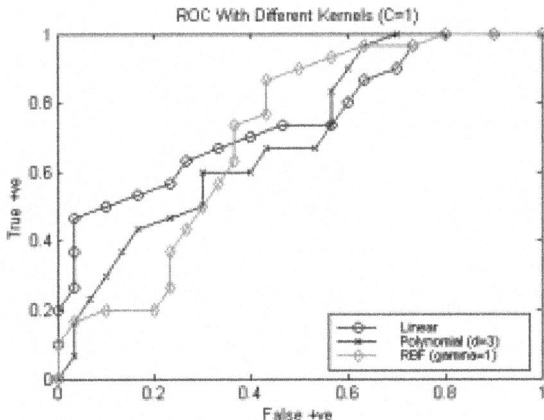

Table 3. ROC areas for a various number of selected features (Adapted from Begg et al., 2005, ©IEEE 2005)

Kernel	Parameters		No. of Features			
		C	3	6	12	24
Linear		1	0.83	0.85	0.84	0.79
Poly	d=3	1	0.83	0.84	0.83	0.64
RBF	g=1.0	1	0.89	0.93	0.94	0.81

Sensitivity, Specificity and ROC Plots

Figure 6 displays ROC plots of linear, polynomial (third order polynomial, d=3) and RBF (width, g=1) kernels when all the features were used as inputs to the SVM classifier. Both Linear and RBF kernels showed better performance compared to the polynomial kernel (ROC_{area}: linear=0.78, RBF=0.82, polynomial=.64).

Feature Selection

A feature selection algorithm was applied to the test results to identify features that provide the most contribution in separation results (ROC_{area}). In this method, one feature

Figure 7. Graph displaying the relationship between ROC area and the number of features selected for linear, polynomial (degree=3) and RBF (g=1) classifiers (Adapted from Begg et al., 2005, ©IEEE 2005)

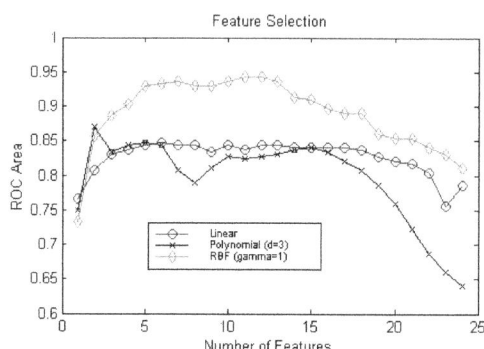

was added at a time that provided the most separation results between the two groups and, subsequently, the classification performance was determined with the new features. When the classifiers were trained with the selected fewer features (Table 3), for example with 6 selected features their classification performance were better (ROC_{area}: Linear=0.85, RBF=0.93, Polynomial=.84) than their performance outcomes utilizing all the input features (24). This is also reflected in the shape of the ROC plots presented in Figure 7 that plot ROC_{area} as a function of features selected by the algorithm. Overall, the RBF kernel displayed better performance (max ROC_{area}=0.94) relative to Linear and Polynomial kernels. It is also important to note from these graphs that all three kernels attained their maximum performance with only a small number of features. Once maximum ROC_{area} was achieved, it was found to be fairly unaffected by further increase in the number of gait features — eventually, the graph showed a downward trend indicating that too many features in this data would not be helpful. In fact, once the maximum accuracy is obtained adding further features that are not good representative of the separation of two classes could have detrimental effect on the classification performance. Similar dependence of classification performance on features has been reported elsewhere. For example, in a movement classification task using features extracted from electroencephalographic signals, Yom-Tov and Inbar (2002) demonstrated that only a small number of features (10-20) out of 1,000 features were necessary to achieve maximum performance and the classification rate decreased for features >20. Similar conclusion has also been drawn using features taken from visual-field location measurements in Glaucoma diagnosis application using various machine classifiers (Chan et al., 2002).

Figure 8. 2D scatter-plot using two features showing decision boundary on the test data for the case of a 3ʳᵈ order polynomial kernel (C=1) (Adapted from Begg et al., 2005, ©IEEE 2005)

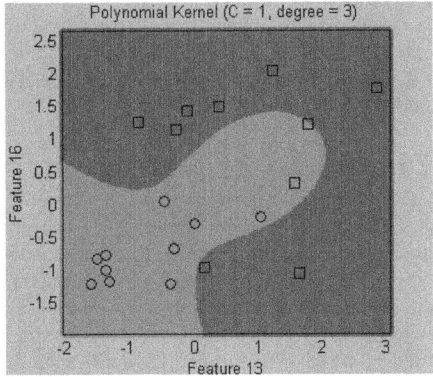

Figure 9. 2D scatter-plot of the test data, and the effect of threshold change on the decision boundaries and classification results, polynomial kernel classifier (Adapted from Begg et al., 2005, ©IEEE 2005)

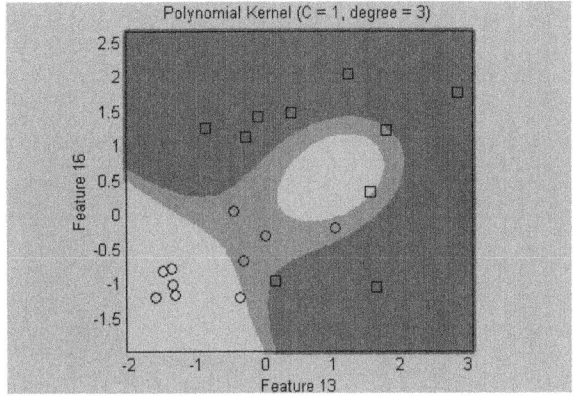

Decision Boundaries

Figure 8 displays visual representation of the data as a function of the two selected features (feature numbers 16 & 13 extracted from the MFC histogram and Poincare plots). These features were selected when the feature selection algorithm was applied. In this case, a separating hyperplane (decision surface) calculated by a third-order polynomial kernel is superimposed on the test data to illustrate the correct and wrong classification. The threshold level can also affect the classification outcomes. Figure 9 illustrates this by showing the effect of the threshold level on the decision boundaries and hence, its influence on the classification accuracies.

Regularization Parameter

Regularization parameter (C) can affect the performance of an SVM classifier. Figure 10 shows ROC_{area} plots of the linear kernel as a function of C. As can be seen from this plot, for this classifier, there is a range of C values that would produce an optimum performance (ROC area=.83). However, for C values outside of this range, the performance can be seen to be less than the optimum value. C is the penalty parameter for misclassification and has to be carefully selected to achieve maximum classification accuracy. Optimal value of C could be different for different number of features and has to be selected by trial and error method. One way of tackling this would be to plot a graph showing the reliance of classifier performance on C parameters and then selecting the optimal C value from the graph. When compared across different C's as well as the number of features for this application, RBF kernel was found to perform the best in recognizing MFC gait-patterns (ROC_{area}=0.95), suggesting that RBF is able to better exploit the features compared to linear and polynomial classifiers (Begg et al., 2005).

Figure 10. Plot of ROC-area as a function of regularization parameter, C (Adapted from Begg et al., 2005, ©IEEE 2005)

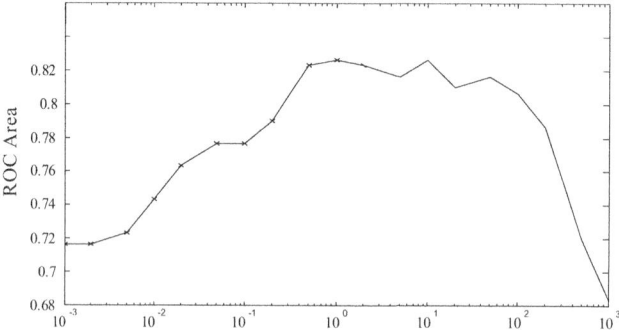

CONCLUDING REMARKS AND FUTURE WORKS

In this chapter, we have described an approach based on support vector machines to classify young/old gait types from their characteristics. All the features used to develop the SVM classifier were based on one single gait parameter, the Minimum Foot Clearance (MFC) during the swing phase of the gait cycle recorded over multiple gait cycles. Such long-term MFC data were transformed into *Histogram* and *Poincaré* plots for feature extraction purposes. The outcomes suggest that MFC data can provide useful information regarding the steady-state walking characteristics of individuals such that the features could be used to train a machine-learning tool for automated recognition of gait changes due to ageing. Early detection of locomotor impairments using such supervised pattern recognition techniques could provide the opportunity to identify at-risk individuals, and the appropriate corrective measures could be undertaken to avoid the occurrence of a fatal event (e.g., identify potential elderly fallers for falls prevention). Automated gait classification techniques also include neural networks and fuzzy clustering-based techniques (Holzreiter & Kohle, 1993; O'Malley et al., 1997) for applications such as diagnosis of pathological gait. Because of the superior performance demonstrated by the SVM for gait classification tasks (Begg et al., 2005) and also in other biomedical applications, support vector machine appears to have significant potential in automated gait diagnosis. Also, this may be applied for monitoring the progress of treatment or intervention outcomes in clinical and rehabilitation situations via gait classification. There may be a number of ways the gait recognition performance of the classifier can be further improved and expanded. These include:

- Generalization performance of a classifier depends primarily among other factors, on the successes of the selection of good features (i.e., utilizing those input features that represent maximal separation between the classes). A feature selection algorithm was employed here which iteratively searches for features that positively improves at least a reduction in identification results and keeps on including them in the search algorithm. The test results conducted on a number of training and test samples clearly show that with our gait data, a small subset of features selected from all the features is more effective in discriminating gait types due to ageing compared to all the features.

- Classification performance of the SVM depends on the value of the regularization parameter (i.e., C), which needs to be carefully selected through a trial-and-error procedure. A wrong selection of C value can drastically affect the classifier performance (see Figure 10).

- Other computational intelligence techniques (e.g., neural networks, fuzzy logic or genetic algorithm) can be tried to work in cascade with the SVM technique in a hybrid design mode to further improve the performance of the classification outcomes.

- Both histogram and Poincaré plot features appear to be very effective in discriminating the two age groups. This suggests that gait changes with age are reflected in these plots. Future research should include other population groups, such as

people with falls and balance problems to develop gait models for these populations and to examine how these plots would be affected due to previous falls
history.

- Other types of gait features such as kinetics (muscle moments, joint moments, etc.) and electromyography data (timing of muscle activity patterns, etc.) can be included in future classification models involving SVM, which could provide useful additional information for better predictive outcomes.

ACKNOWLEDGMENT

The authors would like to thank Simon Taylor and Brendan Owen for help with data analysis and figures.

REFERENCES

Barton, J. G., & Lees, A. (1997). An application of neural networks for distinguishing gait patterns on the basis of hip-knee joint angle diagrams. *Gait & Posture, 5*, 28-33.

Begg, R. K., Palaniswami, M., & Owen, B. (2005). Support vector machines for automated gait recognition. *IEEE Transactions on Biomedical Engineering, 52*, 828-838.

Best, R. J., Begg, R. K., & Dell'Oro, L. (1999, August 8-13). The probability of tripping during gait. In *Proceedings of the International Society of Biomechanics Conference,* Calgary, Canada.

Brennan, M. Palaniswami, M., & Kamen, P. W. (2001). Do existing measures of Poincare plot geometry reflect nonlinear features of heart rate variability? *IEEE Transactions on Biomedical Engineering, 48*, 1342-1347.

Burges, C. J. C. (1998). A tutorial on support vector machines for pattern recognition. *Knowledge Discovery Data Mining, 2*(2).

Chan, K., Lee, T. W., Sample, P. A., Goldbaum, M. H., Weinreb, R. N., & Sejnowski, T. J. (2002). Comparison of machine learning and traditional classifiers in glaucoma diagnosis. *IEEE Transaction on Biomedical Engineering, 49*, 963-974.

Chau, T. (2001). A review of analytical techniques for gait data. Part 2: Neural network and wavelet methods. *Gait & Posture, 13,* 102-120.

Cortes, C., & Vapnik, V. (1995) Support vector networks. *Machine Learning, 20*, 273-297.

Guyon, I., Boser, B., & Vapnik, V. (1993). Automatic capacity tuning of very large VC-dimension classifiers. In S. J. Hanson, J. D. Cowan, & C. L. Giles (Eds.), *Advances in neural information processing systems* (Vol. 5, pp. 147-155). San Mateo, CA: Morgan Kaufmann,.

Holzreiter S. H., & Kohle, M. E. (1993). Assessment of gait pattern using neural networks. *Journal of Biomechanics, 26*, 645-651.

Karst, G. M., Hageman, P. A., Jones, T. F., & Bunner, S. H. (1999). Reliability of foot trajectory measures within and between testing sessions. *Journal of Gerontology: Medical Sciences, 54*, 343-347.

Kecman, V. (2002). *Learning and soft computing: Support vector machines, neural networks and fuzzy logic models*. Cambridge, MA: MIT Press.

Lee, L., & Grimson, W. E. L. (2002). Gait analysis for recognition and classification. In *Proceedings of the Fifth International Conference on Automatic Face and Gesture Recognition.* IEEE Computer Society.

O'Malley, M. J., Abel, M. F., Damiano, D. L., & Vaughan, C. L. (1997). Fuzzy clustering of children with cerebral palsy based on temporal-distance gait parameters. *IEEE Transactions on Rehabilitation Engineering, 5,* 300-309.

Owings, T. M., Pavol, M. J., Foley, K. T., & Grabiner, M. D. (1999). Exercise: Is it a solution to falls by older adults? *Journal of Applied Biomechanics, 15,* 56-63.

Princea, F., Corriveaua, H., Héberta, R., & Winter, D. A. (1997). Gait in the elderly. *Gait & Posture, 5,* 128-135.

Schölkopf, B. Burges, C., & Vapnik, V. (1995). Extracting support data for a given task. In U. M. Fayyad & R. Uthurusamy (Eds.), In *Proceedings of the 1st International Conference on Knowledge Discovery & Data Mining.* Menlo Park, CA: AAAI.

Schöllhorn, W. I. (2004). Applications of artificial neural nets in clinical biomechanics. *Clinical Biomechanics, 19,* 876-898.

Su, F. C., & Wu, W. L. (2000). Design and testing of a genetic algorithm neural network in the assessment of gait patterns. *Medical Engineering and Physics, 22*(1), 67-74.

Su, F., Wu, W., Cheng, Y., & Chou, Y. (2001). Fuzzy clustering of gait patterns of patients after ankle arthrodesis based on kinematic parameters. *Medical Engineering and Physics, 23*(2), 83-90.

Taylor, S. (2003). *Characterising intra- and inter-limb gait stability using minimum foot clearance.* Unpublished Masters Thesis, Victoria University, Melbourne.

Vapnik, V. N. (1995). *The nature of statistical learning theory.* New York: Springer.

Winter, D. (1991). *The biomechanics and motor control of human gait: normal, elderly and pathological.* Waterloo, Canada: University of Waterloo Press.

Wu, W. L., Su, F. C., & Chou, C. K. (1998). Potential of the back propagation neural networks in the assessment of gait patterns in ankle arthrodesis. In E. C. Ifeachor, A. Sperduti, & A. Starita (Eds.), *Neural networks and expert systems in medicine and health care* (pp. 92-100). Singapore: World Scientific Publishing.

Yom-Tov, E., & Inbar, G. F. (2002). Feature selection for the classification of movements from single movement-related potentials. *IEEE Transactions on Neural System and Rehabilitation Engineering, 10,* 170-177.

Zavaljevski, N., Stevens, F. J., & Reifman, J. (2002). Support vector machines with selective kernel scaling for protein classification and identification of key amino acid positions. *Bioinformatics, 18,* 689-696.

Section III

Applications in
Rehabilitation and Sport

Chapter IX

Control of Man-Machine FES Systems

Rahman Davoodi, University of Southern California, USA

Gerald E. Loeb, University of Southern California, USA

ABSTRACT

Movement disabilities due to spinal cord injury (SCI) are usually incomplete, leaving the patients with partially functioning movement system. As a result, functional electrical stimulation (FES) systems for restoration of movement to the paralyzed limbs must operate in parallel with the residual voluntary movements of the patient. In the resulting man-machine system, the central nervous system (CNS) controls the residual voluntary movements while the FES system controls the paralyzed muscles of the same limbs. Clearly, these two control systems must work in synchrony to benefit the patient. In this chapter we will discuss different methods for cooperative control of man-machine FES systems and use a clinical FES system to demonstrate the successful application of these strategies.

BACKGROUND

Advances in rehabilitation technology are producing increasingly complex and sophisticated rehabilitation devices that must interact closely with the residual neural control of patients. The successful operation and optimal performance of the resulting man-machine system depends on a precise coordination and integration between the man and the machine. The same is true when the humans operate machinery such as driving a car, operating an anti-aircraft gunner, or flying an airplane. All of these tasks have had the benefit of over 100 years of development of user interfaces and training methods. To develop a systematic approach for analyzing such man-machine systems, control engineers model the human operator as an element of the automatic control system. Early models of the human operator were linear transfer functions that could be analyzed by linear control theories (Mcruer & Graham, 1967; Kleinman & Perkins, 1974). Later, the models of the human operator were expanded to multi-input multi-output nonlinear systems using optimal control theory, neural networks, and fuzzy logic control (Doman & Anderson, 2000; Zapata, Galvao, & Yoneyama, 1999). The focus of these expanded models was still the manual operation of machinery by well-trained human operators.

Man-machine rehabilitation systems are similar in that the human subject must voluntarily produce commands to operate the rehabilitation device but there are also stark differences. For example, the rehabilitation device is not an independent, well-behaved machine, but an integral part of the limb that is supposed to operate it. For example, the artificial hand is attached to an amputated arm that should produce the command signals to operate it as well as some of its movement through space. Similarly, in a FES reaching system, the intact shoulder joint may have to produce voluntary commands to operate the FES control of the paralyzed elbow joint (Popovic & Popovic, 2001). Instead or in addition, the FES controller may be driven by residual movement of unrelated body parts (e.g., tongue or contralateral arm), electromyography (EMG) of the intact muscles, or electrical activity of the neural cells in the sensorimotor cerebral cortex. Development of mathematical models for the human operator in these man-machine systems is a challenging task and has been studied only in simple, well-defined tasks. For example, Davoodi et al. used fuzzy logic (Davoodi & Andrews, 1998) and neural network (Davoodi et al., 1998; Davoodi, Kamnik, Andrews, & Bajd, 2001) models to simulate the CNS control of the arm forces in FES assisted standing up, which is a highly constrained and relatively simple movement. Until we have a better understanding of the principles underlying the operation of the CNS, the development of a formal analysis that applies to all man-machine FES systems is unlikely. In the absence of formal systematic analysis, creative methods have been used to design coordinated man-machine FES systems that are discussed next.

MAN-MACHINE COORDINATION IN FES

In order to improve performance and patient acceptance, FES systems must provide natural and intuitive interfaces that allow the patients to be in full control of the combined man-machine system. The extent of the man-machine interactions between the patient and FES depends on the level and type of the spinal cord injury. A complete paralysis in the lower thorax paralyses the legs but the arms and trunk muscles are still under

Figure 1. A typical FES system showing two strategies for man-machine coordination

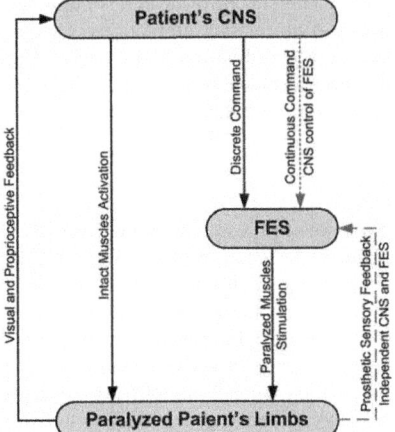

The patient's CNS and FES control intact and paralyzed muscles of the patient's limbs, respectively. Solid lines represent minimal CNS-FES interface that must be present in all FES systems. In CNS control of FES (dotted line) CNS generates continuous command to consciously control the stimulation of the paralyzed muscles. In independently operating CNS and FES (dashed line), FES operates independently but uses prosthetic sensory feedback to coordinate its control actions with CNS.

voluntary control of the patient. A FES system for standing and walking in these patients can be used to control the paralyzed leg muscles while a high degree of residual voluntary control in the upper extremity provide additional help by operating the FES, providing additional arm support to lift part of the body weight, and providing balance to prevent falls. As the level of spinal cord injury moves to higher levels, there will be less residual voluntary movements and more demand on the FES system. Most quadriplegic patients have varying level and degrees of paralysis of the arms because few patients survive complete high cervical transections that paralyze respiratory muscles. Usually, many muscles in the upper arm are at least partially intact while most of the muscles in the lower arm and hand are paralyzed. Even among these patients, however, the underlying neurological status is quite heterogeneous and secondary changes such as disuse atrophy and contractures further complicate the selection and fitting of prosthetic systems.

Interactions between the patient and FES can be divided into two general categories: discrete and continuous. The patient can use discrete commands to start and stop the FES or switch among different FES programs and functions. The continuous interactions are the moment-to-moment interactions between the patient and FES to cooperatively control the dynamics of the limb during movement. At least a few discrete interactions to select states or modes of operation are usually present in all FES systems and provide the patient with high level control, which is important for safety and user acceptance. Based on the nature of continuous interactions between the patient and FES, we can divide the man-machine coordination strategies in FES systems into two categories: CNS control of FES and independently operating CNS and FES. Figure 1 shows these strategies in a typical man-machine FES system.

CNS Control of FES

This strategy relies on the residual capacity of the CNS to control not only some voluntary movements but the FES system as well. The CNS generates continuous command signals to the FES system, which applies the CNS control signals or an open-loop function of those control signals to the paralyzed muscles. The FES, therefore, establishes a bridge that enables the CNS to regain control of the paralyzed muscles.

This coordination strategy is very attractive because it solves some of the most difficult design problems in FES. First, it greatly simplifies the design of the FES controller because the patient's CNS performs motor planning, coordination and learning. The decisions for moment-to-moment control of the muscles are made by the CNS and the FES simply executes them. Second, by unifying the center of command in the CNS, it automatically coordinates the voluntary and FES movements. Therefore, coordination is an important byproduct of this strategy and no specific action is required by the FES. Third, the direct control by the CNS gives the patient a feeling of integration and full control over the limb which is important for the acceptance and use of FES system. The users are usually uncomfortable with FES systems they don't understand or can't control effectively.

On the down side, this strategy relies on the ability of the CNS to control simultaneously multiple FES channels via unnatural interfaces. Humans have inherent limitations in their ability to control several variables consciously and simultaneously. This must be considered when designing man-machine interfaces (Chan & Childress, 1990; Ison, 1988). Performance depends not only on the number of controlled variables, but also on the nature of the interface for recording the control signals. For example, Mesplay and Childress (1988) found that joints that move limb segments with a low inertia such as the wrist have higher capacity for control of a prosthesis than joints such as the knee. Because of these limitations this coordination strategy is limited to FES systems with few degrees of freedom (e.g., activation of quadriceps muscles to control the knee joint angle) and those in which the relationship between muscle action and the resulting motion is simple and easy to understand by the patient (see manual control of FES rowing). In some of the more complex FES systems, it may be possible to bury the complexities of the musculoskeletal system in the FES and provide the patient with a small number of intuitive control "knobs" that are easier to learn and operate. For example the patient might operate a "knob" that represents the flexion/extension movement of a joint and the FES controller uses an inverse model of the limb to calculate the required muscle stimulation patterns for both synergists and antagonists.

From the early days, rehabilitation engineers have been fascinated with the creativity of patients in using their rehabilitation devices and have tried to use their residual voluntary motor control as a source of control. For example, upper limb prostheses have been controlled by distributed forelimb pressure created by voluntary muscle contraction (Curcie, Flint, & Craelius, 2001), voluntary EMG activity of the intact arm muscles (Englehart & Hudgins, 2003), and even by direct connections to the intact muscles via tendon exteriorization cineplasty (Weir, Heckathorne, & Childress, 2001; Doubler & Childress, 1984). Similarly, several FES applications have tried to tap into the residual motor capacity in the CNS as a source of command signal for FES. Most FES systems for restoration of hand grasp use open-loop stimulation patterns to open and close the paralyzed hand. Voluntary movements of the intact shoulder and wrist joints and EMG activity of the intact muscles in the forearm have been used to proportionally control the intensity of the open-loop stimulation patterns to control the degree of hand opening and the grasping force (Hart, Kilgore, & Peckham, 1998; Smith, Mulcahey, & Betz, 1996; Riso, Ignagni, & Keith, 1991; Thorsen, Spadone, & Ferrarin, 2001). The direct voluntary control provides the patient with an intuitive interface to consciously and continuously control the movement of the paralyzed hand via a simple open-loop FES system.

Independently Operating CNS and FES

In this strategy for man-machine coordination, FES and CNS operate independently, each making their own control decisions, but they coordinate their actions to achieve coordinated movement. The patient may provide discrete commands to start and stop FES but does not provide ongoing continuous command during the movement.

The coordination strategy here makes use of the fact that both CNS and FES have the common goal of controlling the same movement. For example, control of a partially paralyzed arm may involve the CNS control of the intact muscles in the upper arm and FES control of the paralyzed muscles in lower arm, but the common objective is to move the whole arm in a synergistic manner to perform useful reach and grasp movements. Similarly, in FES standing, the CNS controls the intact upper body to provide balance and partial support of body weight while the FES controls the paralyzed lower extremity to provide most of the weight support, but they have a common objective.

The CNS uses proprioception in the intact muscles and visual feedback to monitor the FES generated movements. This sensory information helps the CNS to react appropriately as the FES program unfolds and to learn to operate in synchrony with the FES toward achieving the common objective. The independently operating FES needs prosthetic feedback sensors to monitor the voluntary movements of the subject and take appropriate actions to coordinate actively with the CNS (see automatic control of FES rowing).

In this strategy, the patient does not have direct and conscious control of the muscle stimulation in the FES system. As the main consequence, the FES control will have more responsibility because it has to make moment to moment decisions about the control of the paralyzed muscles and it has to coordinate actively its control actions with the CNS. This will require a more sophisticated FES control system that may be more difficult to design and implement. Further, because the CNS does not continuously control the FES, the patient may feel less in control and less integrated with the semiautonomous behavior of the FES. The patient, therefore, has to rely and trust the FES to properly perform its functions. This may be a difficult psychological barrier for the patient to overcome, especially in applications such as standing and walking where the patient's safety depends on reliable operation of FES. On the other hand, an independent FES does not need continuous voluntary command signals from the patient. This relieves the patient from paying attention to generating these voluntary command signals. Also, because there is no need for continuous command signals from the patient, the FES control system can be arbitrarily complex without burdening the patient or being limited by the patient's abilities (see automatic control of FES rowing).

In most of the early applications, FES controllers were open-loop and did not actively coordinate their control actions with the CNS. For example, a common method for FES standing and walking (Marsolais & Kobetic, 1987; Graupe, 1994; Bajd et al., 1981) used open-loop FES to control the paralyzed leg muscles while the patient voluntarily controlled the upper body musculature. The open-loop FES provided a pre-programmed stimulation pattern without actively trying to coordinate with the CNS. The CNS decided when to trigger the open-loop FES stimulation patterns via thumb switches. It also controlled the upper body to carry part of the body weight and provide balance to prevent falls. Therefore, the overall coordination of the movement was provided solely by the CNS, which simply triggered a specific FES program. Closed-loop FES controllers introduced later to these FES applications used the sensory feedback of the movement to coordinate actively their actions with the CNS. For example, Mulder et al. (1992) used a switching curve FES controller that tried to make the knee joint follow normal standing up trajectories. The FES controller stimulated the knee joint muscles based on the error between the actual and desired knee joint trajectories. Riener et al (Riener & Fuhr, 1998) used a closed-loop FES controller for standing up that operated independently of CNS, but coordinated its control actions by monitoring the voluntary actions of the upper body. The FES controller used the measured posture and upper body forces of the patient to calculate and produce the necessary moments in the paralyzed leg joints to complement the voluntary actions of the upper body.

Miller et al. (1989) and later Grill and Pechham (1998) developed a closed-loop FES control of elbow joint for overhead reaching. The FES operated independently of the CNS (i.e., the CNS had no conscious control over FES operation). But FES coordinated its actions to CNS by measuring the voluntary movements of the intact shoulder joint and extending the elbow joint whenever the shoulder movement indicated that the subject was attempting to reach above the shoulder level. Popovic and Popovic (2001) developed a closed-loop FES reaching controller that detected the voluntary movement of the shoulder and tried to move the paralyzed elbow joint according to normal reaching synergies. Their argument was that the patients will find such biomimetic interfaces more intuitive and easier to learn because they automatically mimic the movements they have lost to injury. A similar strategy was suggested by Davoodi and Loeb (2003) to restore the reaching movement to partially paralyzed arms of quadriplegic patients. The FES control of the paralyzed lower arm would detect the intended movement of the patient from their voluntary shoulder joint movements and would move the lower arm according to normal reaching synergies intended by the patient. In their control strategy, the FES controller has a biomimetic design that tries to fill the gap created by the spinal cord injury by similar CNS control circuits. The role of the FES controller here is similar to the role of the spinal cord, which mediates between the voluntary control centers in the brain and the complex musculoskeletal system. These spinal control circuits are believed to have two important functions: local regulation of joint stiffness to reject computational noise and external disturbances and reduction of the number of degrees of freedom in the musculoskeletal system so that the voluntary control centers in the brain see a system with fewer degrees of freedom that can be operated with less control effort (Loeb, 2001; Loeb, He, & Levine, 1989). Similarly, FES systems can be designed to handle some of the low-level control functions to relieve the CNS to deal with higher level control functions. Further, these FES systems can reduce the number of independent control variables in the paralyzed limb to a smaller set of more intuitive variables that are easier to control by the CNS.

CASE STUDY:
FES ASSISTED INDOOR ROWING EXERCISE

People with paralysis need to exercise to stay fit and prevent cardiovascular diseases, which is the major cause of death in this population (Washburn & Figoni, 1999). Because of their sedentary lifestyle, paralyzed patients need physical activity and exercise even more than the normal population but they cannot use most standard exercise equipment. As a result, exercise by these patients is limited to upper body activities that they can still perform such as wheelchair propulsion and arm cranking, which have limited benefits for cardiovascular fitness (Raymond, Davis, Climstein, & Sutton, 1999). To provide paralyzed patients with a total body exercise machine that combines upper and lower body exercise, a standard indoor rowing machine (Concept II Inc., USA) was modified for use by the paralyzed patients (Davoodi, Andrews, Wheeler, & Lederer, 2002b). The main modification was the addition of an electrical stimulator to activate the paralyzed muscles of the knee joint. Electrical stimulation of the quadriceps and hamstring muscles can extend and flex the legs, respectively. Appropriate combination of the leg extension and flexion with the voluntary pull of the handle by the intact upper body enables the paralyzed patient to perform rowing exercise. The details of the electrical stimulator and modifications to provide safety, seating stability, patient interface, and feedback position sensors can be found elsewhere (Davoodi et al., 2002b).

The Coordination Problem in FES Rowing

Rowing exercise is a cyclic movement in which a rowing cycle is repeated as long as the subject wishes to exercise. Each rowing cycle is composed of several phases that must be performed in an orderly manner (Figure 2). In normal rowing, the orderly execution of these phases and the smooth transition from one cycle to another is coordinated by the CNS. In FES rowing, the upper body can still be controlled by the CNS but not the paralyzed lower extremities, whose active movement must be restored by a FES controller. The coordination problem, therefore, is to synchronize the actions of the CNS and FES to perform coordinated rowing exercise. Two approaches for achieving such man-machine coordination are shown in Figure 3. In the first method, the coordination is

Figure 2. A complete indoor rowing cycle that is performed repeatedly to exercise. In normal rowing, the CNS controls the muscles in the upper and lower extremities

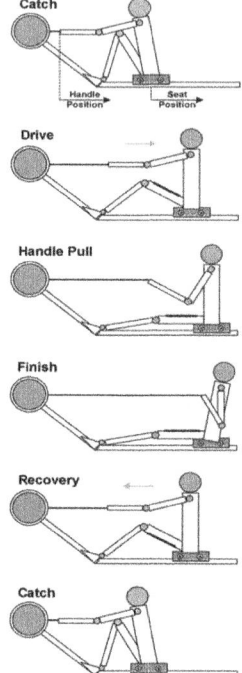

In rowing by paralyzed patients, CNS controls upper body muscles while FES controls the paralyzed muscles in the legs. The thick lines on the thigh show the active muscles in each phase of the rowing cycle. The patient uses two control buttons in the handlebar to provide command signals to the FES. Positions of the seat and handle are measured by two optical encoders that provide the FES controller with prosthetic sensory feedback.

achieved by unifying the center of command in the CNS so that the patient's CNS is in full control of the combined man-machine system. In the second method, the CNS and FES operate independently but coordinate their actions to achieve the common objective.

Coordination by CNS Control of FES

In this method for coordination of FES rowing, the operation of the whole system is under the complete control of the CNS. As a result, in addition to voluntarily controlling the upper body part of the rowing maneuver, the user directly controls the timing and level of electrical stimulation to the paralyzed leg muscles. To enable the patient to control the electrical stimulation of the paralyzed leg muscles, two control buttons were attached to the right and left side of the handlebar (Davoodi et al., 2002b). By pressing on these buttons the patient could activate the knee extensor muscles that would extend the legs and move the seat backward, or the knee flexor muscles that would flex the legs and move the seat forward. In addition to controlling the upper body to pull the handle, the CNS must learn the new skills to control the timing of the button presses in the handle. Clearly, the timing and duration of the button presses must be coordinated with the voluntary upper body movements to be effective.

Figure 3. Two control strategies for man-machine coordination in FES rowing

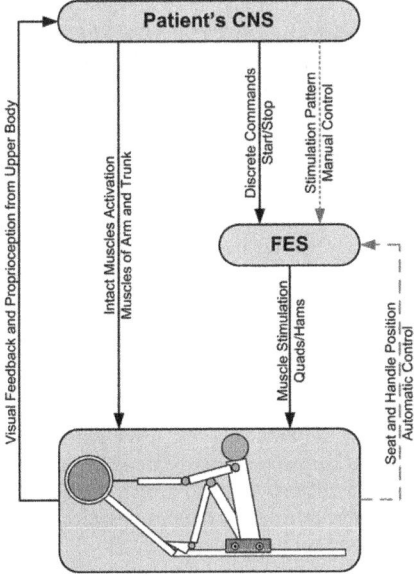

In manual control (dotted line), the timing and pattern of stimulation to the paralyzed knee muscles is directly controlled by a continuous command from the CNS. FES control here is open-loop and simply executes the commands from the CNS. In automatic control (dashed line), the continuous control of CNS over paralyzed muscles is removed. Instead, prosthetic sensory feedback of the seat and handle positions are used by a closed-loop FES controller to determine the timing and pattern of the stimulation to the knee muscles.

The rowing exercise starts in the Catch position where the subject presses on the right button to apply a constant level stimulation to the knee extensors on both legs. The knees extend and the seat moves back along the track. During the latter part of this phase, when the knees are completely extended, the subject pulls on the handle. While pulling the handle, FES is maintained on the knee extensors to keep the legs fully extended. When the handle is fully withdrawn the subject releases the right button, presses the left button and slackens off the handlebar and the damper cord retracts. These actions cause: constant level stimulation to be applied to the knee flexors, stimulation is removed from knee extensors, and the seat can return to the starting position. The cycle repeats when the subject returns to the starting Catch position.

Coordination by Independently Operating CNS and FES

In this strategy, CNS and FES operate independently to control the upper and lower extremity parts of the rowing maneuver, respectively. But the two controllers synchronize their actions by each monitoring and reacting to the control actions of the other. The CNS uses visual feedback of the leg movements to monitor FES control actions. In addition, the intact proprioception sensors in the upper body can sense the upper body reaction forces that are generated by the FES controlled leg movements. These reaction forces are due to dynamic coupling among the joints in the human body that enable muscles about a single joint to affect the movement of many other joints. The combined visual and proprioceptive feedback helps the CNS to monitor, react, and adapt to FES control system. To provide the FES control system with similar capabilities to monitor and react to the voluntary CNS control actions, two sensors were used to measure the positions of the seat and the handlebar. The closed-loop FES controller itself had two levels of hierarchy (Davoodi, Andrews, & Wheeler, 2002a). In the higher level a finite state controller applied a set of state transition rules to the input sensory information to identify the current state (phase) in the rowing cycle and dispatched the appropriate low-level controller for that state. The state-dedicated, low-level controllers in turn controlled the level of electrical stimulation to the paralyzed muscles as long as the system remained in that state.

The rowing exercise starts in the Catch position. Pressing the start button propels the system into the automatic mode. First, the low-level controller applies maximal constant-level stimulation to the knee extensors to extend the knee and move the seat backward. In full knee extension, the low-level controller continues to apply constant level stimulation to the knee extensors to keep the knees extended while the subject pulls on the handle. When the high-level controller detects that the knees are fully extended and the handle is fully withdrawn, it instructs the low-level controller to apply maximum constant-level stimulation to the knee flexors to flex the knees. As a result, the seat moves forward while at the same time the subject voluntarily extends the arms. Simultaneous occurrence of the full knee flexion and full arm extension signals the completion of one rowing cycle. At this point, the system immediately enters the next rowing cycle, which can be repeated until the subject decides to stop exercising.

Figure 4. Trajectories and electrical stimulation patterns for two typical cycles of normal rowing (top), and FES rowing with manual (middle) and automatic (bottom) FES controllers by a paraplegic volunteer

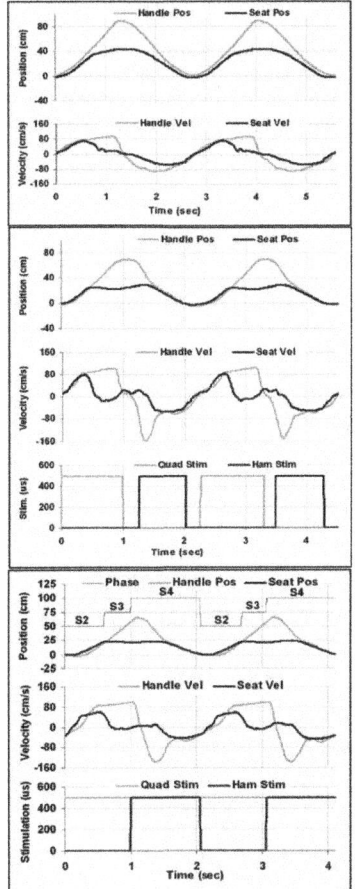

The trajectories for FES rowing are almost the same, but slightly different from normal mainly due to the difference in the size of the subjects and safety constraints in FES rowing. In manual control, the patient learns to control the timing of the muscle stimulation. The automatic controller detects different phases of the rowing cycle (S2, S3, and S4) and dispatches state-dedicated control strategies to deliver the muscle stimulation.

Comparison of Control Strategies for FES Rowing

The strategies for man-machine coordination in FES rowing were evaluated by two paraplegic volunteers (Davoodi et al., 2002a; Davoodi et al., 2002b). The results for one of the paraplegic volunteers (51 years old, 70.7 kg, T4 complete, and 31 years after injury) are reviewed here.

Both strategies were easy to learn and the user quickly learned to perform rowing exercise in a few minutes. The arbitrary rule for the manual control strategy (i.e., pressing right/left switches extends/flexes the knee joint) was easy to learn and to employ. The automatic control strategy was even easier to operate because the user did not have to

Figure 5. Seat-versus-handle position graphs for two minutes of continuous rowing for normal rowing (top), and FES rowing with manual (middle) and automatic (bottom) FES controllers by a paraplegic volunteer

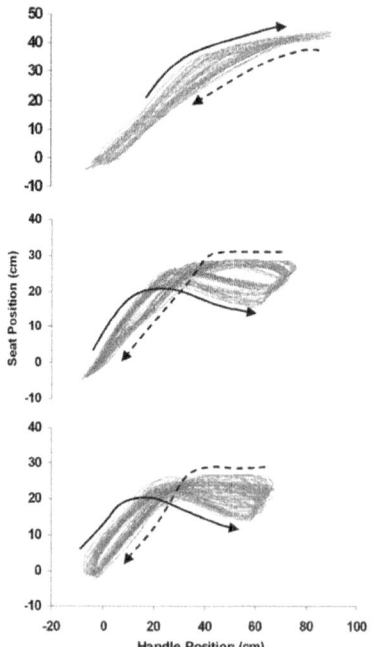

Solid and dashed arrows represent the Drive and Recovery phases of the rowing cycle

pay attention to continuously pressing the control switches. The results of normal and FES rowing by the two man-machine coordination strategies are shown in Figures 4-5. Figure 4 shows the trajectories of the seat and handle and the electrical stimulation of the knee joint muscles. The trajectories are slightly different from normal rowing but the cycles are almost identical and the subject could repeat them successfully to perform FES rowing exercise. Figure 5 shows seat-versus-handle position graphs over two-minute continuous rowing. These graphs are indicators of the coordination between the CNS control of the upper body and FES control of the lower extremity which was achieved consistently during the exercise.

When the CNS is in control of the combined man-machine system, the user is in full control of the movement. As a result, the user not only controls the initiation and termination of the rowing exercise but the moment-to-moment control of the electrical stimulation of the paralyzed muscles. Because the control decisions are made by the CNS, the FES control is simple and therefore easy to design and implement. The FES system simply executes the commands by the CNS and does not have to make the decisions to control the complicated musculoskeletal system. An important consequence of this control strategy is that it automatically achieves man-machine coordination by unifying the control in the CNS. The success of this control strategy depends, however, on the capacity of the CNS to learn new and sometimes unnatural skills and the patient's ability and willingness to pay attention to executing those skills consistently, at least initially. In FES rowing, the user must learn the timing of pressing two buttons that had easy-to-understand consequences of moving the seat forward or backward. This was easy to learn by the users, but learning will be difficult when the number of controls increase (to independently control more muscles and joints in the legs) or when the relation between the control actions and the controlled variables are more complicated (such as the relation between the muscle stimulation and the resulting movement). To reduce the mental workload of the patient and to improve consistency, the patient's skills in operating the FES may be transferred to an automatic controller. Such automatic controllers could be handcrafted (Davoodi et al., 2002a) or trained by machine learning techniques (Kirkwood & Andrews, 1989; Kostov, Andrews, Popovic, Stein, & Armstrong, 1995) to operate the FES controller in place of the patient.

The FES control strategy that operates independently of the CNS is more difficult to design and implement because it needs sensors to monitor the voluntary movements of the patients and it has to make decisions for moment-to-moment control of the paralyzed muscles in coordination with the voluntary movements. The successful operation of the two independent CNS and FES controllers depends on each controller performing its task and relying on the other controller to do its part. This may be difficult for the user, especially when his/her safety depend on the proper operation of the man-machine system. This mutual trust and cooperation will be improved if the user feels he/she has adequate control of the whole system. In FES rowing, the user can start and stop exercise at any time, providing a degree of overall control. Further, because the FES moves are conditional on completion of the voluntary parts of the rowing maneuver, the pace of the rowing exercise can also be controlled by the subject. For example, at the end of each rowing cycle, the rower must voluntarily extend his/her arm to bring the handlebar forward before the FES controller starts a new cycle by stimulating the knee extensors. With this level of user control combined with the fact that the failure of the FES system does not endanger the user safety, it was easy for the subject to accept and use the FES rowing system. But creating a high level of trust and therefore coordinated control between the user and the FES system for standing and walking will certainly be more difficult. The users of FES assisted standing usually overload their upper body, at least initially. Over time, they might learn to trust the FES and put more weight on the legs, but this will depend on the reliable and robust operation of the FES.

The independently operating FES controller can employ sophisticated control strategies while keeping the CNS-FES interface simple and without worrying about command generation capabilities of the patients. For example, the FES controller for rowing employed simple on/off control (Davoodi et al., 2002a), switching curve control (Davoodi & Andrews, 2003), and fuzzy logic control (Davoodi & Andrews, 2004) to improve the performance of FES rowing while keeping the CNS-FES interactions exactly the same. In principle, this permits more sophisticated multijoint FES control systems to be implemented without worrying about the limitations of the subject in operating it.

FUTURE TRENDS

Movement rehabilitation systems will increase in complexity to provide more functions and better performance. Increasingly, the limitations of intelligent control systems for man-machine coordination will limit the clinical applications of these systems. Researchers in FES have already acknowledged and are motivated to address these shortcomings. One major obstacle will be the range of different control schemes needed to address the heterogeneous patient population and the clinical expertise required to select and fit a controller for each patient.

There is a great advantage in using the residual control capacity of the patient's CNS and it will continue to play an important role in coordination and integration of FES and other rehabilitation devices. In the past, CNS commands have been relayed to FES indirectly via residual voluntary limb movements, pressing control buttons and switches, and EMG activity of the intact muscles. Future FES systems may be able to employ a more direct interface to the CNS via chronically implanted electrodes in the motor cortex of the patient (Schwartz, 2004; Donoghue, 2002). These command signals are expected to be richer in information and may enable more natural and easier to operate FES systems but will also require sophisticated algorithms to decipher these command signals (Schwartz, Taylor, & Tillery, 2001). Furthermore, the FES systems operated by these CNS command signals will be more likely to benefit from FES controllers that mimic more closely the biological systems they are trying to replace (Davoodi et al., 2003; Loeb, Brown, & Cheng, 1999). Such biomimetic FES controllers should be more natural and intuitive for the patients to learn and operate.

More intelligent FES control strategies will reduce reliance on the residual control capacity in CNS, much as the highly evolved circuitry of the spinal cord presumably simplifies the control of limbs by the brain. For this to happen, however, the FES controllers will probably need to employ learning and adaptation techniques to coordinate their actions automatically with the CNS. Because we cannot model the behavior of the CNS, such adaptive FES controllers must be able to learn from their trial and error interactions with the patients. Machine-learning techniques such as reinforcement learning and evolutionary computation are particularly appropriate for such learning tasks (Davoodi et al., 1998; Davoodi & Andrews, 1999).

One particularly challenging problem that must be addressed in the future is the dynamics of learning in the CNS and FES controllers as they interact to control the same limb. In current FES systems the learning occurs only in the CNS. The FES engineer can observe the results of CNS learning and can modify the parameters of the FES system for better performance. But there are no tools to predict the CNS learning and, therefore, to incorporate it in a systematic approach for the design of FES control systems. The FES system on the other hand usually has fixed parameters and has no automatic learning capability. A fixed-parameter FES provides the CNS with a predictable system that might be easier to operate but the potentials of such FES systems are limited. An ideal scenario is a man-machine system in which CNS and FES both learn the optimal control strategies by trial and error interactions with each other and the paralyzed limb.

CONCLUSION

Effective man-machine coordination is essential for successful operation of rehabilitation systems, especially given the sophisticated interactions required to reanimate a paralyzed limb by FES. In this chapter we introduced strategies for man-machine coordination that focused on continuous interactions between the patient and FES system to control the dynamics of the limb movement cooperatively. FES-assisted rowing exercise provides a useful clinical case study to demonstrate and compare different man-machine coordination strategies. The coordination strategies that rely on the residual control capacity in the CNS to operate FES and coordinate the overall movement are attractive because they simplify the design and implementation of the FES system. But they require the patient's active and attentive participation in control and are therefore limited to simpler movements with few degrees of freedom. Reduction of the burden on the patient and extension to more complex movements require the use of intelligent closed-loop FES controllers that operate independently but automatically coordinate their actions with the CNS.

REFERENCES

Bajd, T., Kralj, A., Sega, J., Turk, R., Benko, H., & Strojnik, P. (1981). Use of a two-channel functional electrical stimulator to stand paraplegic patients. *Physical Therapy, 61,* 526-527.

Chan, R. B., & Childress, D. S. (1990). On information-transmission in human-machine systems - channel capacity and optimal filtering. *IEEE Transactions on Systems Man and Cybernetics, 20,* 1136-1145.

Curcie, D. J., Flint, J. A., & Craelius, W. (2001). Biomimetic finger control by filtering of distributed forelimb pressures. *IEEE Transactions on Neural Systems and Rehabilitation Engineering, 9,* 69-75.

Davoodi, R., & Andrews, B. J. (1998). Computer simulation of FES standing up in paraplegia: A self-adaptive fuzzy controller with reinforcement learning. *IEEE Transactions on Rehabilitation Engineering, 6,* 151-161.

Davoodi, R., & Andrews, B. J. (1999). Optimal control of FES-assisted standing up in paraplegia using genetic algorithms. *Medical Engineering & Physics, 21,* 609-617.

Davoodi, R., & Andrews, B. J. (2003). Switching curve control of functional electrical stimulation assisted rowing exercise in paraplegia. *Medical & Biological Engineering & Computing, 41,* 183-189.

Davoodi, R., & Andrews, B. J. (2004). Fuzzy logic control of FES rowing exercise in paraplegia. *IEEE Transactions on Biomedical Engineering, 51,* 541-543.

Davoodi, R., Andrews, B. J., & Wheeler, G. D. (2002a). Automatic finite state control of FES-assisted indoor rowing exercise after spinal cord injury. *Neuromodulation, 5,* 248-255.

Davoodi, R., Andrews, B. J., Wheeler, G. D., & Lederer, R. (2002b). Development of an indoor rowing machine with manual FES controller for total body exercise in paraplegia. *IEEE Transactions on Neural Systems and Rehabilitation Engineering, 10,* 197-203.

Davoodi, R., Kamnik, R., Andrews, B. J., & Bajd, T. (2001). Predicting the voluntary arm forces in FES-assisted standing up using neural networks. *Biological Cybernetics, 85,* 133-143.

Davoodi, R., & Loeb, G. E. (2003). A biomimetic strategy for control of FES reaching. In *Proceedings of the 25th Annual International Conference of the Engineering in Medicine and Biology Society.*

Doman, D. B., & Anderson, M. R. (2000). A fixed-order optimal control model of human operator response. *Automatica, 36,* 409-418.

Donoghue, J. P. (2002). Connecting cortex to machines: Recent advances in brain interfaces. *Nature Neuroscience, 5,* 1085-1088.

Doubler, J. A., & Childress, D. S. (1984). Design and evaluation of a prosthesis control system based on the concept of extended physiological proprioception. *Journal of Rehabilitation Research & Development, 21,* 19-31.

Englehart, K., & Hudgins, B. (2003). A robust, real-time control scheme for multifunction myoelectric control. *IEEE Transactions on Biomedical Engineering, 50,* 848-854.

Graupe, D. (1994). *Functional electrical stimulation for ambulation by paraplegics.* Malabar, FL: Krieger Publishing.

Grill, J. H., & Peckham, P. H. (1998). Functional neuromuscular stimulation for combined control of elbow extension and hand grasp in C5 and C6 quadriplegics. *IEEE Transaction on Rehabilitation Engineering, 6,* 190-199.

Hart, R. L., Kilgore, K. L., & Peckham, P. H. (1998). A comparison between control methods for implanted FES hand-grasp systems. *IEEE Transaction on Rehabilitation Engineering, 6,* 208-218.

Ison, K. T. (1988). Assessment of control ability in the physically-handicapped. *Journal of Biomedical Engineering, 10,* 430-437.

Kirkwood, C. A., & Andrews, B. J. (1989). Finite State Control of FES Systems: application of AI inductive learning techniques. In *Proceedings of 11th Annual International Conference of IEEE Engineering in Medicine and Biology Society* (pp. 1020-1021).

Kleinman, D. L., & Perkins, T. R. (1974). Modeling human performance in a time-varying anti-aircraft tracking loop. *IEEE Transaction on Automatic Control, AC-19,* 297-306.

Kostov, A., Andrews, B. J., Popovic, D. B., Stein, R. B., & Armstrong, W. W. (1995). Machine learning in control of functional electrical stimulation systems for loco-motion. *IEEE Transaction on Biomedical Engineering, 42,* 541-551.

Loeb, G. E. (2001). Learning from the spinal cord. *Journal of Physiology, 533,* 111-117.

Loeb, G. E., Brown, I. E., & Cheng, E. J. (1999). A hierarchical foundation for models of sensorimotor control. *Experimental Brain Research, 126,* 1-18.

Loeb, G. E., He, J., & Levine, W. S. (1989). Spinal cord circuits: Are they mirrors of musculoskeletal mechanics? *Journal of Motor Behavior, 21,* 473-491.

Marsolais, E. B., & Kobetic, R. (1987). Functional electrical stimulation for walking in paraplegia. *Journal of Bone Joint and Surgery [Am], 69,* 728-733.

Mcruer, D. T., & Graham, D. (1967). Manual control of single-loop systems: Part I. *Journal of the Franklin Institute, 283,* 1-29.

Mesplay, K. P., & Childress, D. S. (1988). Capacity of the human operator to move joints as control inputs to prostheses. In *Proceedings of the ASME Conference on Modeling and Control Issues in Biomechanical Systems* (pp. 17-25).

Miller, L. J., Peckham, P. H., & Keith, M. W. (1989). Elbow extension in the C5 quadriplegic using functional neuromuscular stimulation. *IEEE Transaction on Biomedical Engineering, 36,* 771-780.

Mulder, A. J., Veltink, P. H., & Boom, H. B. (1992). On/off control in FES-induced standing up. A model study and experiments. *Medical & Biological Engineering & Computing, 30,* 205-212.

Popovic, M., & Popovic, D. (2001). Cloning biological synergies improves control of elbow neuroprosthesis. *IEEE Engineering in Medicine and Biology Magazine, 20,* 74-81.

Raymond, J., Davis, G. M., Climstein, M., & Sutton, J. R. (1999). Cardiorespiratory responses to arm cranking and electrical stimulation leg cycling in people with paraplegia. *Medicine & Science in Sports & Exercise, 31,* 822-828.

Riener, R., & Fuhr, T. (1998). Patient-Driven control of FES-supported standing up: A simulation study. *IEEE Transaction in Rehabilitation Engineering, 6,* 113-124.

Riso, R. R., Ignagni, A. R., & Keith, M. W. (1991). Cognitive feedback for use with FES upper extremity neuroprostheses. *IEEE Transaction on Biomedical Engineering, 38,* 29-38.

Schwartz, A. B. (2004). Cortical neural prosthetics. *Annual Review of Neuroscience, 27,* 487-507.

Schwartz, A. B., Taylor, D. M., & Tillery, S. I. H. (2001). Extraction algorithms for cortical control of arm prosthetics. *Current Opinion in Neurobiology, 11,* 701-707.

Smith, B. T., Mulcahey, M. J., & Betz, R. R. (1996). Development of an upper extremity FES system for individuals with C4 tetraplegia. *IEEE Transaction on Rehabilitation Engineering, 4,* 264-270.

Thorsen, R., Spadone, R., & Ferrarin, M. (2001). A pilot study of myoelectrically controlled FES of upper extremity. *IEEE Transactions on Neural Systems and Rehabilitation Engineering, 9,* 161-168.

Washburn, R. A., & Figoni, S. F. (1999). High density lipoprotein cholesterol in individuals with spinal cord injury: the potential role of physical activity. *Spinal Cord, 37,* 685-695.

Weir, R. F., Heckathorne, C. W., & Childress, D. S. (2001). Cineplasty as a control input for externally powered prosthetic components. *Journal of Rehabilitation Research and Development, 38,* 357-363.

Zapata, G. O. A., Galvao, R. K. H., & Yoneyama, T. (1999). Extracting fuzzy control rules from experimental human operator data. *Ieee Transactions on Systems Man and Cybernetics Part B-Cybernetics, 29,* 398-406.

Chapter X

Evolutionary Methods for Analysis of Human Movement

Rahman Davoodi, University of Southern California, USA

Gerald E. Loeb, University of Southern California, USA

ABSTRACT

A movement rehabilitation therapist must first diagnose the cause of disability and then prescribe therapies that specifically target the dysfunctional unit of the movement system. Objective diagnosis and prescription are difficult, however, because human movement is the result of complicated interactions among complex and highly nonlinear elements. Treatment based on limited observations may target the wrong element of the movement system. Researchers in central nervous system (CNS) control of human movement and functional electrical stimulation (FES) restoration of movement to paralyzed limbs face similar challenges in objective analysis of the integrated movement system. In this chapter, we will present evolutionary methods as powerful new tools for analysis and rehabilitation of human movement. These methods have been modeled after the same biological processes that have been optimized for the control of human movement in the process of biological evolution. Therefore, it is logical to think that these methods, if applied properly, could help us understand the control of human movement and repair it when it is disabled. A case study demonstrates the potential of evolutionary methods in movement analysis and rehabilitation.

BACKGROUND

As long as the movement system in humans functions properly, its operation seems simple and effortless. After diseases or injuries impair human movement, however, we realize how difficult it is to rehabilitate or restore function. We often find it difficult even to understand the exact cause of the impairment. The difficulty arises from the fact that human movement is produced by complex interactions among complex elements such as muscles, skeletal linkage, sensors, and the many CNS circuits. Our limited understanding of these movement components and the interactions among them makes it difficult to properly diagnose and treat movement disabilities. Often clinicians have to make therapy decisions based on limited observations and examinations. Such decisions are usually subjective and may target the wrong parts of the movement system. Improvements in diagnosis and treatment therefore require more sophisticated tools for measurement and analysis. The need will only increase with the growing sophistication and complexity of the rehabilitative interventions. Consider, for example, the use of robots in movement rehabilitation (Jezernik, Colombo, Keller, Frueh, & Morari, 2003). Because the robot cannot exercise clinical judgment during the course of the interaction with the patient, the prescribing clinician must specify the details of the treatment and contingencies for unexpected outcomes. In FES control of movement, paralyzed muscles are contracted by applying electrical currents. Over the years, sophisticated hardware has been developed to interface safely with the biological tissue and stimulate individual muscles (Loeb, Peck, Moore, & Hood, 2001; Smith et al., 1998), but still missing is an equally sophisticated control system to coordinate the activation of the muscles. Development of such control systems is hampered by our limited understanding of the musculoskeletal systems and the way they interact with and adapt to each other.

Various methods have been used in the past to analyze and repair human motor behavior. Neurophysiologists have used electrophysiological recordings to study the role of elements such as the muscles, sensors, and the CNS circuits. But these measurements alone cannot elucidate the interactions between these elements and the mechanical system that actually creates the movements. Rehabilitation of movement deals with the movement in its entirety and, therefore, requires the analysis of these interactions. Clinical motion analysis focuses on the movements themselves and their underlying kinetics. Movement patterns of the patient can be compared with the norm to identify abnormalities and monitor the effect of rehabilitative interventions over time. More sophisticated inferential analyses (such as the estimation of individual muscle forces) may be used to identify the cause of a movement deficit. Motion analysis, however, is limited to analyzing existing movement in healthy and moderately disabled subjects. It provides little guidance when movement is entirely lost or severely limited. For example, an FES engineer is interested in novel movement patterns that are customized to the weaker muscles of the paralyzed patient and the proper strategy for the control of muscles to realize them. A therapist might be interested in novel movement trajectories that are customized to the capabilities of a disabled patient. Such customized trajectories then may be used to provide feedback to patients in physical therapy sessions (Colborne, Olney, & Griffin, 1993; Wu, 1997). But the answers to these questions are not unique because the musculoskeletal system is highly redundant and there are many ways to accomplish a movement task. The task of the therapist, therefore, is to find the best solution among many possibilities — a task that can be formulated as an optimization problem.

The controller or the trajectory of the movement is first described as a function of a set of tunable parameters. Then an optimization algorithm is used to find the optimal values of the tunable parameters to minimize an objective function related to the quality of movement such as trajectory errors or the energy expenditure. But these movement optimization problems are difficult to solve with traditional optimization algorithms (van Soest & Casius, 2003). The objective functions typically have many local optima that could easily trap the optimization algorithms that start with a random point in the search space and move down the first hill they come across. The typically discontinuous nature of these objective functions is detrimental to optimization algorithms that depend on the calculation of the derivatives to determine the search direction. A further complicating factor is the unavailability of the objective function and the constraints as explicit functions of the optimization variables. For example, if one wishes to optimize a controller to successfully perform a movement, it is not clear how the success/failure of a movement can be described as explicit function of the control parameters. Because of the unavailability of explicit objective functions, they must be evaluated by time-consuming simulations. Finally, the search space is very large even for simple problems. As the result of these complexities, the traditional optimization methods are not applicable, inefficient, or at best limited to very simple problems. It is clear that the optimization problems that arise in movement rehabilitation require search algorithms that are efficient in searching large spaces and can robustly handle objective functions that may be multimodal, discontinuous, and not available in explicit form. One of the main approaches for such difficult optimization problems is evolutionary optimization that may not be as efficient as the conventional methods, but can solve problems that could not be solved before. Availability of such robust techniques is especially important for movement science because most movement optimization problems are hard to formulate and solve using conventional optimization methods.

EVOLUTIONARY COMPUTATION

Evolutionary computation is a set of optimization and learning algorithms that are loosely based on the mechanics of natural evolution, which include the very processes that directed the development of neuromusculoskeletal systems in the first place. These computations can simulate the very slow process of natural evolution within a computer program that can produce increasingly fitter solutions to the learning and optimization problems in a short period of time. Researchers in various fields have been fascinated by these methods mainly because they can be applied to a variety of problems that are otherwise difficult to solve using conventional methods. The first evolutionary methods, known as genetic algorithms, were introduced by Holland (1962). Genetic programming (Koza, 1994) is an important new variation. The main difference between the two algorithms is the parameters they tune. In genetic algorithm, an individual solution is a bit string of 0s and 1s that code the tuning variables of the optimization problem. For example, if each tuning variable is coded by an 8-bit binary number, an optimization problem with 5 tuning parameters will be represented by a bit string whose length is 40.

In genetic programming, the individual solution is a computer program rather than a bit string. So, each individual represents one way of putting smaller primitive program elements together to form a more complex program. Genetic programming, therefore, finds the optimal configuration for assembling the program from smaller program elements rather than just optimizing the values of the parameters for a solution whose basic form is fixed. The ability to optimize the architecture in addition to the parameters is a powerful concept that has been applied to problems such as assembling optimal electric circuits from a number of simple circuit elements (Koza, Keane, & Streeter, 2004). It may have applications in designing neural circuits in the CNS from a collection of simple elements such as sensory neurons, interneurons, motor neurons, and synapses among them.

All variations of the evolutionary methods follow similar procedures to produce increasingly fitter solutions to an optimization or learning problem. First, a generation containing a large population of randomly selected individuals is built. Each individual encodes all the tuning parameters and, therefore, represents a complete solution to the optimization or learning problem. The individual solutions in the first generation are then tried on the problem and a measure of their fitness in solving the problem is calculated. The next step is to produce a new generation of individuals by applying genetic operators such as reproduction, crossover, and mutation. Reproduction produces a new generation from the individuals in the old generation where individuals with higher fitness have

Figure 1. Structure of the genetic algorithm optimization method

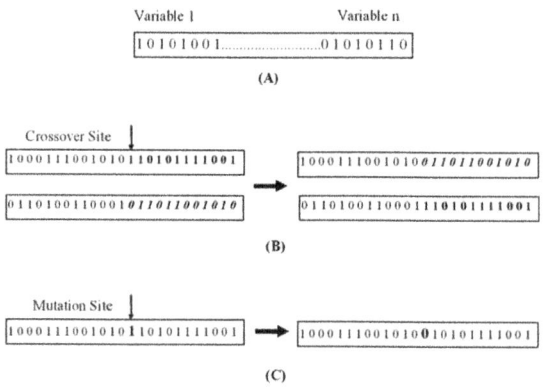

In (A) a chromosome is shown containing the optimization variables coded by binary strings. The crossover and mutation operations are demonstrated in (B) and (C). The chromosomes before and after operation are shown in the left and right, respectively.

better chance of selection. Crossover results in randomized, yet structured, information exchange between individuals to produce new patterns while mutation maintains diversity in the population. The process of reproduction is repeated until no further improvement in the fitness can be made (convergence) or an optimization threshold is met. A sample coding of the tuning parameters and the genetic operations for mutation and crossover are shown in Figure 1. Many variations to these operations have been developed to improve the efficiency of the evolutionary computation or customize it to a specific problem. For a review of different evolutionary methods and techniques see Back, Hammel, and Schwefel (1997), Goldberg (1989), and Koza (1992).

Evolutionary computations must be programmed in a computer code and executed to simulate the evolutionary process. In the past, the users had to write their own programs to implement the evolutionary computations (Davoodi & Andrews, 1999). The basic algorithms are straightforward and can be easily programmed in any programming language. These user implementations were very useful in helping researchers gain a deeper understand of the evolutionary methods and devise new variations to improve their efficiency. Now, thanks to the availability of the new general purpose software (e.g., Genetic Algorithm and Direct Search Toolbox, Mathworks Inc., USA), most users do not have to write their own code. But they still must formulate their problem by proper selection and encoding of the tuning parameters, the formulation of the fitness function and the selection of other parameters such as the mutation and crossover rates. The programs of the evolutionary computation then must be run or simulated in a computer. The simulations contain two types of computation: application of evolutionary operations to produce new generations and evaluation of individuals on the optimization problem. The latter is usually the most time-consuming part of the computation, particularly for complex systems such as those responsible for sensorimotor behaviors. For example, evaluation of each individual may require the simulation of a sit-to-stand movement. State variables from the resulting movement can be used to calculate the fitness of the individual according to predetermined performance criteria. Because each generation contains a large number of individuals, the evaluation of fitness in each generation and in a number of generations could easily take hours or even days on a single computer. Unlike conventional optimization methods, however, the computations of evolutionary methods can be performed in parallel (van Soest et al., 2003). For example, a master computer can assign a number of networked computers to each evaluate one individual. Then it can apply the evolutionary operations (which are computationally cheaper) to select the "winners" and use their features to produce a new generation of individuals for evaluation in the networked computers.

APPLICATIONS OF EVOLUTIONARY METHODS IN MOVEMENT SCIENCE

Evolutionary methods have been applied to various problems in movement science such as studies of the mechanisms underlying the control of movement and the design of prosthetic control systems to repair and restore the lost movement. Rehabilitation of movement is another area that may benefit from the application of evolutionary methods.

Biomechanics and Motor Control of Movement

Control of movement in humans and animals is the result of natural evolution. Different designs for musculoskeletal components and the CNS control of movement have been tried and improved over many generations based on their fitness to help the animal or the human survive. Motor control theorists are interested in the biological principles underlying the control of movement, but these are influenced by both the performance criteria and the processes of incremental development, both phylogenetic and ontogenetic. Evolutionary methods fit naturally with these problems because they can simulate processes similar to those that formed the musculoskeletal and CNS systems in the first place. We can hope that they are more likely to produce control systems that are similar to the CNS than more conventional engineering methods such as those that are typically employed in robotics. In order to design a motor or any other control system using conventional methods, one first has to select a control architecture and then use an analytical or learning algorithm to optimize its parameters. However, the architecture of the biological control system itself is unknown. An important advantage of the evolutionary methods is their ability to evolve both the control architecture and its parameters (Koza, Keane, Yu, Mydlowec, & Bennett, 2000; Bennett, Koza, Yu, & Mydlowec, 2000). Therefore, they can be used first to explore the range of control circuits and then to optimize their parameters to perform specific movements.

Several recent studies have explored the use of evolutionary methods in biomechanics and motor control studies. Cerveri et al. used evolutionary optimization to estimate spinal joint centers from motion data (Cerveri, Pedotti, & Ferrigno, 2004). The evolutionary methods enabled them to estimate the joint centers for joints whose range of motion was limited. Raikova and Aladjov (Raikova & Aladjov, 2002, 2004) used an evolutionary optimization method to gain better insight into motor control of the elbow muscles. They compared evolutionary and conventional optimization methods in finding muscle forces that correspond to known elbow joint moments. In their view, the main advantage of the evolutionary methods was its ability to handle a more complex model that allowed them to include motor unit properties in their muscle models. As a result, in addition to optimal muscle forces, they could predict corresponding motor unit firing patterns that could shed light on CNS control strategies. Ijspeert et al. have used evolutionary methods and biomechanical simulations to investigate the neural mechanisms for control of swimming in Lamprey and gait in Salamander (Ijspeert, 2001; Ijspeert & Kodjabachian, 1999). Their goal was to explore the type of neural circuitry that can produce the observed movement patterns in these animals. They used a staged evolution to optimize different parts of the control system that produced the observed movement patterns. Schaal and Sternard (1993) used evolutionary methods to learn the control of juggling a ball. They were able to find successful control strategies, but the learning speed and the quality of the resulting solutions depended on the formulation of the problem. The formulation of the optimization problem, especially the fitness function, is therefore critical. The optimal solutions discovered by the evolutionary optimization will match the CNS control strategies only if the fitness function uses the same criteria that have been employed by natural evolution.

Design of Prosthetic Control Systems

Prosthetic control systems are used to restore movements that are lost due to injury or disease. For example, spinal cord injury paralyzes those muscles whose motor neurons are located below the level of injury. Except for motor neurons damaged at the site of the injury itself, the remaining motor neurons continue to innervate the muscle fibers of the paralyzed muscles. Their large, myelinated axons are readily stimulated by electrical pulses applied in the ventral horn, the peripheral nerve or the muscle itself. FES is a technique for delivering such signals but it has to solve the same control problems as the CNS (i.e., how to control the activation of the muscles to produce effective, efficient and stable limb movement). The biological and prosthetic solutions can be expected to have some similarities, but there are also reasons why they might be different:

- The physiological properties of muscles that have undergone disuse atrophy and chronic electrical stimulation are different than normal muscles and electrical stimulation produces unphysiological patterns of recruitment and frequency modulation in the motor units.
- Prosthetic sensors for feedback are likely to be different and relatively impoverished compared to the diverse and numerous somatosensory receptors.
- The command signals that can be recorded to drive the prosthetic controller will be very different from those actually produced by the CNS.
- The set of tasks to be performed is a highly limited set of those for which the biological system was designed.

Thus, the prosthetic control systems do not have to be exactly similar to the CNS to be practical. Therefore, both conventional and biomimetic control systems can be designed using evolutionary methods to restore limited but essential repertoire of movements to the patient.

Another property that makes the evolutionary methods more attractive for the design of prosthetic control systems is their ability to solve reinforcement type control problems (Juang, Lin & Lin, 2000; Whitley, Dominic, Das, & Anderson, 1993; Davoodi et al., 1999). In designing the prosthetic control systems, the detailed knowledge of the desired control signals such as the muscle stimulation patterns are not known. Therefore, one cannot use the supervised learning methods to design control systems to produce the desired stimulation patterns. In most cases, the more distant knowledge of the desired behavior such as the joint torques, or joint trajectories is also unknown. As the result, computational methods such as those in Robotics (e.g., computed torque method) may not be applicable. What is more common, however, is the knowledge of the final outcome such as the success or failure of the movement. These indicators of desired movement are delayed reinforcement because they will be known only after the completion of the movement. Reinforcement learning methods have been developed to solve such control problems but scaling them to large real-world problems have proven to be difficult (Davoodi & Andrews, 1998). Evolutionary methods are a more robust alternative and can solve similar control problems.

Davoodi et al. (1999) used evolutionary methods to automatically tune the parameters of closed-loop FES control systems for restoration of standing in paraplegic patients. This movement is controlled by active voluntary movement of the upper body and FES control of the paralyzed leg muscles. Their goal was to improve the movement parameters related to the safety and energetics of the FES-controlled movement. The main advantage of the evolutionary methods was their ability to optimize the parameters of the FES control system that was interacting with the residual voluntary movements, a problem intractable to conventional optimization methods. The optimal controller synchronized the FES system with the volitional movements of the patient based on minimal information such as the success/failure of standing up or desired leg kinematics. They suggested that the evolutionary method can be applied directly to the patients but because the number of movement trials for optimization was high, the controllers must be pre-trained using a computer model scaled to the individual patient. Hussein and Granat (2001) later used evolutionary methods to optimize the parameters of a neural network controller for FES standing-up and sitting-down. Their objective was to minimize the upper limb support, the trajectory error, and the average muscle stimulation. They reported a robust performance in the presence of unpredictable changes or alteration that could happen during standing-up and sitting-down maneuvers. Evolutionary methods have also been used to derive the voluntary command signals from the EMG activity of the forearm muscles (Peleg, Braiman, Yom-Tov, & Inbar, 2002), extraction of angular position from sensory electroneurographic signals of the muscle afferents (Cavallaro, Micera, Dario, Jensen, & Sinkjaer, 2003), and classification of the patient's intended movement from electroencephalographic recordings (Yom-Tov & Inbar, 2001). These could be used as voluntary command signals and as sensory feedback in neural prostheses.

Movement Therapy and Rehabilitation

It will be easier to repair a complex system if we know how it worked in first place. Similarly, the knowledge we gain by the evolutionary exploration of the CNS control mechanisms can help us to improve movement rehabilitation techniques. Currently, movement therapy decisions are based on limited observations and measurements from parts of the complicated movement system. The therapist often does not have a clear understanding of the integrated movement system and, therefore, cannot properly diagnose and treat the real underlying cause of the disability. The trial-and-error methods that are commonly adopted by the therapists are inefficient at best and ineffective or even deleterious at worst. New methods are needed to diagnose objectively and definitively the mechanism responsible for the movement disability and recommend targeted movement therapies, perhaps in combination with orthotic or prosthetic aids. One area of potential application is the use of evolutionary methods to customize the movement therapy to the limited capabilities of the patients. For example, evolutionary methods can be used to explore feasible movement patterns that are customized to the specific muscle force profile of the patient and targets specific impairment. These trajectories then can be used to prescribe movement therapies to improve the existing functions or design FES control systems to restore the lost movement. To our knowledge, the use of evolutionary methods in this area has not been explored yet. The following case study will demonstrate the potential uses of evolutionary methods in movement rehabilitation.

CASE STUDY:
FEASIBLE SIT-TO-STAND TRAJECTORIES

Sit-to-stand movement is a demanding maneuver whose successful completion requires both adequate joint moments and proper coordination of multi-joint movements (Kelly, Dainis, & Wood, 1976; Riley, Schenkman, Mann, & Hodge, 1991; Riley, Krebs, & Popat, 1997). As the functional capacity of the muscles in the lower extremity diminishes, the CNS modifies the multi-joint coordination strategy to successfully perform the sit-to-stand maneuver. Such modifications have been observed in elderly who tend to flex the trunk more before standing to reduce the burden on their knee joint muscles (Papa & Cappozzo, 2000). Such gradual adaptation is not possible, however, when the subject suddenly looses significant muscle capacity and the ability to stand up unaided. This could happen due to ailments affecting one or more muscles in the lower extremity, such as muscle atrophy, surgery, paralysis, etc. To restore unaided sit-to-stand to these patients, it is necessary to determine whether standing up with the diminished muscle capacity is still possible and what patient-specific joint trajectories must be used for multi-joint coordination. Once a feasible strategy is determined, it can be used as a guide in exercise therapy for movement rehabilitation. These customized trajectories can also be used in designing FES controllers to restore sit-to-stand movement to the paralyzed patients. Of course, these are the sorts of judgments that experience therapists make intuitively. What is proposed here is to embed their expert knowledge and informal kinetic analysis into computer software to produce more objective and consistent results.

To explore such feasible trajectories, Pai and Patton (1997) used a single-link inverted pendulum model of sit-to-stand movement to find the set of velocity-position combinations for the center of mass that maintained balance during seat take-off. They showed that with a proper velocity during the take-off, the balance could be maintained even if the center of mass is outside the base of support. In the following case study, genetic algorithm optimization method is used to explore feasible trajectories for a more complex multi-joint model of sit-to-stand movement.

Biomechanical Model of Sit-to-Stand Movement

A planar eight-segment model was used to model the sit-to-stand movement (Figure 2). The lower extremity was modeled by a foot, a leg, and a thigh segment. The trunk was modeled by a single segment; and the upper extremity was modeled by two upper arms and two forearms allowing independent movement of the left and right arms. Frictionless pin joints were used to connect the segments together. Ankle, knee and hip joints were actuated by torque producing muscles while the arms were allowed to move freely and were not actuated. The mass center of each segment lied on the line connecting the two adjacent joints and did not change during the motion. Typical male proportions were used to determine anatomical parameters (Winter, 1990).

The foot was attached to the ground by two pin joints under the heel and metatarsal head. The direction of the vertical reaction forces developed in these two pin joints were used to detect whether the sit-to-stand maneuver was dynamically stable and therefore feasible. The movement was considered as feasible as long as the vertical forces applied to the foot at these two joints were in upward direction (i.e., the foot at these two locations was pressing on the ground). SD/FAST (Symbolic Dynamics Inc., USA) was used to

Figure 2. Biomechanical model of sit-to-stand movement

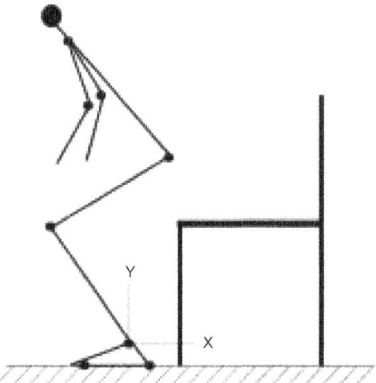

Ankle, knee and hip joints are actuated by torque-generating muscles while the arm joints are free to move and are not actuated. Two joints connecting the foot to the ground are used for stability detection. The model represents typical male proportions for a subject with mass = 70 kg and height = 1.75 m.

formulate and automatically derive the equations of motion for the model, which were then solved by a 4th-order variable step Runge-Kutta numerical integration method.

Parameterization of Sit-to-Stand Trajectories

To optimize the sit-to-stand trajectories, they must first be parameterized. Sit-to-stand maneuver is comprised of the movements in the hip, knee and the ankle joints, but there are distinct patterns and phases to these movements in normal sit-to-stand movement. For example, the forward movement of the trunk starts the maneuver while the other joints start moving at appropriate times later in the maneuver. It is the timing of the joint movements and their patterns that create different sit-to-stand movement patterns.

Similar to normal standing up, the movement was assumed to start with the trunk moving forward while the subject is on the chair. Other joints could start moving forward at anytime to lift the subject off the chair. The trunk motion during sit-to-stand can be divided into two phases. Initially it flexes forward, but later in the movement it reverses direction to straighten the trunk by extending the hip joint. To model these two phase, the trunk motion as a function of time was represented by two polynomials. The tuning parameters were the fractions of time for each trajectory section and joint position in via point (the transition point from the first to the second trajectory). The movement of the knee joint was modeled by a single polynomial to allow extension only movement. The tuning parameter was the time at which the knee joint starts its movement. Two polynomials modeled the two phases of the ankle joint movement: forward movement of the shank and reversal to vertical position. The tuning parameters were the initiation time, the fraction of time for each trajectory and the ankle position in via point. In total five polynomials were used to parameterize the movement of the hip, knee and ankle joints. All the trajectories were represented with 5th-order polynomials with zero initial and final velocities. The total time of the sit-to-stand maneuver was fixed at 2.5 sec.

Formulation of Genetic Algorithm Optimization

The five parameterized trajectories of the hip, knee, and ankle joints had six tuning parameters that were optimized using genetic algorithms. Each parameter was coded by an 8-bit binary number resulting in a chromosome of length 48. Therefore, each chromosome contained all the tuning parameters of the joint trajectories and specified a unique sit-to-stand movement pattern. The genetic algorithm randomly selected the parameters of all the chromosomes (100) in the first generation, applied them to the model subject and measured their fitness. The objective is to increase the measure of fitness from generation to generation until it can not be improved anymore (i.e., it converges to the optimal solution). The process of reproduction to form the next generation used tournament selection method, crossover probability of 0.8 and mutation rate of 0.05.

The measure of fitness for each chromosome was the inverse of the objective function. Maximization of fitness, therefore, results in minimization of the objective function. The objective function was defined as:

$$J = w_1 LoB + w_2 \tau_{total} + w_3 \tau_{max}^{Ankle} + w_4 \tau_{max}^{Knee} + w_5 \tau_{max}^{Hip}$$

where J is the objective function, w_1, w_2, w_3, w_4, w_5 are the weights of individual criteria, LoB is the number of time steps during a maneuver the model loses its balance, $\tau_{max}^{Ankle}, \tau_{max}^{Knee}, \tau_{max}^{Hip}$ are the maximum joint torques during standing up, and τ_{total} is the square sum of joint torques during standing up. The latter is an indication of the total muscle energy expenditure during the movement. Each criterion is normalized before inclusion in the objective function. Therefore, by changing the relative weights of individual criteria, one can change its influence in the final solution.

Feasible Sit-to-Stand Trajectories

The optimal sit-to-stand trajectories depend on the criteria included in the fitness function and their relative weights. For example, the relative weight of the maximal knee joint torques can be increased if the subject has diminished knee joint torque. This will result in sit-to-stand trajectories that minimize the demand on the knee joint muscles. The recommended knee joint torques then could be compared to the available knee joint torque to examine whether the subject has the muscle strength to execute the recommended movement. If not, the recommended trajectories may be used as a guide in movement therapy to target those muscles that need strengthening.

To demonstrate the effect of the selected criteria on optimal trajectories, we performed four genetic algorithm optimizations. The goal of the first optimization (OPT1) was to find trajectories that were always dynamically balanced. This required the minimization of the single criteria, LoB, that was achieved by setting all the weights in the objective function to zero except for w_1 which was set to 1. The performance of the genetic algorithm in exploring the search space to find dynamically balanced movements is shown in Figure 3.

Figure 3. The trend of the objective function for the best solution in each generation in OPT1

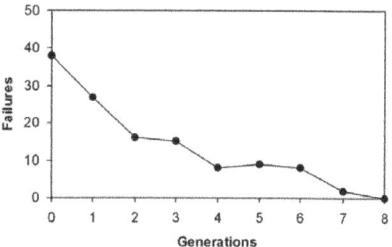

The goal of the optimization was to minimize the number of time steps during which the model failed to be stable (Failures)

Figure 4. Stick figures (intervals of 100 ms) representing the optimal movement pattern in OPT1

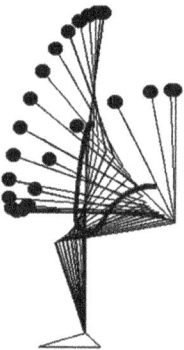

The trajectory of the center of mass is also shown. The arms are not drawn for clarity.

Figure 5. Trajectories of the center of mass in normal and sit-to-stand movements optimized by genetic algorithms (OPT1-4)

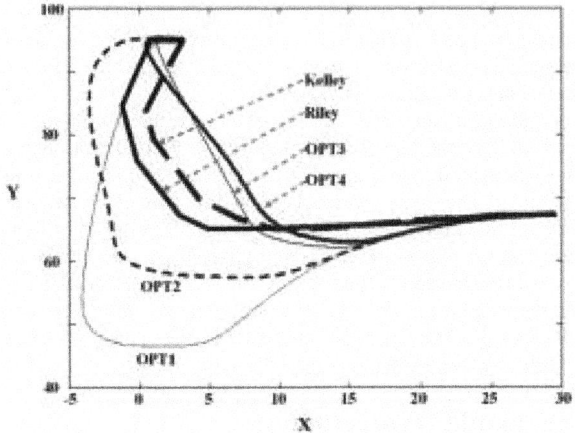

"Kelly" and "Riley" are trajectories of the center of mass in normal subjects adopted from Kelly et al. (1976) and Riley et al. (1991), respectively. The position of the center of mass is measured in the coordinate system shown in Figure 2.

Figure 6. Angular velocities of the hip, knee, and ankle joints for sit-to-stand movements optimized by genetic algorithm (OPT1-4)

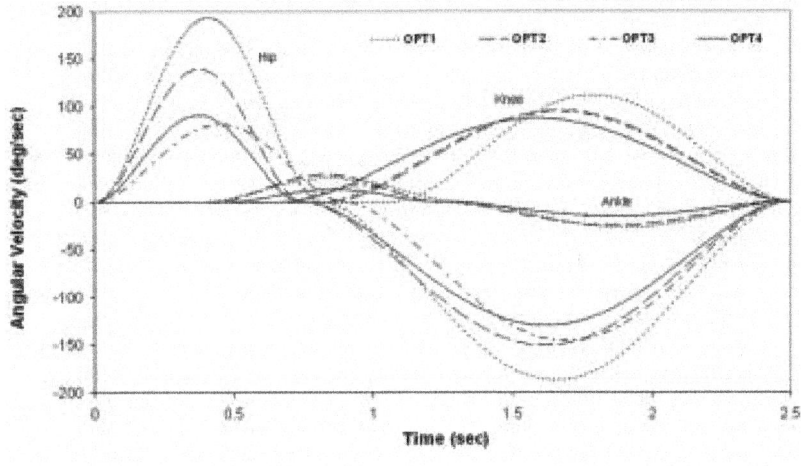

The first generation was selected randomly, resulting in candidate solutions that were all unbalanced. Even the best candidate solution in the first generation was out of balance 38 times out of the total 250 time steps. But the genetic algorithm optimization managed to converge to the optimal solution after 8 generations (800 sit-to-stand movements) and find a motion pattern that was dynamically balanced over the whole duration of the movement. The stick figure animation of the optimal solution and the trajectory of the center of mass are shown in Figure 4.

The goal of the second optimization (OPT2) was to minimize the maximum required torque in hip, knee, and ankle joints. This would simulate overall weakness in all lower extremity muscles, for example, due to old age. Therefore the weights of these three criteria in the objective function, w_3, w_4, w_5, were set to 1 while the rest of the weights were set to zero. In the third optimization (OPT3), the sum of all joint torques, which is an indicator of total muscle energy expenditure, was minimized by setting w_2 to 1 and the rest of the weights to zero. In the fourth and final optimization (OPT4), all the criteria were included by setting all the weights to 1. The genetic algorithm was able to converge to optimal solution in all cases. The trajectories of the center of mass and lower extremity joints for the optimal solutions are shown in Figure 5 and 6.

Discussion of Feasible Trajectories

Unlike, traditional optimization methods, the genetic algorithm optimization did not require the mathematical formulations of the equations of motion or the constraints in terms of the optimization parameters. These requirements usually limit the application of the optimization method to simple models (Pai et al., 1997). The genetic algorithm developed here works with a simulated subject that can include complex properties of the musculoskeletal system. For example, discontinuous and non-differentiable elements can be easily included in the simulation model. Therefore, the genetic algorithm broadens the range and complexity of the motion optimization problems that can be solved.

Selection of the appropriate criteria with proper weights is probably the most critical factor for the success of genetic algorithm because it represents the goal of the optimization. For example, in OPT1 the objective was to maximize the dynamic stability irrespective of the required joint moments. The optimal movement, as a result, employs excessive forward flexion of the trunk that improves stability but also requires high hip joint moments. But in OPT2, where the maximal joint moments are minimized, the trajectory of the center of mass was closer to the normal movement and required lower joint moments. The optimal trajectories will be similar to the norm only if the CNS actually optimized the movement and we knew what the optimization criteria were and how each criterion should be weighed in the fitness function. Fortunately, this knowledge is not necessary for rehabilitation applications because we are interested only in feasible movements that respect the known limitations of a disabled patient.

The feasible movements in OPT1 employ excessive forward movement of the trunk to move the center of mass over the base of support and build momentum before the seat take-off. This strategy requires higher hip joint torques because later in the movement, the forward movement of the trunk must be halted by the extensor muscles at the hip joint. But at the same time this dynamic maneuver reduces the demand on the knee and ankle joint muscles. Apparently, the momentum buildup in the trunk is transferred to the other segments as suggested by Riley et al. (1991). The implication of this result for movement rehabilitation is that an altered movement pattern can make use of the higher strength in one or more joints to compensate for the weakness in other joints. If these modified strategies can be identified by the genetic algorithm, they can be used in targeted rehabilitation of movement. As an example procedure, a musculoskeletal model of the impaired subject can be used by the genetic algorithm to find optimal movement patterns to prevent, for example, loss of balance. These movement patterns then could be used to rehabilitate the subject. The therapists can strengthen the weak muscles so that the recommended movement patterns by the genetic algorithm become feasible for the patient (Hughes, Myers, & Schenkman, 1996). If the patient has the required muscle strengths, but still does not perform the maneuver, the recommended movement patterns can be used as a training guide in movement therapy sessions. For example, during the movement therapy sessions, audio and/or visual signals could signal the amount of deviation from the desired trajectories. Such augmented feedback has been used by Wu (1997) to train elderly patients to maintain upright standing posture and by Colborne et al. (1993) in rehabilitation of hemiplegic patients. Alternatively, robotic rehabilitators (Hesse, Schmidt, Werner, & Bardeleben, 2003) can be used to assist the patient to perform the recommended movement.

The technique introduced here could also produce feasible trajectories for FES control of movement. In the past, trajectories of movements performed by healthy subjects have been used as the desired trajectory in FES control systems (Dolan, Andrews, & Veltink, 1998; Mulder, Veltink, & Boom, 1992). But these trajectories may not be feasible for patients with limited physical or sensory capabilities (Acosta & Kirsch, 2000). An adequate understanding of each patient's capabilities and their feasible movements is a prerequisite for designing feasible FES control systems.

FUTURE TRENDS

Human movement is a complicated system whose study and treatment require sophisticated techniques, such as evolutionary computation. But successful application of evolutionary methods in movement science requires expertise in movement science and computer modeling and simulation of movement and evolutionary computation that are currently difficult to master. The required effort and expertise, however, will be lower and more researchers will be drawn in the field as the usability of software for evolutionary computation and musculoskeletal modeling improves. Eventually, these tools will be embedded into software packages that can be operated by therapists. It is these types of applications that drive a collective efforts by the musculoskeletal modeling community to develop new sophisticated software for musculoskeletal modeling (Davoodi, Urata, Todorov, & Loeb, 2004). With the availability of the more sophisticated software tools, more complicated problems will be tackled. Evolutionary methods alone or in combination with other intelligent systems may be used for simultaneous identification of the architecture and parameters of the control circuits in the CNS or a prosthetic system. These investigations will further our understanding of the human movement in its entirety and help us to develop new therapies to repair and restore movement after disease or injury.

CONCLUSION

We have learned a great deal about human movement and its treatment by experimental measurements and quantitative analysis of its individual elements. To further our understanding of human movement and to develop new movement therapies and rehabilitation techniques, such motor behaviors must be studied in their entirety as an integral of the components and interactions among them. Evolutionary methods introduced in this chapter are intelligent systems that can further our understanding of human movement and help us devise new treatments. The main motivations for the use of evolutionary methods are that they resemble the biological processes that led to the development of natural human movement, and they can solve optimization and learning problems that cannot be solved with existing methods. The basic evolutionary computations are simple and straightforward to implement, but the computation time for complex biomechanical systems may be prohibitive. Evolutionary computation is well suited to parallel processing that can reduce the computation time significantly. Despite their great potential, evolutionary methods have yet to be applied extensively to studying or treating human movement disorders. This should change as new methods and software tools for evolutionary computation and biomechanical modeling and simulation become readily available.

REFERENCES

Acosta, A. M., & Kirsch, R. F. (2000). Feasibility of restoring shoulder function in high tetraplegia. *Assistive Technology, 12,* 154.

Back, T., Hammel, U., & Schwefel, H. P. (1997). Evolutionary computation: comments on the history and current state. *IEEE Transactions on Evolutionary Computation, 1,* 3-16.

Bennett, F. H., Koza, J. R., Yu, J., & Mydlowec, W. (2000, April 17-19). Automatic synthesis, placement, and routing of an amplifier circuit by means of genetic programming. In *Proceedings of Evolvable Systems: from Biology to Hardware, Third International Conference, ICES 2000.* Edinburgh, Scotland, UK, April 17-19 (Vol. 1801, pp. 1-10). Springer.

Cavallaro, E., Micera, S., Dario, P., Jensen, W., & Sinkjaer, T. (2003). On the intersubject generalization ability in extracting kinematic information from afferent nervous signals. *IEEE Transactions on Biomedical Engineering, 50,* 1063-1073.

Cerveri, P., Pedotti, A., & Ferrigno, G. (2004). Evolutionary optimization for robust hierarchical computation of the rotation centres of kinematic chains from reduced ranges of motion the lower spine case. *Journal of Biomechanics, 37,* 1881-1890.

Colborne, G. R., Olney, S. J., & Griffin, M. P. (1993). Feedback of ankle joint angle and soleus electromyography in the rehabilitation of hemiplegic gait. *Archives of Physical Medicine and Rehabilitation, 74,* 1100-1106.

Davoodi, R., & Andrews, B. J. (1998). Computer simulation of FES standing up in paraplegia: A self-adaptive fuzzy controller with reinforcement learning. *IEEE Transaction on Rehabilitation Engineering, 6,* 151-161.

Davoodi, R., & Andrews, B. J. (1999). Optimal control of FES-assisted standing up in paraplegia using genetic algorithms. *Medical Engineering & Physics, 21,* 609-617.

Davoodi, R., Urata, C., Todorov, E., & Loeb, G. E. (2004). Development of clinician-friendly software for musculoskeletal modeling and control. In *Proceedings of the 26th Annual International Conference of the Engineering in Medicine and Biology Society*.

Dolan, M. J., Andrews, B. J., & Veltink, P. H. (1998). Switching curve controller for FES-assisted standing up and sitting down. *IEEE Transaction on Rehabilitation Engineering, 6,* 167-171.

Goldberg, D. E. (1989). *Genetic algorithms in search, optimization and machine learning.* New York: Addison-Wesley.

Hesse, S., Schmidt, H., Werner, C., & Bardeleben, A. (2003). Upper and lower extremity robotic devices for rehabilitation and for studying motor control. *Current Opinion in Neurology, 16,* 705-710.

Holland, J. H. (1962). Outline for a logical theory of adaptive systems. *Journal of Association for Computing Machinery, 3,* 297-314.

Hughes, M. A., Myers, B. S., & Schenkman, M. L. (1996). The role of strength in rising from a chair in the functionally impaired elderly. *Journal of Biomechanics, 29,* 1509-1513.

Hussein, S. E., & Granat, M. H. (2001). A neueo-genetic model for standing up and sitting down in paraplegia. In *Proceedings of International Functional Electrical Stimulation Society Conference*, Cleveland, OH (pp. 294-296).

Ijspeert, A. J. (2001). A connectionist central pattern generator for the aquatic and terrestrial gaits of a simulated salamander. *Biological Cybernetics, 84,* 331-348.

Ijspeert, A. J., & Kodjabachian, J. (1999). Evolution and development of a central pattern generator for the swimming of a lamprey. *Artificial Life, 5,* 247-269.

Jezernik, S., Colombo, G., Keller, T., Frueh, H., & Morari, M. (2003). Robotic orthosis Lokomat: A rehabilitation and research tool. *Neuromodulation, 6,* 108-115.

Juang, C. F., Lin, J. Y., & Lin, C. T. (2000). Genetic reinforcement learning through symbiotic evolution for fuzzy controller design. *IEEE Transactions on Systems Man and Cybernetics Part B-Cybernetics, 30,* 290-302.

Kelly, D. L., Dainis, A., & Wood, G. K. (1976). Mechanics and muscular dynamics of rising from a seated position. In P. V. Komi (Ed.), *Biomechanics* (pp. 127-134). Baltimore: University Park.

Koza, J. R. (1992). *Genetic programming: On the programming of computers by means of natural selection.* Cambridge, MA: MIT.

Koza, J. R. (1994). Genetic programming as a means for programming computers by natural-selection. *Statistics and Computing, 4,* 87-112.

Koza, J. R., Keane, M. A., & Streeter, M. J. (2004). Routine automated synthesis of five patented analog circuits using genetic programming. *Soft Computing, 8,* 318-324.

Koza, J. R., Keane, M. A., Yu, J., Mydlowec, W., & Bennett, F. H. (2000, April). Automatic synthesis of both the topology and parameters for a controller for a three-lag plant with a five-second delay using genetic programming. In Cagnoni, Stafano et al. (Eds.) *Proceedings of the Real-World Applications of Evolutionary Computing. EvoWorkshops 2000. EvoIASP, Evo SCONDI, EvoTel, EvoSTIM, EvoRob, and EvoFlight,* Edinburgh, Scotland, UK, (Vol. 1803, pp.168-177). Berlin,Germany: Springer-Verlag.

Loeb, G. E., Peck, R. A., Moore, W. H., & Hood, K. (2001). BION system for distributed neural prosthetic interfaces. *Medical Engineering & Physics, 23*, 9-18.

Mulder, A. J., Veltink, P. H., & Boom, H. B. (1992). On/off control in FES-induced standing up. A model study and experiments. *Medical & Biological Engineering & Computing, 30*, 205-212.

Pai, Y. C., & Patton, J. (1997). Center of mass velocity-position predictions for balance control. *Journal of Biomechanics, 30*, 347-354.

Papa, E., & Cappozzo, A. (2000). Sit-to-stand motor strategies investigated in able-bodied young and elderly subjects. *Journal of Biomechanics, 33*, 1113-1122.

Peleg, D., Braiman, E., Yom-Tov, E., & Inbar, G. E. (2002). Classification of finger activation for use in a robotic prosthesis arm. *IEEE Transactions on Neural Systems and Rehabilitation Engineering, 10*, 290-293.

Raikova, R. T., & Aladjov, H. T. (2002). Hierarchical genetic algorithm versus static optimization - investigation of elbow flexion and extension movements. *Journal of Biomechanics, 35*, 1123-1135.

Raikova, R. T., & Aladjov, H. T. (2004). Simulation of the motor units control during a fast elbow flexion in the sagittal plane. *Journal of Electromyography and Kinesiology, 14*, 227-238.

Riley, P. O., Krebs, D. E., & Popat, R. A. (1997). Biomechanical analysis of failed sit-to-stand. *IEEE Transaction on Rehabilitation Engineering, 5*, 353-359.

Riley, P. O., Schenkman, M. L., Mann, R. W., & Hodge, W. A. (1991). Mechanics of a constrained chair-rise. *Journal of Biomechanics, 24*, 77-85.

Schaal, S., & Sternad, D. (1993). Learning passive motor control strategies with genetic algorithms. In *1992 lectures in complex systems* (pp. 913-918). New York: Adison-Wesley.

Smith. et al. (1998). An externally powered, multichannel, implantable stimulator-telemeter for control of paralyzed muscle. *IEEE Transaction on Biomedical Engineering, 45*, 463-475.

van Soest, A. J. K., & Casius, L. J. R. R. (2003). The merits of a parallel genetic algorithm in solving hard optimization problems. *Journal of Biomechanical Engineering-Transactions of the ASME, 125*, 141-146.

Whitley, D., Dominic, S., Das, R., & Anderson, C. W. (1993). Genetic reinforcement learning for neurocontrol problems. *Machine Learning, 13*, 259-284.

Winter, D. A. (1990). *Biomechanics and motor control of human movement* (2nd ed.) New York: John Wiley & Sons.

Wu, G. (1997). Real-time feedback of body center of gravity for postural training of elderly patients with peripheral neuropathy. *IEEE Transaction on Rehabilitation Engineering, 5*, 399-402.

Yom-Tov, E., & Inbar, G. F. (2001). Selection of relevant features for classification of movements from single movement-related potentials using a genetic algorithm. In *Proceedings of the Annual IEEE Engineering in Medicine and Biology Society Conference*, Istanbul, Turkey.

Chapter XI

Dynamic Pattern Recognition in Sport by Means of Artificial Neural Networks

Jürgen Perl, Johannes Gutenberg University, Germany

Peter Dauscher, Johannes Gutenberg University, Germany

ABSTRACT

Behavioural processes like those in sports, motor activities or rehabilitation are often the object of optimization methods. Such processes are often characterized by a complex structure. Measurements considering them may produce a huge amount of data. It is an interesting challenge not only to store these data, but also to transform them into useful information. Artificial Neural Networks turn out to be an appropriate tool to transform abstract numbers into informative patterns that help to understand complex behavioural phenomena. The contribution presents some basic ideas of neural network approaches and several examples of application. The aim is to give an impression of how neural methods can be used, especially in the field of sport.

INTRODUCTION

If dynamic behavioural processes (e.g., those dealt with in biomechanics, game analysis or medical therapy) are to be analysed, statistical methods are helpful, but not sufficient. In order to understand the dynamics of a process, time-dependent development has to be taken into consideration, quantitatively as well as qualitatively. This means that in addition to quantitative characteristic parameters in particular qualitative information like patterns are necessary for an appropriated identification and characterisation of processes. The problem, however, is that we can automatically record huge amounts of data from processes, but then we have to find out the particularly relevant information within the data. This is the point where Artificial Neural Networks can become extremely helpful. They are able to learn from data, compressing them to useful information. They can recognize patterns indicating characteristic properties of the process. They are also helpful in detecting hidden information and striking features that cannot be seen from data or even video pictures.

The contribution starts with a brief introduction to the main ideas and concepts of Artificial Neural Networks and then gives a number of applications that demonstrate their usability.

BACKGROUND

Neural Networks

Artificial Neural Networks are presumably the best known biologically inspired approach within the field of artificial intelligence and machine learning. By developing abstract models of biological neural structures and performing them on standard computers or special hardware, researchers have been trying to reproduce some of the cognitive abilities one can observe in higher animals and, of course, in humans.

First approaches of neuron-like circuitry go back to the early 1940s (McCulloch & Pitts, 1943). A simple kind of learning behaviour was implemented in the model by the psychologist Donald Hebb (1949).

The best known model of these pioneer years is the so-called *perceptron* (Rosenblatt, 1958), which was able to learn a binary classification of patterns. However, it was shown in the 1960's that Rosenblatt's *perceptrons* were able to implement only a rather restricted class of classifications (Minsky & Papert, 1969). Even simple concepts like *"either ... or"* cannot be learnt by this approach.

This (legitimate) fundamental criticism meant a major set-back for Artificial Neural Network research followed by a period sometimes referred to as "neural winter." The shortcomings of the *perceptrons* were not overcome until the mid-1980s, when a learning algorithm was developed that was applicable to an Artificial Neural Network consisting of several layers of neurons (Rumelhart & McClelland, 1986). The connection of multiple layers of adaptive nonlinear neurons allows for the ability to learn complex relationships. Recurrence is another interesting aspect of neural networks. In recurrent networks, signals already processed by some of the neurons can be fed back into the system, which induces a potentially complex dynamics. In the Hopfield approach (Hopfield, 1982) this dynamics has a well-controllable attractor structure corresponding to a number of

Figure 1a. Supervised learning: Input I and desired output O$_{des}$ are presented to a learning algorithm

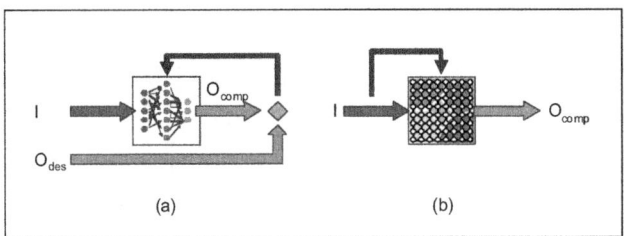

(a)   (b)

The connections of the neurons are adapted according to the difference of desired output O$_{des}$ and computed output O$_{comp}$. Figure 1b. Unsupervised learning: there is no desired output; input data are classified according to similarity of the input data points.

distinct patterns that can be recognized. A short summary of the approaches described above can be found in Ritter et al. (1992).

The Artificial Neural Networks models described so far mainly deal with the problem of reproducing a given input-output-behaviour. This kind of learning is known as "supervised learning." Every input instance is combined with the desired output. A comparison between the desired and the computed output is used to correct the interconnections between the single neurons (which is illustrated in Figure 1a).

Another kind of learning is called "unsupervised." Inputs are presented to the neural network and the network finds a classification (based on given similarity measures) without desired output values being provided (see Figure 1b).

Which approach is chosen mainly depends on the problem to be tackled. If a given input-output behaviour is to be achieved, supervised methods mostly appear to be more appropriate. If, however, unknown structures in data are to be discovered, unsupervised methods are to be preferred. Neural Networks have been applied to various kinds of data analysis problems. A number of approaches are depicted in Schöllhorn (2004) and Holzreiter (2005). Self-organizing maps, the topic of this chapter, belong to this second class of unsupervised methods.

The Kohonen Feature Map

The best known self-organizing map is the Kohonen feature map (Kohonen, 1982). We will now sketch in brief the principles of this approach.

Let us consider a number d of variables which can take on values from a given range. For instance, we could observe temperature, humidity, wind strength and barometric pressure at a given time.

An observation of such a combination of values forms a data point, which can be conceived as a point in the d-dimensional space. If $d>3$, it is hard for us to recognize structures in a set of data points.

Figure 2. A high-dimensional input data point (left) is mapped onto a lower-dimensional lattice of neurons (right)

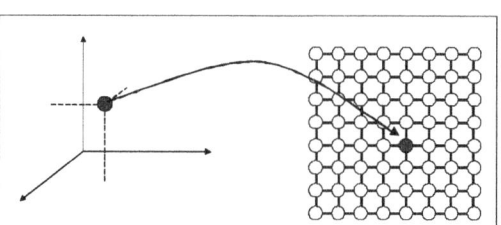

The aim of Kohonen feature maps (KFM) is to map such data points to a lower-dimensional structure (typically one- or two-dimensional, sometimes slightly higher-dimensional) in order to be able to recognize structures either visually or by further computer-based methods.

In the case of KFM, this lower-dimensional structure consists of a set of neurons arranged in a lattice of the respective dimensionality.

An arbitrary mapping, of course, is not sufficient to recognize structures in a set of data points. There are two basic requirements which are fulfilled by the KFM:

- Regions of the d-dimensional space containing a large number of data points shall be mapped to more neurons than regions containing few data points or no data points at all.
- Two data points having a low distance in the d-dimensional space should be mapped (largely) to the same neuron or to two adjacent neurons in the lattice.

Figure 3. Visualization of the adaptation to a new data point according to the neighbourhood-structure

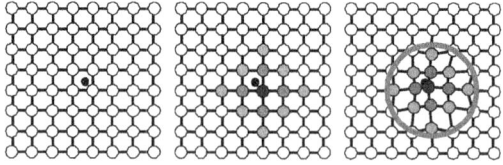

The vector of the "winner neuron" is changed most. Its direct neighbours are changed slightly less, and so on.

Figure 4. Spreading of the lattice in the data space during a learning process (from Goodson, 2004)

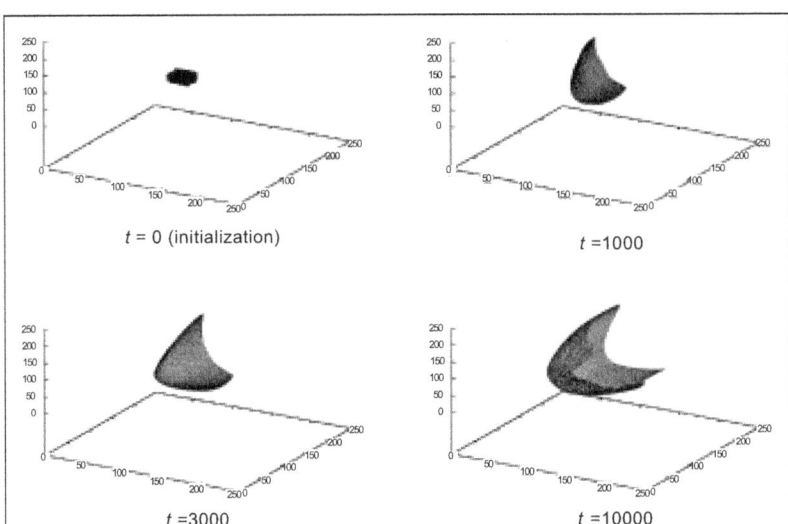

Such a mapping is implemented by assigning a d-dimensional vector to each neuron. A data point is mapped to the neuron with the vector which is nearest to the data point (i.e., which has the least distance in the d-dimensional data space). This neuron is often referred to as the "winner neuron" or the "best matching unit." During a learning period, data points are presented to the network and the vectors are adapted taking into account the neighbourhood-structure of the neurons as sketched in Figure 3.

If one considers the vectors assigned to each neuron, the rectangular lattice can be thought to be embedded in the higher-dimensional space. The learning process corresponds to a spreading of the lattice in the high-dimensional data space (see Figure 4).

A more formal description of Kohonen feature maps can be found in Ritter et al. (1989) or in Dauscher and Uthmann (2004).

In sport science, applications of neural networks can mainly be found in the areas of sport game analysis, rehabilitation, and movement pattern analysis. Some examples are given in the regarding section of this contribution. Those examples have in common that they are characterized by time-depending behavioural processes, which can be classified in an easy way by means of Artificial Neural Networks (see Figure 5). During the training with the process data, the network arranges the information by mapping similar process patterns to neighboured neurons. Each neuron then represents a type of process, and each cluster represents a class of similar process types.

Figure 5. During the net training, classes of similar process patterns are mapped to clusters of neighboured neurons, where each neuron then represents a type of process

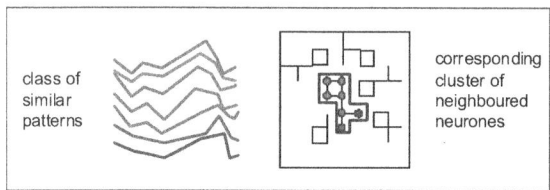

EXTENSIONS OF THE KFM CONCEPT

There are several extensions of the KFM concept dealing with some weaknesses of the approach which are undesirable in some applications.

The Growing Neural Gas Approach

In Figure 4 one can observe how the embedded lattice adapts to a distribution of the data points forming a 2-dimensional surface within the data space. The way the lattice is embedded is therefore rather natural. If the distribution of the data has another dimensionality than the lattice, however, this may cause problems. Especially if the data point distribution is higher-dimensional than the lattice, the lattice folds in a rather arbitrary way in order to adapt to the distribution. In every-day life, this would correspond to the attempt to fold a sheet of paper (2-dimensional) to fill a box (3-dimensional). If the sheet is very large, the attempt might be sufficiently successful.

Figure 6. Adaptation to a data point distribution of locally different dimensionalities obtained (a) by a rectangular lattice of a Kohonen feature map (b) by a growing neural gas model

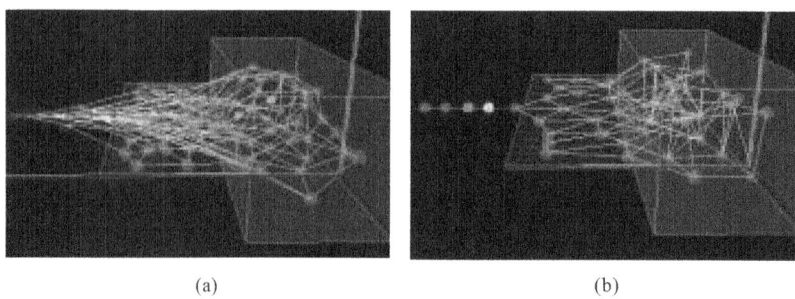

(a)   (b)

However, the way the sheet is folded or crumpled is rather arbitrary. If the sheet is rather small, the filling of the box will be extremely incomplete.

The latter case is shown on the right-hand side in Figure 6a for a Kohonen feature map. Although the data points are equally distributed within the cuboid box, the lattice does not reflect this distribution appropriately.

Furthermore, neurons are "wasted" by the Kohonen Algorithm in regions of lower dimensionality (Figure 6a, left-hand side).

In order to overcome these weaknesses of the Kohonen approach, (Fritzke, 1995, 1997) presented the growing neural gas model, an extension that adapts not only the localization of the neurons to the distribution but also the neighbourhood structure (also referred to as topology of the network).

Starting with a relatively small number of neurons, the model allows the network to grow (up to a given limit) in those regions where a necessity of additional neurons is detected.

The result of this extension is shown in Figure 6b. One can see that the dimensionality of the lattice is 1 on the left side, 2 in the middle and 3 on the right side, corresponding to the respective local dimensionality of the data point distribution.

There are numerous applications of growing neural gas models including, for example, practical applications in the field of robotics as well as theoretical investigations concerning net-based learning analyses.

The DyCoN Approach

Although the KFM-approach is most helpful in the field of process analysis, there are two specific problems that sometimes make it difficult to work with.

On the one hand, the demand for training data is very high and often cannot be satisfied by usual tests. The reason is that the impact of a single training input is extremely small, while the number of patterns that have to be learned is large. Therefore, a lot of training steps are necessary in order to reach a sufficient and significant distribution of information. Normally, a number of about 15,000 to 30,000 training steps are necessary, depending on the complexity and variability of the patterns.

On the other hand, a training cannot be continued or started again once it has been finished. The reason is that the learning behaviour of a conventional KFM is externally controlled, using predefined and inflexible functions for controlling the learning process. The result is that at the end of the training the net is frozen in a final state and cannot be reactivated for further training anymore, but just used as a tool. Therefore, a conventional KFM can be used neither for continuous learning nor for dynamic adaptation to time-dependent processes.

In order to avoid these disadvantages, a particular type of a continuously learning KFM-based network has been developed by our working group. The concept of DyCoN (i.e., **dy**namically **co**ntrolled **n**etwork (c.f. Perl, 2001)) allows for adapting the neurons individually to the learning contents and thus enables continuous learning as well as learning in separate phases. This way, for example, a DyCoN can be prepared using Monte-Carlo-generated data in a first phase. In a second phase, this reference network can be specifically coined using original data.

This approach has successfully been applied in a lot of cases in medicine, rehabilitation, and sport and even in fraud detection in car insurance.

Figure 7. DyCoN-based analyses of rehabilitation processes

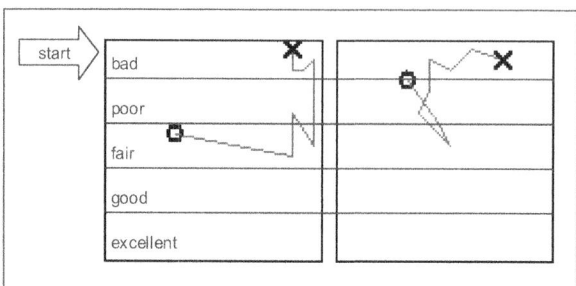

A first example may demonstrate how to use the DyCoN-approach in case of process analysis (also see Perl, 2001; Rebel, 2004).

Processes from rehabilitation are often documented by time series consisting of weekly records of attributes. Transferred to nets, these weekly attribute vectors form the input to the neurons, while the time series of those vectors is represented by a corresponding trajectory connecting the involved neurons. The problem is, however, that normally the number of recorded weeks is small (between 5 and 10) and the number of attributes per week is comparatively high (about 10 to 30). This leads to the rather bad situation of a small number of highly complex data, which normally prevents statistical, as well as net-based, analyses.

A DyCoN can be prepared step-by-step, using Monte-Carlo-generated data as described above, until the necessary patterns are represented sufficiently well.

Finally, the net can be interpreted semantically (e.g., by marking areas of good or bad conditions as has been done in Figure 7). It shows two simplified instances of a DyCoN trained with rehabilitation data: the left one tested with a successful process, while the right one shows the trajectory of a more unsuccessful attempt.

The second example deals with fraud detection (e.g., in insurance). An insurance case normally is characterized by a number of attributes, the values of which are corresponding with each other in a complex way. They form patterns, which are hard to detect but significant for classifying a case to be "good" or "bad." Insofar, networks of the KFM-type are perfectly useable for that pattern recognition. The problem is, however, that the strategies in the area of defraud change, not least in response to detection methods. A conventional KFM reacts on changing patterns only by reduced precision of recognition, without giving any information about what has changed. In contrast, the DyCoN approach allows for continuous learning and adapting to the changing pattern distribution. The results of such an adaptive learning process can be evidence of change clusters. Clusters which indicate "schools" of fraud not only increase or decrease but also disappear completely or appear initially.

Finally, DyCoN can be most useful in the field of motor analysis, if the aim is not only to describe perfect standard patterns, but also to detect striking features in the time-depending changes of motor patterns (also see Perl, 2004a). This approach is helpful in the case of detecting fatigue effects as well as in the case of disabled athletes or persons undergoing a rehabilitation process.

Summed up, three main advantages of the DyCoN approach are the small number of necessary original data, the very low preparing effort and the transparency of the results.

While normally the number of necessary original data sets ranges from about 15,000 to 30,000, DyCoN only needs some hundreds of data sets, supposing they contain enough information for Monte Carlo-generation.

Different from concurrent approaches, neither the internal structure of the case data nor their logical rules have to be analysed and fed to rule systems but are prepared by the net training process automatically.

The information contained in a neuron can be checked. Therefore, it is possible to calibrate the net to the current problem situation, using corresponding original data. Furthermore, the recognition of a pattern, which can be done by the net by highlighting the corresponding neuron, can be checked and evaluated. Finally, dynamic changes of the pattern landscape, which might indicate changing structures or distributions in the problem area, can be recognised and taken into consideration.

APPLICATIONS IN DATA ANALYSIS AND VISUALIZATION

The RoboCup Approach

An interesting approach to motion analysis in team sport has been made by Wünstel et al. (2001). The underlying question was whether typical patterns of soccer player movements could be detected using a Kohonen feature map.

Instead of using player position data from real soccer games (which are rather intricate to acquire), surrogate data stemming from RoboCup 2D-Simulation League logfiles were used.

The heart of RoboCup 2D-Simulation League is a virtual soccer playground. The physics of the ball as well as the player movements are roughly modelled and simulated by the so-called "Soccer Server." The team players' "intelligence" (i.e., their behaviour in a given situation)

is implemented by competing groups of programmers (coming from a number of different universities and other institutions around the world). All the movements of the virtual players as well as that of the ball are logged in a so-called "logfile." Such a logfile (stemming from a match between Carnegie Mellon University vs. Mainz Rolling Brains) provided the data for the analyses performed by Wünstel et al. (2001). It is clear that the dynamics of a RoboCup Simulation League match is not equal to that of real soccer in all aspects. However, having a look at the player movements (using an appropriate visualization tool) reveals already a certain amount of game complexity. At least for a non-expert, some soccer-like patterns are observable.

In order to usefully apply Kohonen feature maps to the thus generated data, an appropriate pre-processing has to be performed. It consists mainly of two steps:

1. The absolute positions at two subsequent timesteps of the players were trans-
    formed to relative motion vectors. In other investigations, the relative position to
    the ball was also computed.
2. These data were further aggregated over several timesteps forming high dimen-
    sional vectors describing segments of motion of a fixed length. It is to be noted that
    this is similar to time delay embedding methods in nonlinear data analysis methods
    as described, for example, in Kantz and Schreiber (2004).

These high-dimensional vectors (stemming from both teams) were now used to train
a Kohonen feature map. After the training process, data of each team were analyzed
separately using the map. The number of vectors that belong to a neuron, is visualized
appropriately by a kind of scatter-plot (see Figure 8). Different kinds of motion patterns
correspond to different "clusters" (i.e., groups of adjacent neurons marked by white lines
in Figure 8 and labelled by Roman numbers). The frequency distribution of the data points
over these clusters can reveal, for example, superiority of a team or the fact that a skill
(like dribbling) is not implemented in the virtual players of a team.

In Figure 8, for example, one can see that cluster (I) is remarkably more populated
for the MRB team than for the CMU team. A further investigation showed that this cluster
corresponds to backward movement of the player. The density difference in this cluster
is likely to occur due to the superiority of the CMU team in this match.

A more detailed overview about the methods used by Wünstel et al. (2001) can also
be found in Uthmann and Dauscher (2004).

It is to be remarked, however, that the theoretical applicability of such methods for
data analysis is only one aspect. In order to make these methods easily accessible for
sport scientists and especially sport practitioners, it is necessary to embed these
methods into an appropriate software framework. One approach is to combine the self-

*Figure 8. Visualization of the behaviours of two players of different teams in a RoboCup
Simulation League match (adapted from Wünstel, 1999)*

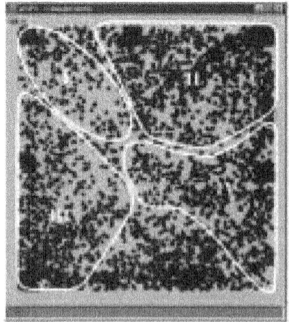 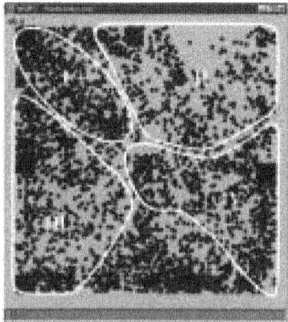

(a) Carnegie Mellon University (CMU) team player; (b) Mainz Rolling Brains (MRB) team player

organizing map with database techniques on the one hand and interactive graphical user interfaces on the other hand (Boll, 1999; Jouraai, 2003).

Such techniques enable the user to store data in one or more database tables and then to choose a set of observables interactively which are used then for the adaptation of a self-organizing map. Clusters representing a given situation or movement can be identified, coloured and labelled appropriately. This cluster structure can be stored in a database again (see Figure 9).

Pattern-Recognition in Sport Science

Besides a lot of different applications, the method of pattern analysis can successfully be used in the field of complex processes like human behaviour or, more specifically, human movement. From the viewpoint of mathematics or computer science, such a pattern is just a high-dimensional vector of numbers encoding, for instance, biomechanical quantities like coordinates, and so can be handled using mathematical or dataprocessing methods.

The problem of analysing single patterns of the same type, like fingerprints or physiognomies, is first to know what the characteristic types are. If the types are known and fixed, stochastic clustering methods are helpful in order to identify patterns as members of clusters that represent the types. If the types are not known or changing over time, net-based approaches that detect and adjust appropriate cluster structures by means of dynamic learning are more useful.

One of our first approaches in the area of sport game analysis was net-based, rally-analysis in squash. Without getting into technical details of data pre-processing, the results were like those presented in Figure 10. First of all, the squash data trained the network in a very specific way, making the structure of the game and the main types of rallies transparent. A meta-pattern of marked neurons represents the basic player-

Figure 9. Cluster structure in a self-organizing map indicated by connecting edges and coloured clusters

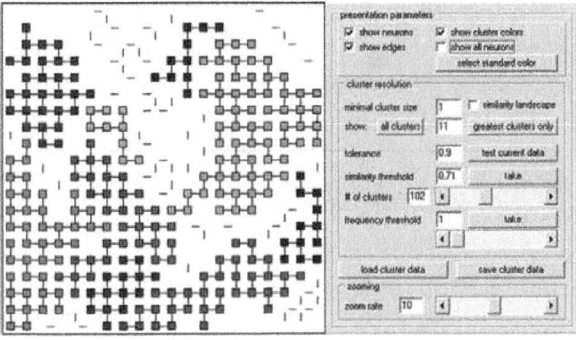

Figure 10. Left graphic: Blue marked neurons represent the meta-pattern of squash rallies. Right graphic: Individual distribution of rally frequencies, indicated by the diameters of the red marked neurons.

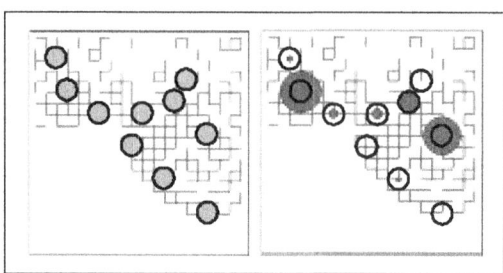

Figure 11. Two games of player B, showing different strategic patterns, which are not player specific, but game specific

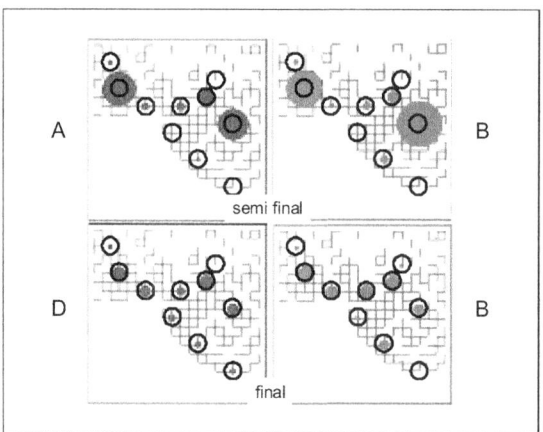

independent playing behaviour (Figure 10, left graphic, blue marked neurons). The player-specific behaviour then is more or less a sub-pattern of that meta-pattern (right graphic, red marks), which characterizes his/her individual strategy. While the marked neurons themselves stay unchanged, the diameters of the graphical representations of the neurons vary depending on the players and so indicate the player-dependent frequencies of the corresponding rallies.

Surprisingly enough, the individual patterns turned out not to be athlete-specific, but to be specific to the pair of opponents. Even in a high level tournament (as the data are from) the players did not (or could not) put through their individual strategy.

Figure 11 demonstrates that player B has quite different rally frequencies in the game against player A, compared to the game against player D. In turn, during a game the two opponents reach something like a common playing rhythm, which in Figure 11 is indicated by rather identical frequency distributions in both examples (also see Perl, 2001, 2002; Lames & Perl, 1999).

The last example shows some (unpublished) results obtained by J.P. considering post-operative rehabilitation of risk-patients. After the operation, psychological items were recorded from the patients for 7 days (3 times a day, 10 values). Among others, these items regarded their pain and their anxiety in order to analyse and improve the rehabilitation process. The problem was that the answers were quite subjective, with the values depending on the complex situation and feeling. Therefore, the combination of a rather high number of items and the small number of repetitions seemed to prevent the detection of characteristic patterns. The net-based method we used was similar to the rehabilitation process from above (see Figure 7). From the very few recorded data, the amount of data necessary for DyCoN-training was Monte Carlo-generated and the net was trained. The data representation then was cut into segments of one-day length, where the three neurons that represented the respective three daily data sets were connected by a "daily trajectory" (see Figure 12).

The coloured areas in the net shown in Figure 12 contain neurons of equal or similar meaning or patient states. This way, the 10-dimensional data sets could be reduced to 2-dimensional neuron positions that can much easier be followed and interpreted. Figure 13 shows one typical result, where during the third, fourth, and fifth day the processes are quite similar (positive during the days and reset to a negative state during the respective nights). During the sixth and the seventh day the processes run inverse, indicating a dramatic change to a negative state.

In comparison, Figure 14 shows the example of another patient with an obviously quite different rehabilitation process. The day-night rhythm starts to change already

Figure 12. Example of a daily trajectory, starting in the red neuron bottom-left and ending in the yellow neuron top-left

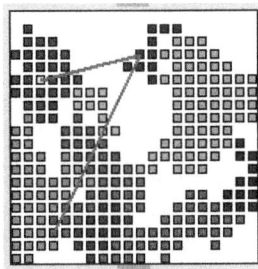

Figure 13. Daily trajectories, in the days 3, 4 and 5 running "from right to left," and in the days 6 and 7 running "from left to right"

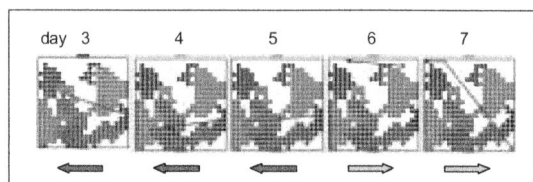

The change of trajectory directions between the fifth and the sixth day indicates a change to a negative state

Figure 14. Trajectories of a process that is starting to stabilize after the fourth day

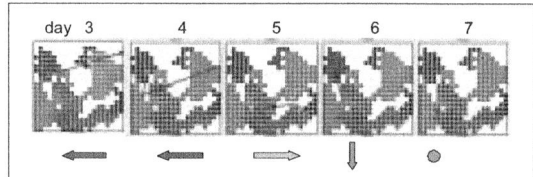

after the fourth day. But in contrast to Figure 13, this change indicates the start of stabilization that is finished after the seventh day in a positive state.

As is demonstrated in the example above, the problem of analysing human behavioural processes is that they form time-dependent structures of situations, which differ inter-individually and intra-individually, depending on types and contexts of activities. Processes can be described as time-dependent trajectories of single patterns of situations, where the trajectories are patterns on a higher level (with the situation patterns as components) and so can be analysed using the methods mentioned above. This approach is particularly useful if the complex situation patterns can be replaced by their much simpler type specifications. In this way, the high-dimensional original process can be replaced by a low-dimensional representing trajectory, which is much easier to handle and to analyse.

Although a net-based pre-processing of patterns improves the trajectory handling, the problem remains if and how trajectories can be compared for intra- and inter-individual analysis. It turns out that due to technical reasons similarity of movement-trajectories is difficult to evaluate, as will be discussed in the following section. Therefore, the idea of taking such trajectories as meta-patterns and again using a net for clustering can get to work, but needs some preparation by means of conventional data processing methods, which are known, for instance, from voice recognition.

Process Analysis in Sport Science

Processes in sport can be described as time series of patterns, which can as well characterize situations (e.g., positions on the playground or angles of articulations) as activities (e.g., moving of players or angle speeds). As has been discussed before, patterns can be learned and recognized by means of Artificial Neural Networks of the KFM-type.

Therefore, networks like KFM or DyCoN can help to analyse processes in sport, as has been done in several approaches (see Lames & Perl, 1999; Schöllhorn & Perl, 2002; Schöllhorn et al. 2002).

In the following, two examples will be presented that belong to the area of biomechanics/motor control and help to understand what the problems are and how networks can help to handle them.

The first example deals with running. The aim was to compare different kinds of running (e.g., jogging and sprinting) intra- as well as inter-individually (see Lippolt et al. 2004). The articulations of the runner's legs were marked, and the position-, velocity- and acceleration-attributes were recorded automatically. A total of 20 attributes were recorded per time step (1/100 sec.). We fed a properly pre-trained net with the process representing time series of that 20-dimensional attribute vectors and got the 2-dimensional trajectories of the running process (see Figure 15, left graphic). One first result was that the number of attributes could be reduced dramatically without losing important information (see Figure 15, right graphic).

On the basis of those trajectories it seemed much easier to calculate intra- and inter-individual comparisons. It seemed even possible to understand the trajectories as high-dimensional patterns and compare them by means of an appropriate neural network. This in principle is a good idea, which for similar problems works perfectly.

It turned out, however, that the trajectories of running processes contain a number of measurement artefacts, in particular regarding the synchronisation of starting and ending points of a step. The beginning and the end of a step neither intra-individually nor inter-individually are well-defined quantities, but depend on the type of running and

Figure 15. Trajectories of one step of a running process, beginning with "x" and ending with "o"

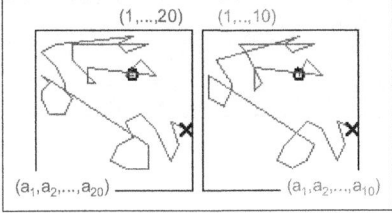

The left trajectory uses the complete set of 20 attributes, while the right trajectory only uses the 10 most relevant of the attributes

Figure 16. Similarity of the main parts of two trajectories: The left one reduced by its (green) tail, the right one reduced by its (blue) starting part

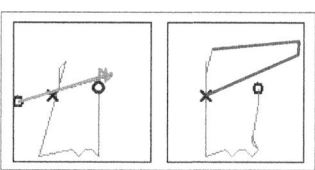

the type of runner. Hence, the recorded intervals do not necessarily match each other and so were not comparable. Therefore, a pre-processing was necessary in order to reduce the complete sets of recorded data to their comparable main parts (see Figure 16).

In Figure 16 the situation is sketched for the comparison of two trajectories, where the left one got cut off at its tail, while the right one got cut off at its starting part. The remaining main parts then show a high degree of similarity (see Perl, 2004b). Additional problems can result from the relation between leg length, step width, and velocity and require different pre-processing like stretching or scaling. Similar problems are well-known, for example, from the field of voice recognition (Mehler, 1994).

Using ergometers, as has been done in the following last example of rowing, prevents problems of this kind. Every data were recorded automatically by means of sensors, and the synchronisation was defined by the ergometer's internal clock. The three phases of interest of the rowing process were the drive phase acceleration, the drive phase deceleration, and the recovery phase acceleration (also see Figure 17).

The results from the net-based trajectory analysis were as exemplarily presented in Figure 18 Inter-individual comparisons showed a high degree of similarity, which was not too surprising because of the normalizing factor of the ergometer. Intra-individual

Figure 17. The three phases of interest of the rowing process

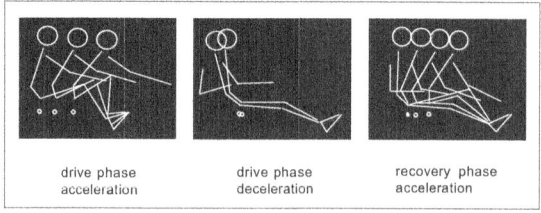

Figure 18. Trajectories of sequences of strokes

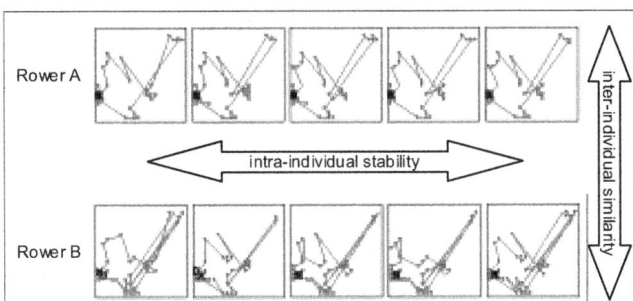

comparisons, however, showed significant differences. One aspect that is focused on in Figure 18 is the stability: Rower A shows a high degree of stability, while rower B does not.

For a better understanding of the trajectories, Figure 19 shows the correspondences between the different parts of a trajectory and the three phases of the stroke. It can be seen at first sight that the drive phase acceleration is the phase of instability of rower B, followed by the drive phase deceleration. In contrast, the recovery phase acceleration seems to be rather stable, although clearly different from the recovery phase acceleration of rower A.

A more detailed analysis, however, showed that the small systematic difference between the recovery phase acceleration of rower A and B meant a significant difference in the respective techniques: Rower B always added a hand rotation that was too small to be striking and so could not be recognized by video-based analysis. Nevertheless, it costs some 1/100 seconds each, which in the result can sum up significantly. Once detected from the trajectories, the hand rotation could be recognized even from the video pictures and also analysed from the recorded data.

Figure 19. Correspondences between trajectory and stroke phases

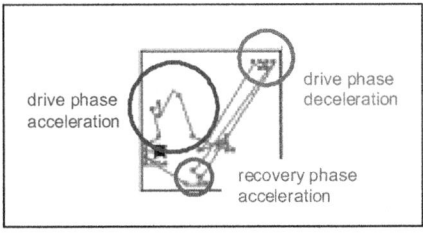

Figure 20. Stroke trajectories for different rowing velocities

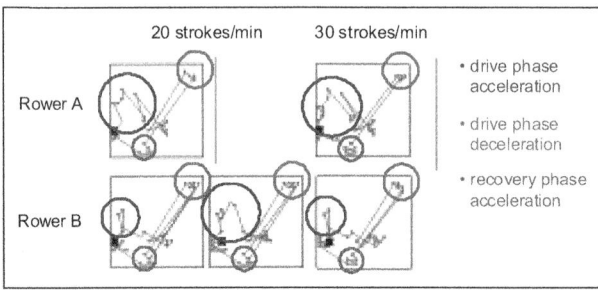

Finally, it turned out that there are significant differences with regard to the velocities (see Figure 20). While the strokes of rower A were more or less independent of the rowing velocity, the strokes of rower B varied with the velocity. In particular in cases of low velocity, the strokes seem to become a bit unstable. (Note that both athletes are top level national rowers.)

We asked athletes as well as coaches about this phenomenon. The congruent interpretation was that in cases of high velocity motor control worked automatically, while in cases of low velocity, intentional elements disturbed the internally controlled rhythm.

CONCLUSION

As has been shown, there are several problems where net-based analyses can help to detect and better understand effects and phenomena, which are normally hidden in a huge amount of complex data. Therefore, an appropriate combination of data recording, numerical analysis and video analysis on the one hand with net-based pattern recognition on the other hand seems to meet various demands of motor analysis in a quite satisfying way.

It should be noted, however, that an appropriate pre-processing of the raw data is often necessary in order to successfully use the networks for pattern analysis. Therefore, according to our experience, sport scientists or practitioners should collaborate with experienced colleagues from computer science.

REFERENCES

Boll, M. (1999). *Analyse von Verhaltensprozessen mit Hilfe Neuronaler Netze.* Master's thesis, Mainz: Johannes Gutenberg-Universität, Institute of Computer Science.
Fritzke, B. (1995). A growing neural gas network learns topologies. In G. Tesauro, D. S. Touretzky & T. K. Leen (Eds.), *Advances in neural information processing systems 7* (pp. 625-632). Cambridge, MA: MIT.

Fritzke, B. (1997). A self-organizing network that can follow non-stationary distributions. In *Proceedings of ICANN97, International Conference on Artificial Neural Networks*, (pp. 613-618). Berlin; Heidelberg; New York: Springer.

Goodson, C. (2004). *Classification of visual sensor data for an autonomous robot*. Bachelor thesis, Mainz: Johannes Gutenberg Universität, Institute of Computer Science.

Hebb, D. O. (1949). *The organization of behavior*. New York: John Wiley.

Holzreiter, S. (2005). Autolabeling 3D tracks using neural networks. *Clinical Biomechanics, 20*(1), 1-8.

Hopfield, J. J. (1982). Neural networks and physical systems with emergent collective computational abilities. In *Proceedings of the National Academy of Sciences* (Vol. 79, pp. 2554-2558).

Jouraai, M. (2003). *Kopplung von Selbstorganisierenden Karten mit Datenbanken zur Prozeßanalyse*. Master's thesis, Mainz: Johannes Gutenberg-Universität, Institute of Computer Science.

Kantz, H., & Schreiber, T. (2004). *Nonlinear time series analysis*. Cambridge, MA: Cambridge University.

Kohonen, T. (1981). Automatic formation of topological maps of patterns in a self-organizing system. In E. Oja & O. Simula (Eds.), In *Proceedings of 2SCIA, Scand. Conference on Image Analysis* (pp. 214-220). Helsinki: Suomen Hahmontunnistustutkimuksen.

Kohonen T. (1982). Self-organized formation of topologically correct feature maps. *Biological Cybernetics, 43*, 59-69.

Kohonen T. (1995). *Self-organizing maps*. Berlin; Heidelberg; New York: Springer.

Lames, M., & Perl, J. (1999). Identifikation von Ballwechseltypen mit Neuronalen Netzen. In K. Roth, Th. Pauer & K. Reichle (Eds.), *Dimensionen und visionen des sports. Schriften der dvs, 108*, 133.

Lippolt, T., Schöllhorn, W. I., Perl, J., Bohn, C., Schaper, H., & Hillebrand, Th. (2004). Differenzielles Training im Leichtathletischen Sprint – Koordinierung von Sprintkoordinationsübungen mit Simulation und Optimierung eines Trainingsprozesses. In Bundesinstitut für Sportwissenschaft (Ed.), *BISp Jahrbuch 2003* (pp. 267-274). Bergheim: Druckpunkt Offset.

McCulloch, W. S., & Pitts, W. (1943). A logical calculus of the idea immanent in nervous activity. *Bulletin of Mathematical Biophysics, 5*, 115-133

Mehler, F. (1994). *Selbstorganisierende Karten in Spracherkennungssystemen*. PhD thesis. Mainz: Johannes Gutenberg-Universität, Institute of Computer Science.

Minsky, M., & Papert, S. (1969). *Perceptrons: An introduction to computational geometry*. Cambridge, MA: MIT.

Perl, J. (2001). DyCoN: Ein dynamisch gesteuertes neuronales netz zur modellierung und analyse von prozessen im sport. In J. Perl (Ed.), *Sport & Informatik VIII* (pp. 85-98). Köln: Strauß.

Perl, J. (2002b). Game analysis and control by means of continuously learning networks. *International Journal of Performance Analysis of Sport, 2*, 21-35.

Perl, J. (2004a). Artificial neural networks in motor control research. *Clinical Biomechanics, 19*(9), 873-875. Retrieved from http://authors.elsevier.com/sd/article/S0268003304000774

Perl, J. (2004b). A neural network approach to movement pattern analysis. *Human Movement Science*, *23*, 605-620. Retrieved from http://authors.elsevier.com/sd/article/S0167945704000818

Rebel, M. (2004). Wenn der Kopf in Die Knie geht – Analyse von Rehabilitationsverläufen nach Kreuzbandrekonstruktionen. *Schriften zur Sportwissenschaft, 51.* Bonn: Kovac.

Ritter, H., Martinetz, T., & Schulten, K. (1992). *Neural computation and self-organizing maps: An introduction.* New York: Addison-Wesley.

Rosenblatt, F. (1958). The perceptron: A probabilistic model for information storage. *Psychological Review, 65*, 386-408.

Rumelhart, D., & McClelland, J. (1986). *Parallel distributed processing. Vol.1+2.* Cambridge, MA: MIT.

Schöllhorn, W., & Perl, J. (2002). Prozessanalysen in der Bewegungs- und Sportspielforschung. *Spectrum der Sportwissenschaften, 14*(1), 30-52.

Schöllhorn, W. I., Schaper, H., Kimmeskamp, S., & Milani, T. L. (2002). Inter- and intra-individual differentiation of dynamic foot pressure patterns by means of artificial neural nets. *Gait & Posture, 16*, 159.

Schöllhorn, W. (2004). Applications of artificial neural nets in clinical biomechanics. *Clinical Biomechanics, 19*(9), 876-898.

Uthmann, T., & Dauscher, P. (2005). Analysis of motor control and behavior in multi agent systems by means of artificial neural networks. *Clinical Biomechanics, 20*(2), 119-125.

Wünstel, M., Polani, D., Uthmann, T., & Perl, J. (2001). Behavior classification with self-organizing maps. In P. Stone et al. (Eds.), *RoboCup 2000*, LNAI 2019 (pp. 108-118), Berlin; Heidelberg; New York: Springer.

Section IV

Computational Modelling for Predicting Movement Forces

Chapter XII

Estimation of Muscle Forces About the Ankle During Gait in Healthy and Neurologically Impaired Subjects

Daniel N. Bassett, University of Delaware, USA

Joseph D. Gardinier, University of Delaware, USA

Kurt T. Manal, University of Delaware, USA

Thomas S. Buchanan, University of Delaware, USA

ABSTRACT

This chapter describes a biomechanical model of the forces about the ankle joint applicable to both unimpaired and neurologically impaired subjects. EMGs and joint kinematics are used as inputs and muscle forces are the outputs. A hybrid modeling approach that uses both forward and inverse dynamics is employed and physiological parameters for the model are tuned for each subject using optimization procedures. The forward dynamics part of the model takes muscle activation and uses Hill-type models of muscle contraction dynamics to estimate muscle forces and the corresponding joint moments. Inverse dynamics is used to calibrate the forward dynamics model predictions of joint moments. In this chapter we will describe how to implement an EMG-driven hybrid forward and inverse dynamics model of the ankle that can be used in healthy and neurologically impaired people.

INTRODUCTION

It is difficult to estimate muscle forces during human movements. Such forces cannot be measured directly apart from implantation of force sensors into the muscles — a painful procedure that is difficult to use on multiple muscles and would likely cause unnatural changes in the way such muscles are activated (e.g., limping). Although biomechanical models are commonly used to analyze human motion, estimation of individual muscle forces remains problematic as it is a difficult modeling task to do with reasonable accuracy. The main reason for this difficulty is that the musculoskeletal system is redundant: there are an infinite number of combinations of muscle forces that could create a desired limb movement or, in most cases, even a single joint moment.

Nevertheless, there are two approaches that have been used with reasonable success. The first uses a mathematical approach, calculating muscle forces based on optimization of muscle forces according to a user-defined cost function (e.g., Neptune, 1999; Anderson & Pandy, 2001). The other approach uses information about how the nervous system activates the muscles, as recorded from electromyograms (EMGs) and from this the muscle forces are estimated (e.g., Lloyd & Besier, 2002; Buchanan et al., 2004). The later approach will be the focus of this chapter.

Accurate estimation of muscle forces is important to study the neural control of movement, design of advanced prosthetics and to provide clinical tools for rehabilitation. In this chapter the focus will be on applications pertaining to the control of the human ankle and will conclude with an application to gait in a subject who has had a stroke.

MODEL FLOW

There are two fundamentally different approaches employed to study human movement and kinetics: forward and inverse dynamics. Each has strengths and weaknesses and we shall use a hybrid method that takes advantage of the strengths of each method (Figure 1).

Inverse dynamics relies on information about position and ground reaction forces. By simple application of fundamental rigid-body dynamics (e.g., Newton's Laws), joint moments can be calculated from the kinematic and external force data. From joint moment and musculoskeletal geometry one might try to calculate muscle forces. However, the body is a redundant system and multiple muscles contribute to each joint moment. Therefore, the force distribution among the muscles that span a joint cannot be readily determined. In our application, inverse dynamics will only be used to determine joint moments.

Forward dynamics depends on neural command, which can be estimated from electromyograms. EMG is measured directly from each of the interested muscles. *Muscle activation dynamics* is the process by which the EMG (measured in millivolts) is converted to muscle activation, $a(t)$, a dimensionless time varying signal between zero and one (Figure 1). In combination with joint position and velocity data, muscle activation is used to estimate muscle force by means of a process called *muscle contraction dynamics*. The total joint moment due to the muscle forces is easily computed from forces and corresponding muscle-tendon moment arms. In a typical forward dynamic analysis, from here joint kinematics are determined. However, joint accelerations are not easily calculated. Inertial properties of each body segment, external forces and moments must be known to calculate joint accelerations from joint moments. For this reason, in our modeling, forward dynamics will only be used to obtain joint moment.

The hybrid method, described in this chapter, uses the inverse dynamics calculation of joint moment as a benchmark to which the forward dynamics joint moment is optimized. These two joint moments are compared and the differences between them are minimized by adjusting parameters in the model. These model parameters will be described in more detail in subsequent sections. Every time a parameter is changed, a new forward dynamics solution is determined, achieving a new joint moment. If the comparison with the inverse dynamics joint moments is better, the improvement is noted, otherwise the change is discarded. The cycle continues until the joint moments match. The muscle forces are the final output of interest and from these, other important forces can be determined, such as tendon strain, joint compressive forces and ligament loads.

Figure 1. Theoretical flow of a hybrid forward and inverse dynamics muscle force estimating model, made possible by means of optimization

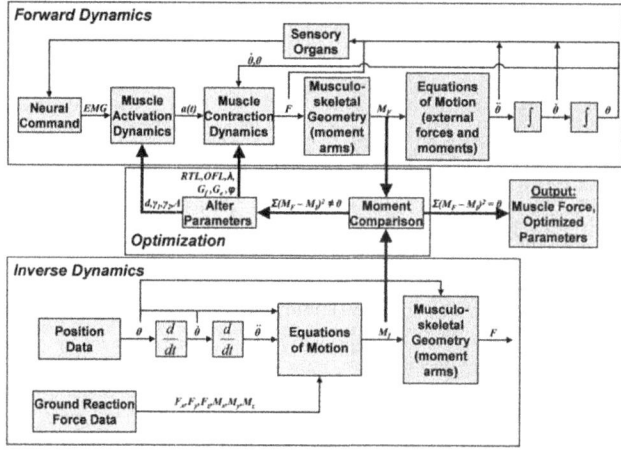

Legend: EMG = *electromyograms;* a(t) = *muscle activation;* F = *muscle force;* M_F = *total joint moment from forward dynamics;* M_I = *total joint moment from inverse dynamics;* q *(and derivatives)* = *joint angle, joint angular velocity, joint angular acceleration;* d = *electromechanical delay;* g_1 *and* g_2 = *recursive filter parameters;* A = *shape factor;* RTL = *resting tendon length;* OFL = *optimal fiber length;* G_f *and* G_e = *flexor and extensor gain coefficients;* f = *pennation angle;* $F_x, F_y, F_z, M_x, M_y, M_z$ = *ground reaction forces and moments in all three directions.*

Optimization of the hybrid method is referred to as model calibration or tuning. Once the model has been calibrated, the inverse dynamics part of the model can be removed and the model can be used to predict joint moments and muscle forces for novel trials. Furthermore, the model's accuracy in predicting joint moments can be determined at any time by comparing with the inverse dynamics solutions.

FORWARD DYNAMICS

Forward dynamics modeling takes an approach similar to that of the human neuromuscular control system. To produce movement the brain must trigger nerve fibers, which in turn activate the muscles. As the muscles produce moments about a joint, physical movement occurs. Similarly in forward dynamics, estimated neural command data are used to calculate muscle forces, which produce a torque about a joint and consequently motion. The measure of this motion is the angle between the involved body segments, represented by θ in Figure 1.

A further parallel between forward dynamics and human biomechanical properties exists. Position, orientation, and motion of body segments do not solely depend on the current innervation of the muscle fibers. They are also affected by the previous history of applied forces (Zajac, 1993).

Error in forward dynamics is primarily due to its subjectivity. The variability of EMG signals makes the estimation of muscle activation difficult. Determination of muscle-tendon moment arms can be very difficult, especially as joint angles change. Finally, considerable error is introduced in finding position from joint torque. Using a hybrid method reduces the inherent errors of forward dynamics and eliminates integration to obtain joint position.

Data Inputs

The input in the forward dynamics loop of Figure 1 is EMG. However, joint moment cannot be determined from EMG alone. In the flow chart, two feedback loops are present. The first, involving sensory organs, sends information back to the nervous system. Given that the first input is EMG, which is an estimation of neural command, the model is already accounting for this feedback loop. Therefore, an explicit formulation of the sensory loop is not included in the model.

The second loop supplies muscle contraction dynamics with muscle-tendon length (l_{mt}) and muscle-tendon velocity (v_{mt}), which are estimated from joint position and velocity (Delp & Loan, 1995). To reduce error introduction, forward dynamics is used only up to joint moment calculation, giving no position or velocity to feedback. In inverse dynamics position data are an input. Therefore, the data are available and the feedback can be replaced by an input.

The last necessary inputs are tendon moment arms. These are difficult to measure in cadavers and even more difficult to estimate *in vivo*, although some promising methods have been developed using MRI-based methods (Arnold et al., 2000). Moment arms (r_{ma}) can be estimated from models of muscle paths and joint kinematics. Moment arm is the partial derivative of muscle length with respect to joint angle (An et al., 1984). The accuracy of this approach depends upon the accuracy of the musculoskeletal geometry in the model.

Figure 2. Flow of EMG processing

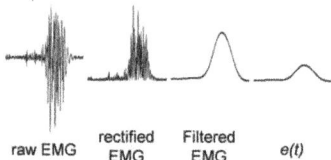

raw EMG   rectified EMG   Filtered EMG   *e(t)*

The raw EMG signal has already been high-pass filtered. The offset has been eliminated from the rectified EMG data. Filtered EMG has been run through a Butterworth filter double-pass. EMG activation has been normalized to a maximum EMG value.

Muscle Activation Dynamics

EMG is not a measure of muscle force or of muscle activation. It is the measure of the electrical signal passing over the muscle inducing contraction. For it to be a useful tool it must be transformed into muscle activation. Raw EMG is simply voltage such as that shown in Figure 2. Muscle activation, on the other hand, is a value between zero and one. It accounts for the electromechanical delay in muscles as they contract, and is non-linearly related to EMG, being a function of its recent activation history.

EMG Processing

The raw EMG signal must be processed before it can be used in the model. First, low-frequency noise, such as signal drift or skin movement, should be removed from the signal by applying a high-pass filter in the 5-30 Hz range. Next the signal should be rectified. The result of rectification is values all greater than zero. The rectified EMG signal, at this point, still contains high-frequency noise like electrical interference or artifacts due to motion. To isolate the actual muscle activity, a low-pass filter with a cut-off frequency between 3 and 10 Hz should be used.

Finally, the EMG signal must be normalized. Normalization allows for comparison between muscles by making the activation value a percentage of maximum activation (Fuglsang-Frederiksen, 2000; Merletti & Lo Conte, 1997). Though there are other methods (Enoka, 2002; Mathiassen et al., 1995; Keen et al., 1994), the value achieved during maximum voluntary contraction (MVC) is the method we use, as done by Manal et al. (2003) and Woods and Bigland-Ritchie (1983). Different trials for each muscle to produce an MVC value should be performed to avoid normalized data exceeding a value of one. Once the normalization is performed, EMG processing is complete, delivering *e(t)*.

Neural Activation

The muscle does not contract at the instant in time the motor unit is triggered. There is a time delay (*d*) during which the muscle is preparing to produce force. Once the muscle begins contracting, the tension ramps up to a peak, which is not coincident with the peak EMG (Herzog et al., 1998; Guimarães et al., 1995; Corcos et al., 1992).

The transformation is not a simple phase shift. A muscle twitch is the smallest and quickest contraction. When a muscle is triggered by a nerve fiber, it twitches and produces force. If triggered again, more force is added to the force from the first twitch. If the triggering frequency is relatively low, the second force is added to the already decaying first force. If the triggering frequency is high enough and sustained, the forces form a smooth ramping curve achieving a fused tetanus contraction (Windmaier et al., 2004; Macefield et al., 1996; Macefield et al., 1991; Luciano et al., 1978). Therefore, neural activation must account for both electromechanical delay and previous activation history.

Zajac (1989) suggested that a relationship between $e(t)$ and muscle activation ($a(t)$) could be modeled with a first order differential equation. Others reported that a second order relationship provides a better depiction of activation dynamics. Thelen (1994) used a first order recursive filter to obtain activation data and later others have used recursive filters based on descritized forms of second order differential equations (Lloyd & Bessier, 2002; Buchanan et al., 2004) as follows:

$$u(t) = \alpha \cdot e(t - d) - \beta_1 \cdot u(t - 1) - \beta_2 \cdot u(t - 2) \qquad (1)$$

Here, t is time, d is the electromechanical delay, α is the gain coefficient for the muscle, γ_1 and γ_2 are the recursive filter coefficients. For this recursive filter to be stable (satisfy the Nyquist stability criteria), Buchanan et al. (2004) showed that following conditions must hold true:

$$\beta_1 = \gamma_1 + \gamma_2$$
$$\beta_2 = \gamma_1 \cdot \gamma_2$$
$$|\gamma_1| < 1 \; and \; |\gamma_2| < 1 \qquad (2)$$

$$\alpha - \beta_1 - \beta_2 = 1 \qquad (3)$$

Equations (2) and (3) define the three coefficients of (1) (α, β_1, β_2) by using only two parameters (γ_1 and γ_2). The recursive components make each time step of EMG activation effect the subsequent two. Electromechanical delay assists in the synchronization between muscle activation and force output. The values of these three parameters cannot be determined experimentally, so a mathematical approach (i.e., optimization analysis) must be used to determine them.

Muscle Activation

Muscle activation dynamics is sometimes considered completed once neural activation has been computed. This implies a linear relationship between isometric EMG and steady-state muscle force. However, this is not necessarily the case. Woods and Bigland-Ritchie (1983) showed that many muscles are non-linearly related to force for the first 30% of activation. Manal and Buchanan (2003) have developed a model of the transformation from neural activation to muscle activation that characterizes the low-level, non-linear region of this relationship as a logarithmic function and the later 70% as a linear function. This relationship can be described by a single parameter, A, that

Figure 3. Graph of one-parameter non-linearization of neural activation to muscle activation

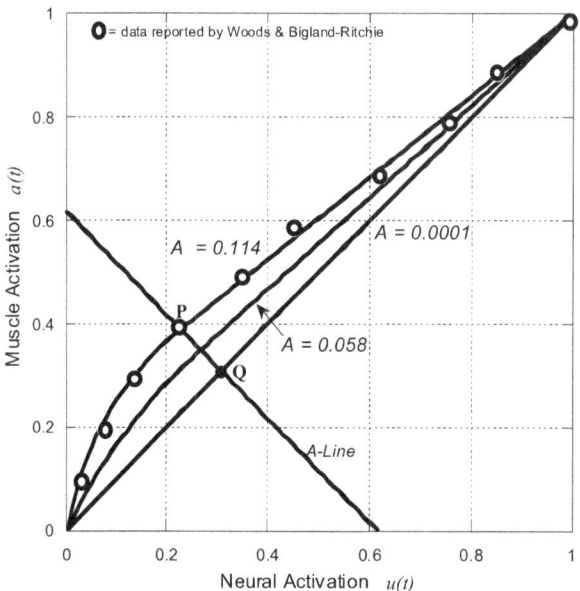

Based on the value of A, the relationship between u(t) and a(t) is defined, and shown to match the results found by Woods (1983).
Legend: A = shape factor measured from P to Q along the A-Line; P = point defined by A; Q = starting point from which A is measured, defined by Manal (2003) as (0.3085, 0.3085).

characterizes the degree of non-linearity (Figure 3). The governing equations are as follows:

$$a(t) = d \cdot Ln(c \cdot u(t) + 1) \qquad 0 \le u(t) < Q_x$$
$$a(t) = m \cdot u(t) + b \qquad\qquad Q_x \le u(t) \le 1 \tag{5}$$

Equation (5) is the mathematical definition of Figure 3, where Q_x is the x coordinate of point Q (0.3085); $d, c, m,$ and b = constants determined from the value of A. This model introduces only one unknown parameter to the forward dynamics process and produces curves similar to experimental results.

Muscle Contraction Dynamics

Muscle contraction dynamics is the process by which force is computed from muscle activation. A Hill-type model is generally used for modeling systems with more

Figure 4. Hill-type model

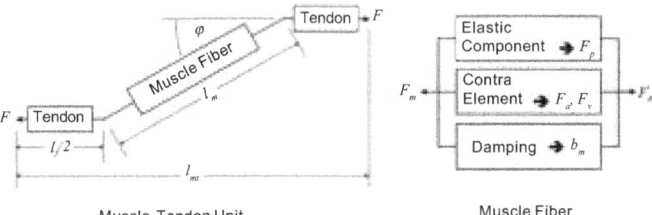

Muscle-Tendon Unit   Muscle Fiber

Muscle-tendon unit has the muscle fiber in series with the tendons. Therefore, the total muscle-tendon force (F) is passing through both components. The pennation angle (ϕ) is the angle between the lines of action of the tendons and the muscle fiber. l_m is the length of the muscle, l_t is the total length of the tendons (each one is half of the length), and l_{mt} is the muscle-tendon length. The muscle fiber produces the total muscle force (F_m), and is comprised of three parallel components. The contractile element produces the force depending on fiber length and fiber velocity. From fiber length active force (F_A) is generated, and from fiber velocity the velocity-dependent force is generated (F_V). The elastic component gives passive force (F_P), and damping is quantified by b_m.

than one muscle. While more complex muscle models that characterize cross-bridge interactions are available (e.g., Huxley, 1958), they are very computationally taxing for modeling multiple muscles and often introduce many parameters that are difficult to determine. Hill-type models, on the other hand, usually require only one differential equation per muscle, making them the most commonly used approach when modeling multiple muscles.

Figure 4 illustrates the design of the Hill-type model that is used in this chapter. The corresponding mathematical model is:

$$F = F_{max}\left[\tilde{F}_A\left(\tilde{l}_m\right)\cdot \tilde{F}_V\left(\tilde{v}_m\right)\cdot a(t) + \tilde{F}_P\left(\tilde{l}_m\right) + b_m\cdot \tilde{v}_m\right]\cdot \cos(\varphi) \qquad (6)$$

where, F is the muscle-tendon force, $\tilde{l}_m$ is the normalized muscle fiber length, $\tilde{v}_m$ is the normalized muscle fiber velocity. F_A, F_V, and F_P are normalized components of force. We use normalized components of force assuming all the muscles we model exhibit the same properties. To "un-normalize" the force it must be transformed from a percentage of the maximum force (F_{max}) to a quantified estimation of force. Note that muscle activation affects active force and velocity-dependent force, not passive force.

Active and Passive Force

Active and passive forces are related in that they are both dependent on the length of the muscle fiber. The active component (F_A) is due to the actual contraction of a muscle, whereas the passive element (F_P) exerts a resistive force as it is stretched.

Figure 5. Force-length relationship

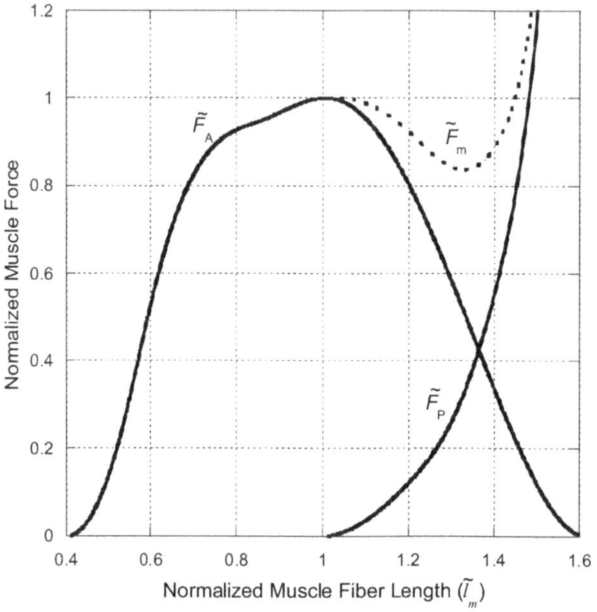

Legend: $\tilde{F}_A$ = *normalized active force curve,* $\tilde{F}_P$ = *normalized passive force curve, and* $\tilde{F}_m$ = *total normalized isometric muscle force due to length.*

The active component can be thought of as a linear motor. Force is created when the myosin pulls the actin filaments. At optimal overlap between the two myofilaments, maximum force is produced (Windmaier et al., 2004; Enoka, 2002; Lieber et al., 1994; Gordon et al., 1966). When the sarcomeres are at this optimal length, the muscle is at optimal fiber length (*OFL*). At greater or lesser lengths, force output will be less than maximal. We model the relationship between force and length in the same way for all interested muscles. Consequently both force and fiber length must be normalized: force by max force, and fiber length by optimal fiber length. This relationship is complex and although it has been modeled as a second order polynomial (Woittiez, 1984), it is better represented as a graphical relationship (Gordon, 1966) (Figure 5).

Passive force is exerted resistively, increasing exponentially at fiber lengths greater than *OFL*. It is due to the intrinsic elasticity of the muscle fibers. Schutte (1993A) modeled the passive force-length relationship with an exponential relationship:

$$\tilde{F}_P = \frac{e^{10(\tilde{l}_m - 1)}}{e^5} \tag{7}$$

Equation (7) is a good approximation of actual data. In modeling it is used to obtain passive force values regardless of fiber length (Buchanan et al., 2004; Lloyd & Besier, 2003).

Under isometric conditions total force of the muscle (F_m) is given by the sum of the normalized active and passive elements of force multiplied by the maximum isometric force:

$$F_m = F_{max} \cdot \left(\tilde{F}_A(\tilde{l}_m) \cdot a(t) + \tilde{F}_P(\tilde{l}_m) \right) \tag{8}$$

Fiber-Velocity-Dependent Force

Hill (1938) reported a relationship between force and velocity. Force production potential decreases due to velocity during muscle shortening, whereas the inverse is true in a lengthening muscle (Epstein, 1998; Kroll, 1988; Edgerton et al., 1986; Hill, 1938).

Figure 6 illustrates the relationship of force-to-fiber velocity. Note that it is not trivial to calculate muscle fiber velocity (v_m). Using an inverted force-velocity curve, Epstein and Herzog (1998) developed two mathematical relationships to compute velocity-dependent force (F_V). Buchanan et al. (2004) and Lloyd and Besier (2003) apply Epstein and Herzog's equations to the traditional curve.

$$\tilde{F}_{V(concentric)} = \frac{d - c \cdot \tilde{v}_m}{d + \tilde{v}_m} \tag{9}$$

Equation (9) is for concentric contractions. The normalized constants c and d can be equated (Lloyd, 2003). The fiber velocity ($\tilde{v}_m$) has been normalized by optimal fiber lengths per second. The equation for eccentric contractions is a bit more involved:

$$\tilde{F}_{V(eccentric)} = F_{ecc} - (F_{ecc} - 1)\frac{d' + c' \cdot \tilde{v}_m}{d' - \tilde{v}_m} \tag{10}$$

Here, c' and d' are also equated normalized constants, and F_{ecc} is the maximum eccentric force multiplier quantified between 1.1 and 1.8 (Epstein & Herzog, 1998).

Figure 6 is a graphical representation of the relationship between force and velocity (Hill, 1938) according to (9) and (10).

$$F_v = F_{max} \cdot \tilde{F}_v \tag{11}$$

Under isometric conditions — when muscle fiber velocity is zero — the normalized muscle force has a value of one, which is when the maximum isometric force (F_{max}) occurs. At all other velocities the maximum force produced is a greater or lesser percentage of F_{max}.

To implement the velocity-dependent component in the muscle force calculation, (11) is inserted into (8).

$$F_m = F_{max} \cdot \left(\tilde{F}_A(\tilde{l}_m) \cdot \tilde{F}_v(\tilde{v}_m) \cdot a(t) + \tilde{F}_P(\tilde{l}_m) \right) \tag{12}$$

Figure 6. Force-velocity relationship

Figure 6. Force-velocity relationship

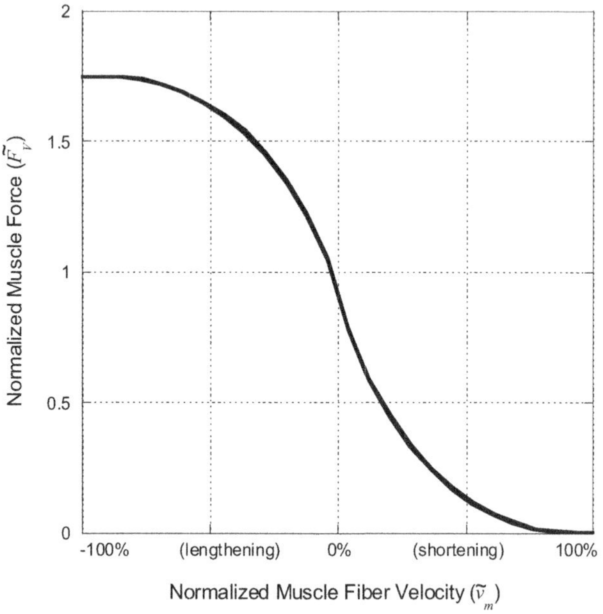

Damping

As seen in Figure 4, a damping component can be modeled parallel to the contractile and elastic elements (Schutte, 1993A). Muscles have intrinsic damping characteristics; hence, this property should be included in a muscle model. The advantage of including passive damping in a model is the stability it provides. It also eliminates the need to invert the force-velocity relationship (Schutte et al., 1993B). To justify the incorporation of the velocity-dependent force, the damping factor must be included in (12) as well:

$$F_m = F_{max} \left[\tilde{F}_A \left(\tilde{l}_m \right) \cdot \tilde{F}_V \left(\tilde{v}_m \right) \cdot a(t) + \tilde{F}_P \left(\tilde{l}_m \right) + b_m \cdot \tilde{v}_m \right] \qquad (13)$$

where b_m is the damping factor. Lloyd and Besier (2003) reported the value b_m to be 0.1.

Pennation Angle

The pennation angle is the angle between the line of action of the tendons and the principle direction of the muscle fibers (Figure 4). This angle varies between muscles, and in some cases it is small enough that its effects can be negligible (Yamaguchi et al., 1990). However, it is best to always account for the pennation angle, especially for muscles in the leg as some have large pennation angles (e.g., soleus). A simple trigonometric relationship exists, and can be modeled by projecting F_m in line with the tendons:

$$F = F_m \cdot \cos(\varphi) \tag{14}$$

The pennation angle ϕ is often modeled as constant, but in reality it changes substantially depending on the fiber length and optimal fiber length (Epstein & Herzog, 1998; Scott & Winter, 1991).

$$\varphi(t) = \sin^{-1}\left(\frac{OFL \cdot \sin(\varphi)}{l_m(t)}\right) \tag{15}$$

Equation 15 characterizes the pennation angle (ϕ (t)) as it changes with respect to time depending on changes in the muscle fiber length ($l_m(t)$). Consistent with other equations that depend on l_m, in equation 15 it is being normalized by the optimal fiber length, OFL. It is possible to experimentally determine values for optimal fiber length in cadaver muscles (e.g., Murray et al., 1995). However, these are tedious to obtain and difficult to scale to individual subjects (Murray et al., 2002). Nevertheless, it is common to use published results taken from cadaver studies such as Yamaguchi et al. (1990) to estimate optimal fiber lengths.

Muscle Fiber Length

Jones (1989) reported a linear relationship between cross-sectional area of a muscle and the maximum force that a muscle can produce. However, Enoka (2002) noted that only muscle fiber length is necessary to obtain muscle force (Kawakami et al., 1994; Narici et al., 1988). This is reflected in (6) which is obtained by inserting (13) in (14).

In equation 6, F_A and F_P are directly dependent on muscle fiber length. The velocity-dependent force depends on l_m because velocity is derived from length. Muscle fiber length also affects the changes in pennation angle (15).

Guimarães et al. (1995) and Huijing (1996) reported that optimal fiber length changed as a function of muscle activation. Lloyd and Besier (2002) mathematically included this relationship in a biomechanical model as follows.

$$OFL(t) = OFL \left[\lambda(1 - a(t)) + 1\right] \tag{16}$$

Where OFL(t) is the optimal fiber length that is changing depending on activation, and λ is the percent change of optimal fiber length. This relationship is important because the measure of l_m is normalized by OFL(t). Anthropometric data can be used for OFL, and λ has been reported to be about 0.15 (Lloyd & Besier, 2002)

The total force a muscle exerts about a joint can be calculated if muscle activation and muscle fiber length (and its derivative) are *known*. However, there is no method available to collect muscle fiber length data *in vivo*. Therefore, equation (6) must be put in terms of l_m and solved. To accomplish this, a forward integration, such as Runge-Kutta-Fehlberg, is employed to obtain muscle fiber length from muscle fiber velocity (Lloyd & Besier, 2002).

Tendon Force, Strain and Stiffness

Calculating the muscle fiber lengths from the only equation available leaves no equations to obtain force from. On the other hand, muscle-tendon force can still be computed from tendon force.

$$F = F_m \cdot \cos(\varphi) = F_t \tag{17}$$

Where, F is the muscle-tendon force, F_m is muscle force, and F_t is the tendon force. Zajac (1989) described the force produced by the muscle passing through the tendon as well (Figure 4). He also described a relationship between normalized tendon force and tendon strain. Where strain is given by:

$$\varepsilon_t = \frac{l_t - RTL}{RTL} \tag{18}$$

Equation (18) is for tendon strain (ε_t), l_t is the tendon length, and RTL is the resting tendon length. The tendon length can be computed by subtracting the muscle fiber length from the muscle-tendon length. Initial values for the resting tendon length were obtained from Delp (1990) and Lloyd and Buchanan (1996). However, the resting tendon length for each muscle was optimized within ±15% of the initial values.

$$\tilde{F}_t = 0 \qquad\qquad \varepsilon \leq 0 \tag{19}$$

$$\tilde{F}_t = 1480.3 \cdot \varepsilon^2 \qquad\qquad 0 < \varepsilon < 0.0127 \tag{20}$$

$$\tilde{F}_t = 37.5 \cdot \varepsilon - 0.2375 \qquad\qquad \varepsilon \geq 0.0127 \tag{21}$$

The relationship between tendon strain and normalized tendon force ($\tilde{F}_t$) is mostly linear for lengths above the RTL value. For lengths below the resting tendon length force is zero (19). It is important to include the non-linear region, because it accounts for the initial loading of the collagen (20). Once the collagen is straightened out the relationship is linear (21). The maximum isometric force falls in this linear region, and corresponds to a tendon strain value of 3.3% (Buchanan et al., 2004; Zajac, 1989).

$$F_t = F_{max} \cdot \tilde{F}_t \tag{22}$$

Equation (22) is the "un-normalization" of tendon force (Zajac, 1989).

Muscle Fiber Velocity

The calculation of fiber velocity (v_m) is essential for the forward integration of (6) to obtain l_m. Runge-Kutta-Fehlberg is applied to $\tilde{v}_m$ in order to obtain muscle fiber length. To calculate velocity Epstein and Herzog (1998) outlined a method. First of all, a new equation for velocity-dependent force is generated:

$$\tilde{F}_V = \frac{\tilde{F}_{vd} - b_m \cdot \tilde{v}_m}{\tilde{F}_A \cdot a(t)} \qquad (23)$$

where:

$$\tilde{F}_{vd} = \frac{\tilde{F}_t}{\cos(\varphi(t))} - \tilde{F}_P \qquad (24)$$

Then (9) and (10) are inserted in (23):

$$\frac{d - c \cdot \tilde{v}_m}{d + \tilde{v}_m} = \frac{\tilde{F}_{vd} - b_m \cdot \tilde{v}_m}{\tilde{F}_A \cdot a(t)} \qquad (25)$$

$$F_{ecc} - (F_{ecc} - 1)\frac{d' + c' \cdot \tilde{v}_m}{d' - \tilde{v}_m} = \frac{\tilde{F}_{vd} - b_m \cdot \tilde{v}_m}{\tilde{F}_A \cdot a(t)} \qquad (26)$$

where, (25) is for concentric contractions, and (26) is for eccentric contractions. Both of which can be manipulated into a quadratic form, and $\tilde{v}_m$ can be solved for by applying the quadratic formula.

Maximum Isometric Force

A reoccurring computation has been "un-normalization" by multiplying the normalized force by F_{max}. However, for the same reasons that we cannot collect muscle force data, it is also not possible to collect maximal force data for all individual muscles. Generally, such values are determined from estimates of the physiological cross-sectional areas which are then multiplied by maximal muscle stress. Such anthropometric data are available to aid in estimations of these values (Delp et al., 1990; Yamaguchi et al., 1990). However, this can be problematic as maximal muscle stress may not be constant (Buchanan, 1994). To avoid this problem, such estimates should not be used for their quantifiable accuracy, but to determine the distribution of force production potential. Scaling factors can be used to obtain subject specific maximum force values, as follows:

$$F_{max} = F_{max-est} \cdot G_f \qquad (27)$$

$$F_{max} = F_{max-est} \cdot G_e \qquad (28)$$

The gain values for the flexors (G_f) and extensors (G_e) are not directly measurable, and must be optimized. $F_{max-est}$ is the maximum isometric force as reported by Delp (1990).

Calculation of Muscle-Tendon Force

For clarity, the following is a recapitulation of how to calculate the force a muscle exerts about a joint, and is done at each time step. All functions of time imply the concerned value is changing with respect to time.

1.   Derive $v_{mt}(t)$ from $l_{mt}(t)$.
2.   Determine $l_m(t)$ from (6) iteratively through forward integration
     (a)   Calculate $OFL(t)$ from OFL, $a(t)$, and λ according to (16).
     (b)   Normalize $l_m(t)$ by $OFL(t)$.
     (c)   Compute $\phi(t)$ from $OFL(t)$, $l_m(t)$ and ϕ according to (15).
     (d)   Calculate $l_t(t)$ from $l_m(t)$ and $l_{mt}(t)$.
     (e)   Compute F_A and F_P from $l_m(t)$ according to Figure 5 and (6).
     (f)   Compute F_t from $l_t(t)$ according to (18) through (22).
     (g)   Compute $v_m(t)$ from F_A, F_P, F_t, and $a(t)$ according to (23) through (26).
     (h)   Obtain F_m from F_A, F_V, F_P and $v_m(t)$ according to (6).
     (i)   Continue iterations until F_m and F_t are equated.
3.   Calculate muscle-tendon force
     (a)   Calculate $l_t(t)$ from $l_m(t)$ and $l_{mt}(t)$.
     (b)   Compute F_t from $l_t(t)$ according to (18) through (21).
     (c)   "Un-normalize" the force by F_{max} and G_f or G_e according to (27) or (28).

Calculation of Joint Moment

Muscles are actuators that create force as they contract; however, exerting this force collinear to the muscle-tendon unit does not create movement of the ankle. Motion occurs due to the moment a muscle creates about a joint:

$$M_m = r_{ma} \cdot F \tag{29}$$

This is the calculation of the joint moment from one muscle (M_m), where, r_{ma} is the moment arm. The moment arms will be positive or negative depending on whether the muscle is a flexor or extensor.

Equation (29) is the calculation of the moment exerted by a single muscle. However, multiple muscles are activating simultaneously about a joint. The calculation of total joint moment is the summation of all the moments about a joint:

$$M_F = \sum M_m \tag{30}$$

where, M_F is the total joint moment that is the final output from forward dynamics calculations.

INVERSE DYNAMICS

Inverse dynamics uses a completely external approach. Referring back to Figure 1, the inputs are ground reaction forces, and position. The ground reaction forces, together with the position data (from which we obtain the joint angles at the specific instant in time), inertia and mass of each body segment, and the equations of motion, enable determination of the joint moments.

Figure 7. Optimization of Hybrid Model flow diagram

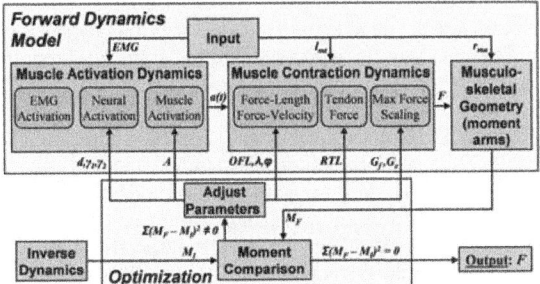

Legend: EMG = electromyograms; a(t) = muscle activation; F = muscle force; M_F = total joint moment from forward dynamics; M_I = total joint moment from inverse dynamics; d = electromechanical delay; γ_1 and γ_2 = recursive filter parameters; A = shape factor; RTL = resting tendon length; OFL = optimal fiber length; G_f and G_e = flexor and extensor gain coefficients; f = pennation angle

OPTIMIZATION

In Figure 1 the optimization is shown comparing the joint moment values from forward and inverse dynamics. This is a good theoretical representation of the hybrid method. However, the actual optimization algorithm can be simpler for a few reasons.

Figure 7 illustrates the simpler algorithm with more detail. The simplification starts with eliminating the feedback loops as described in the data input section for forward dynamics. Also, the inverse dynamics computation does not employ any unknown parameters, and can therefore be performed outside of the optimization algorithm. Finally, the calculation of muscle moment arms can occur before the optimization begins. This implies that both joint moment from inverse dynamics (M_I) and moment arms (r_{ma}) will simply be inputs to the algorithm.

The concept behind optimization is fairly simple. The algorithm is trying to minimize the difference between the forward dynamics calculation of the net joint moment and those obtained from inverse dynamics. Forward dynamics requires several parameters that cannot be easily quantified, and therefore are initially unknown. The optimization algorithm varies each of the unknown parameters within a specified range so that values are constrained to be physiologically relevant. The algorithm continues guessing until the joint moments are equated.

There are several parameters that can be varied in the model. Some terms need to be computed because physiological values cannot be determined: filter coefficients (γ_1, γ_2), gain factors (G_f, G_e), shape factor (A). There are also some parameters for which values reported in the literature have a wide range, such as the resting tendon length, the electromechanical delay, and the optimal fiber length. Note that some of these parameters can be specified to be different for each muscle (e.g., the filter coefficients) and others can be assumed to be the same for all muscles (e.g., the two gain factors, the electrome-chanical delay). The choice of which parameters to be "global" for all muscles or "local" to each muscle will affect the number of parameters be adjusted or tuned in the model. Heine et al. (2003) evaluated the number of parameters required in a static model of the elbow. Their model had seven muscles and they changed the number of parameters from 0 to 57. They found that reasonable predictions could be obtained with as few as one parameter per muscle.

When setting up an optimization algorithm, one might be tempted to include as many parameters as possible. The reasoning makes sense. In fact, the result is a better curve fit of the joint moments. However, the purpose of optimization is to achieve a calibrated model that can be used on novel trials. Including more parameters reduces the predictive ability of the model, because it makes the model more trial specific than subject specific (Heine et al., 2003). Therefore, the fewer parameters that need to be adjusted, the more powerful the model will be.

Table 1. Table of all the parameters that can be optimized

Parameter	Lower Bound	Upper Bound	Reference
Delay (d)	10 msec	100 msec	Corcos (1992)
Filter Coefficients (γ_1, γ_2)	-0.95	0.90	Cohen (2004)
Shape Factor (A)	0.01	0.12	Manal (2003)
Optimal Fiber Length (OFL)	*	*	Delp (1990)
Percent Change of OFL (γ)	0.15*	0.15*	Lloyd (2002)
Pennation Angle (φ)	*	*	Yamaguchi (1990), Delp (1990)
Resting Tendon Length (RTL)	$RTL_i - RTL_i \cdot 0.05$	$RTL_i + RTL_i \cdot 0.05$	Delp (1990), Lloyd (1996)
Gain Factors (G_f, G_e)	$F_{max} \cdot 0.5$	$F_{max} \cdot 8.0$	Lloyd (2003)
Maximum Isometric Force (F_{max})	*	*	Delp (1990)

Those labeled with an asterisk (*) have been consistently reported to optimize to the same value and are therefore fixed—if no value is supplied it is because it is muscle specific and can be found in the literature

Choosing an appropriate optimization algorithm is very critical, and task dependent. Neptune (1999) compares the performance of a sequential programming method (Lawrence et al., 1997), a downhill simplex method (Nelder & Mead, 1965), and a simulated annealing algorithm (Goffe et al., 1994). He found that, in a cycling motion, optimizing 14 muscles by simulated annealing was the best choice. However, for different tasks different optimization algorithms might be better.

Theoretically, the optimization will run until the joint moments from forward dynamics and inverse dynamics are nearly equal. However, this should not be the case. There comes a point at which further refinement reduces effectiveness of calibration. If the calibration is trial specific, the model will not be applicable to novel trials. Neptune (1999) ran his optimization algorithms for 5,000 iterations each. Where one iteration is defined as one new guess of one parameter. The number of iterations an algorithm is run depends on the number of parameters and the number of muscles. It is advisable to start by running a very large number of iterations to determine what type of fit is possible. Running the optimization for other (fewer) numbers of iterations will reveal the best balance by comparing their different predictive abilities.

EXAMPLE

To give an example of modeling muscle forces acting at the ankle joint, the following section will step through the process outlined in the chapter.

A normal neurologically and physiologically unimpaired subject performed gait trials EMG, motion, and force plate data were collected. EMG was collected from the four main contributors to plantar/dorsiflexion of the ankle: tibialis anterior (TA), gastrocnemius lateralis (GL), gastrocnemius medialis (GM), and soleus (Sol) (Figure 8). The data

Figure 8. Ankle joint muscles

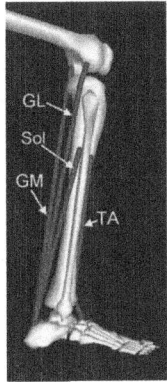

Figure 9. Stroke patient raw EMG

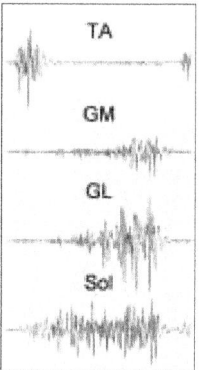

of interest are from heel-strike to toe-off on the force plate. The parameters shown in Table 2 were optimized using a simulated annealing algorithm run for 5,000 iterations.

Figure 9 displays the raw EMG signal, which was input to the forward dynamics model. The other necessary inputs to the model were the muscle-tendon lengths and the moment arms. These were calculated from joint angle.

Both ankle and knee angle were necessary because the gastrocnemii are biarticular muscles, crossing both ankle and knee joints and, therefore, change in length depends on both joint angles. Regarding moment arms, dorsiflexion joint moment compared with inverse dynamics is defined as positive. Therefore, the moment arm sign for the TA,

Figure 10. Joint angles were used to obtain muscle-tendon lengths and moment arms

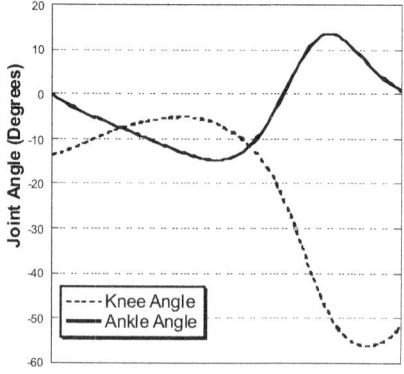

Figure 10. Joint angles were used to obtain muscle-tendon lengths and moment arms (cont.)

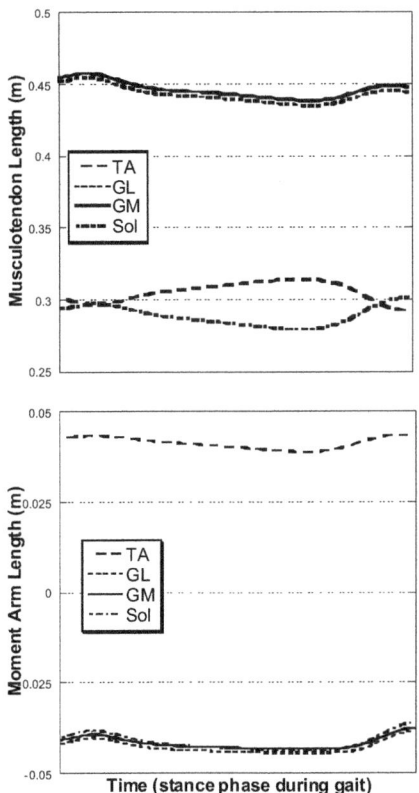

which is the only dorsiflexor modeled, is also positive. The gastrocnemii and soleus are plantarflexors and, therefore, have a negative moment arm to oppose dorsiflexion.

The last necessary inputs to the model are the initial guesses for each of the unknown parameters.

Once all the inputs have been adequately set up, the optimization algorithm can be run.

Figure 11 displays the final result of the joint moment comparison. The criterion for the optimization was for the moments to be equated. However, the end result was an R-squared correlation of 0.98 and a root-mean-squared (RMS) error of 3.57 N-m. These numbers tell us that the optimization was reasonably successful in matching the moments.

Table 2. Initial guesses for optimization algorithm

Parameter	TA	GL	GM	Sol
Delay (d)	50 msec	50 msec	50 msec	40 msec
Filter Coefficients (γ_1, γ_2)	$\gamma_1 = 0.0$ $\gamma_2 = 0.0$	$\gamma_1 = 0.0$ $\gamma_2 = 0.0$	$\gamma_1 = 0.0$ $\gamma_2 = 0.0$	$\gamma_1 = 0.0$ $\gamma_2 = 0.0$
Shape Factor (A)	0.01	0.01	0.01	0.01
Resting Tendon Length (RTL)	0.223 m	0.385 m	0.408 m	0.268 m
Gain Factors (G_f, G_e)	$G_f = 1.0$	$G_e = 1.0$	$G_e = 1.0$	$G_e = 1.0$

Initial resting tendon length is taken from Yamaguchi et al. (1990), and is adjusted based on the initial guess: lower bound is 5 % less and upper bound is 5% greater. Other parameters are adjusted within the bounds described by Table 1.

Figure 11. The joint moment comparison between forward and inverse dynamics for an unimpaired subject

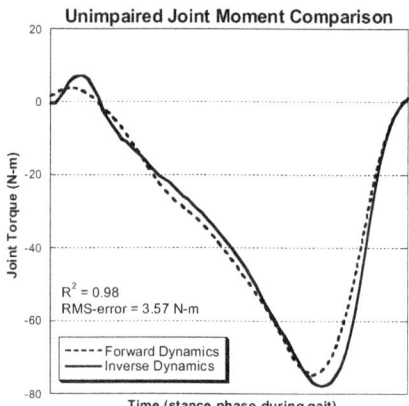

The versatility of the calibration is determined by the ability of the model to predict joint moments in novel trials for which an optimization of unknown parameters is not performed. A successfully calibrated model will predict joint moments with comparable R-squared and RMS error values.

The ultimate goal of optimizing the values of unknown parameters, however, was not to obtain joint moments, as these can be easily obtained from inverse dynamics. The calculation of muscle forces is the strength of the hybrid approach.

Figure 12. Estimation of muscle forces during gait of an unimpaired subject

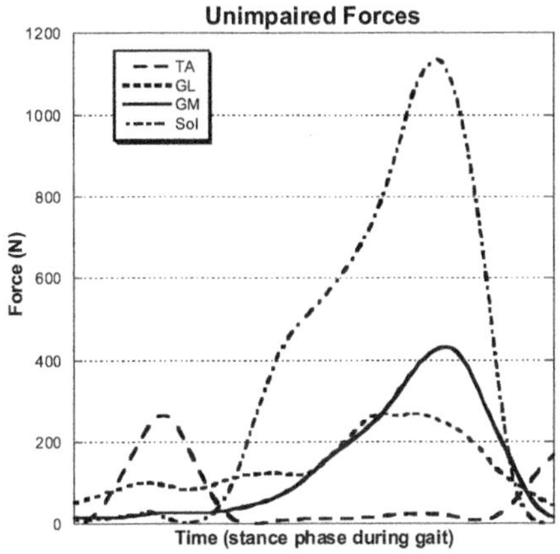

Unimpaired Forces

Table 3. Optimized parameter values

Parameter	TA	GL	GM	Sol
Delay (*d*)	47 msec	37 msec	20 msec	33 msec
Filter Coefficients (γ_1, γ_2)	$\gamma_1 = -0.187$ $\gamma_2 = 0.341$	$\gamma_1 = -0.301$ $\gamma_2 = -0.256$	$\gamma_1 = 0.248$ $\gamma_2 = 0.833$	$\gamma_1 = 0.112$ $\gamma_2 = -0.332$
Shape Factor (*A*)	0.0499	0.0872	0.0054	0.0030
Resting Tendon Length (*RTL*)	0.218 m	0.365 m	0.403 m	0.268 m
Gain Factors (G_f, G_e)	$G_f = 4.83$	$G_e = 2.44$	$G_e = 2.44$	$G_e = 2.44$

Figure 12 displays the muscle forces exhibited by the unimpaired subject during a normal walking trial. Table 3 is a record of the parameters after the optimization. These optimized values can be used in novel trials. One run through the forward dynamics model using these values will give a new estimation of joint torque without the need for optimization or inverse dynamics.

POST-STROKE PATIENT COMPARISON

To better understand the effects of a stroke on the human body and to assist in mobility recovery, it is important to have knowledge of muscle forces and joint moments in a post-stroke subject. Ideally, the goal of rehabilitation of a person who has suffered a stroke is to create the same muscle forces and joint moments as an unimpaired individual exerts. In practice, this level of rehabilitation is not realistic. However, the comparison between post-stroke and unimpaired subjects is a valuable assessment of rehabilitation. It can also be used to measure undesired forces, such as excessive joint compressive force that can lead to osteoarthritis or other joint problems.

We have used our model to estimate muscle forces during gait of a post-stroke subject. Similar to the example with the unimpaired subject the optimization was successful, and the joint moments were matched with an R-squared value of 0.97, and RMS-error of 4.22 N-m (Figure 13), therefore demonstrating the ability of the model to be applied to both normal and neurologically or physiologically impaired subjects.

Figure 14 presents the muscle forces the post-stroke subject exerted during gait. A direct comparison between these muscle forces and the normal muscle force patterns can be made. Note that for the post-stroke patient, the soleus force is much more pronounced in the early part of the trial and, unlike the unimpaired trial, the force profile of the tibialis anterior does not reach a peak. However, the post-stroke forces produced by the gastrocnemii are of comparable magnitude and pattern with the unimpaired muscle forces. Of course, further investigation with more subjects is necessary if generalizable conclusions about differences in these subject populations are desired.

Figure 13. The joint moment comparison between forward and inverse dynamics for a post-stroke subject

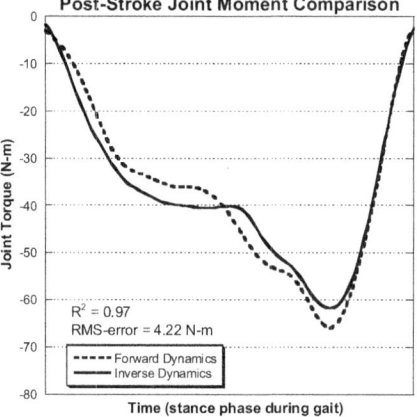

Figure 14. Resulting muscle force during gait of a post-stroke subject

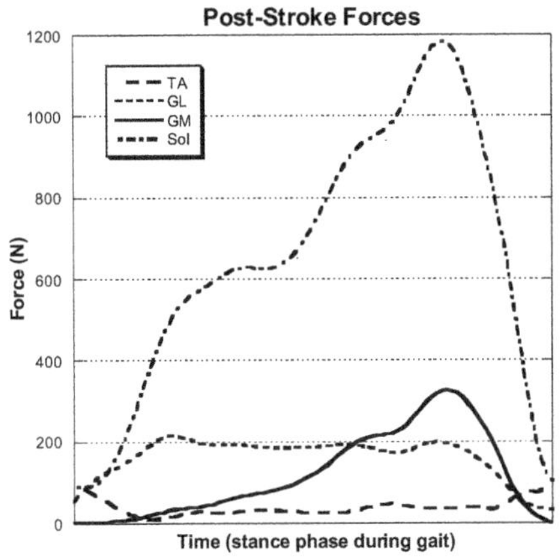

CONCLUSION

A hybrid EMG-driven muscle model has been presented and shown to be applicable to unimpaired and impaired subjects. The forward dynamics portion of the model employs muscle activation dynamics and muscle contraction dynamics, the later of which uses a Hill-type model approach. Forward dynamics requires parameters for which the values are unknown and are, therefore, optimized by using the inverse dynamics calculation of joint moment as a benchmark. The resulting calibration, for both the unimpaired and post-stroke subjects, matched the joint moments with R-squared and RMS error values within an acceptable margin of error. We believe this approach has great potential to the study of muscle forces in healthy and impaired limbs.

EMERGING TRENDS

The methodology presented in this chapter has the potential to develop into an important clinical tool. If the forces in the muscles about a joint can be determined, then it is relatively easy to use this information to estimate forces in other tissues in the joints by using elementary principles of structural mechanics. For example, compressive forces in cartilage can be estimated, which is insightful in studying the progression and rehabilitation of joint diseases such as osteoarthritis. Loads in ligaments can be estimated as well and estimation of muscle and ligament forces can be helpful in studies of ligament-injured patients. For example, it is known that some athletes with complete tears of the anterior cruciate ligament can recover without surgery and return to play while others cannot. Why a person will be in the first group or the second is unknown, but most likely the answer lies in differences in the way the muscles are controlled to compensate for the lost ligament. If this is a skill that can be learned, then rehabilitation of these patients may render surgery unnecessary.

The most exciting developments in this field lie in the innovative use of medical imaging and the increase in computation speed for modeling. Medical imaging data obtained through MRI, ultrasound, or CT scans provide precise information for these models which eliminates the need for much of the parameter adjustment and model tuning. This makes the models subject specific and more accurate. The increase in computational speed, through fast computers and better algorithms for parallel processing, makes it easier to create models that encompass more than a single joint and that are "trained" on multiple trails. The model presented in this chapter takes about 20 minutes to run on a fast (by today's standards), off-the-shelf PC. The possibility of having models that work in real time opens the door to using these models to estimate how to correct inappropriate muscle activity, which could be implemented on-the-fly through functional electrical stimulation in our post-stroke patients.

REFERENCES

An, K. N., Takahashi, K., Harrigan, T. P., & Chao, E. Y. (1984). Determination of muscle orientations and moment arms. *Journal of Biomechanical Engineering, 106*, 280-282.

Anderson, F. C., & Pandy, M. G. (2001). Dynamic optimization of human walking. *Journal of Biomechanical Engineering, 123*(5), 381-90

Arnold, A. S., Salinas, S., Asakawa, D. J., & Delp, S. L. (2000). Accuracy of muscle moment arms estimated from MRI-based musculoskeletal models of the lower extremity. *Computer Aided Surgery, 5*(2), 108-19.

Buchanan, T. S. (1994). Evidence that maximum muscle stress is not a constant: Differences in specific tension in elbow flexors and extensors. *Medical Engineering & Physics, 17*(7), 529-536.

Buchanan, T. S., Lloyd, D. G., Manal, K. T., & Besier, T. F. (2004). Neuromusculoskeletal modeling: Estimation of muscle forces and joint moments and movements from measurements of neural command. *Journal of Applied Biomechanics, 20*(4), 367-395.

Cohen, S. (2004). *The development and evaluation of an emg driven model to estimate ankle moments and muscle forces*. Unpublished master's thesis, University of Delaware, Newark, Delaware, USA.

Corcos, D. M., Gottlieb, G. L., Latash, M. L., Almeida, G. L., & Agarwal, G. C. (1992). Electromechanical delay: An experimental artifact. *Journal of Electromyography and Kinesiology, 2*, 59-68.

Delp, S. L. (1990). A computer-graphics system to analyze and design musculoskeletal reconstructions of the lower limb. (Doctoral dissertation, Stanford University, 1990). *Dissertation Abstracts International, 51*, 5449.

Delp, S. L., & Loan, J. P. (1995). A graphics-based software system to develop and analyze models of musculoskeletal structures. *Computers in Biology and Medicine, 25*(1), 21-34.

Edgerton, V. R., Roy, R. R., Gregor, R. J., & Rugg, S. (1986). Morphological basis of skeletal muscle power output. In N. L. Jones, N., McCartney, & A. J. McComas (Eds.), *Human muscle power*. Champaign, IL: Human Kinetics.

Enoka, R. M. (2002). *Neuromechanics of human movement*. Champaign, IL: Human Kinetics.

Epstein, M., & Herzog, W. (1998). *Theoretical models of skeletal muscle*. New York: Wiley.

Fuglsang-Frederiksen, A. (2000). The utility of interference pattern analysis. *Muscle & Nerve, 23*(1), 18-36.

Goffe, W. L., Ferrier, G. D., & Rogers, J. (1994). Global optimization of statistical functions with simulated annealing. *Journal of Econometrics, 60*, 65-99.

Gordon, A. M., Huxley, A. F., & Julian, F. J. (1966). The variation in isometric tension with sarcomere length in vertebrate muscle fibres. *Journal of Physiology (London), 184*, 170-192.

Guimarães, A. C., Herzog, W., Allinger, T. L., & Zhang, Y. T. (1995). The EMG-force relationship of the cat soleus muscle and its association with contractile conditions during locomotion. *Journal of Experimental Biology, 198*, 975-987.

Heine, R., Manal, K., & Buchanan, T. S. (2003). Using Hill-type muscle models and EMG data in a forward dynamic analysis of joint moment: Evaluation of critical parameters. *Journal of Mechanics in Medicine and Biology, 3*(2), 169-186.

Herzog, W., Sokilosky, J., Zhang, Y. T., & Guimarães, A. C. (1998). EMG-force relation in dynamically contracting cat plantaris muscle. *Journal of Electromyography and Kinesiology, 8*, 147-155.

Hill, A. V. (1938). The heat of shortening and the dynamic constant of muscle. In *Proceedings of the Royal Society of London Series B, 126* (pp. 136-195).

Huijing, P. A. (1996). Important experimental factors for skeletal muscle modelling: Nonlinear changes of muscle length force characteristics as a function of degree of activity. *European Journal of Morphology, 34*, 47-54.

Huxley, A. F. (1958). Muscle structure and theories of contraction. *Progress in Biophysical Chemistry, 7*, 255-318.

Jones, D. A., Rutherford, O. M. & Parker, D. F. (1989). Physiological changes in skeletal muscle as a result of strength training. *Quarterly Journal of Experimental Physiology, 74*, 233-256.

Kawakami, Y., Nakazawa, K. Fujimoto, T. Nozaki, D., Miyashita, M, & Fukunaga, T. (1994). Specific tension of elbows flexor and extensor muscles based on magnetic resonance imaging. *European Journal of Applied Physiology, 68*, 139-147.

Keen, D. A., Yue, G. H., & Enoka, R. M. (1994). Training-related enhancement in the control of motor output in elderly humans. *Journal of Applied Physiology, 77*, 2648-2658.

Kroll, P. G. (1988). The effect of previous contraction condition on subsequent eccentric power production in elbow flexor muscles. (Doctoral dissertation, New York University, 1988). *Dissertation Abstracts International, 49*, 3040.

Lawrence, C. T., Zhou, J. L. & Tits, A. L. (1997). *User's guide for CFSQP Version 2.5: A C code for solving (large scale constrained nonlinear (minimax) optimization problems, generating iterates satisfying all inequality constraints* (Tech. Rep. No. TR-94-16rl). University of Maryland, Institute for Systems Research.

Lieber, R. L., Loren, G. J., & Frieden, J. (1994). In vivo measurement of human wrist extensor muscle sarcomere length changes. *Journal of Neurophysiology, 71*, 874-881.

Lloyd, D. G., & Besier, T. F. (2002). An EMG-driven musculoskeletal model to estimate muscle forces and knee joint moments in vivo. *Journal of Biomechanics, 36*, 765-776.

Lloyd, D. G., & Buchanan T. S. (1996). A model of load sharing between muscles and soft tissues at the human knee during static tasks. *Journal of Biomechanical Engineering, 118*, 367-376.

Luciano, D. S., Vander, A. J., & Sherman, J. H. (1978). *Human function and structure.* New York: McGraw-Hill.

Macefield, V. G., Fuglevand, A. J., & Bigland-Ritchie, B. (1996). Contractile properties of single motor units in human toe extensors assessed by intraneural motor axon stimulation. *Journal of Neurophysiology, 75*, 2509-2519.

Macefield, V. G., Hagbarth, K.-E., Gorman, R. Gandevia, S. C., & Burke, D. (1991). Decline in spindle support to ±-motoneurones during sustained voluntary contractions. *Journal of Physiology (London), 440*, 497-512.

Manal, K. T., & Buchanan, T. S. (2003). A one-parameter neural activation to muscle activation model: estimating isometric joint moments from electromyograms. *Journal of Biomechanics, 36*, 1197-1202.

Manal, K. T., & Gonzalez, R. V., Lloyd, G. L., & Buchanan, T. S. (2002). A real-time EMG-driven virtual arm. *Computers in Biology and Medicine, 32*, 25-36.

Mathiassen, S. E., Winkel, J., & Hägg, G. M. (1995). Normalization of surface EMG amplitude from upper trapezius muscle in ergonomic studies — A review. *Journal of Electromyography and Kinesiology, 5*, 195-226.

Merletti, R., & Lo Conte, L. R. (1997). Surface EMG signal processing during isometric contractions. *Journal of Electromyography and Kinesiology, 7*, 20-25.

Milner-Brown, H. S., Stein, R. B., & Yemm, R. (1973). Changes in firing rate of human motor units during linearly changing voluntary contractions. *Journal of Physiology (London), 228*, 371-390.

Murray, W. M., Buchanan, T. S., & Delp, S. L. (2002). Scaling of peak moment arms of elbow muscles with upper extremity bone dimensions. *Journal of Biomechanics, 35*(1), 19-26.

Murray WM, Delp SL, Buchanan T.S. (1995). Variation of muscle moment arms with elbow and forearm position. *Journal of Biomechanics, 28*(5), 513-25.

Narici, MV., Roi, G. S., & Landoni, L. (1988). Force of Knee extensor and flexor muscles and cross-sectional area determined by nuclear magnetic resonance imaging. *European Journal of Applied Physiology, 57*, 39-44.

Nelder, X. X., & Mead, Y. Y. (1965). A simplex method for function minimization. *The Computer Journal, 7*, 308-313.

Neptune, R. R. (1999). Optimization algorithm performance in determining optimal controls in human movement analyses. *Journal of Biomedical Engineering, 121*, 249-252.

Schutte, L. M. (1993A). Using musculoskeletal models to explore strategies for improving performance in electrical stimulation-induced leg cycle ergometry (Doctoral dissertation, Stanford University, 1993). *Dissertation Abstracts International, 53*, 5847.

Schutte, L. M., Rodgers, M. M, & Zajac, F. E. (1993B). Improving the efficacy of electrical stimulation induced leg cycle ergometry: an analysis based on a dynamic musculoskeletal model. *IEEE Transactions on Rehabilitation Engineering, 1*(2), 109-124.

Scott, S. H., & Winter, D. A. (1991). A comparison of three muscle pennation assumptions and their effect on isometric and isotonic force. *Journal of Biomechanics, 24*, 163-167.

Thelen, D. G., Schultz, A. B., Fassois, S. D., & Ashton-Miller, J. A. (1994). Identification of dynamic myoelectric signal-to-force models during isometric lumber muscle contractions. *Journal of Biomechanics, 27*, 907-919.

Windmaier, E. P., Raff, H., & Strang, K. T. (2004). *Human physiology: The mechanisms of body function.* New York: McGraw-Hill.

Woittiez, R. D., Huijing, P. A., Boom, H. B., & Rozendal, R. H. (1984). A three-dimensional muscle model: a quantified relation between form and function of skeletal muscles. *Journal of Morphology, 182*(1), 95-113.

Woods, J. J., & Bigland-Ritchie, B. (1983). Linear and non-linear surface EMG/force relationships in human muscles. an anatomical/functional argument for the existence of both. *American Journal of Physical Medicine, 62*(6), 287-99.

Yamaguchi, G. T., Sawa, A. G. U., Moran, D. W., Fessler, M. J., & Winters, J. M. (1990). A survey of human musculotendon actuator parameters. *Multiple muscle systems: Biomechanics and movement organization.* New York: Springer-Verlag.

Zajac, F. E. (1989). Muscle and tendon: properties, models, scaling, and application to biomechanics and motor control. *Critical Reviews in Biomedical Engineering, 17*(4), 359-411.

Zajac, F. E. (1993). Muscle coordination of movement: A perspective. *Journal of Biomechanics, 26*(1), 109-124.

Chapter XIII

Computational Modelling in Shoulder Biomechanics

David C. Ackland, The University of Melbourne, Australia

Cheryl J. Goodwin, The University of Texas at Austin, USA

Marcus G. Pandy, The University of Melbourne, Australia

ABSTRACT

The objectives of this chapter are as follows. First, a background in anatomy and biomechanics of the shoulder complex is presented to provide a brief review of the essential functions of the shoulder. Second, important features of practical shoulder models are discussed with reference to capabilities of current computational modelling techniques. Third, techniques in computational modelling of the shoulder complex are compared and contrasted for their effectiveness in representing shoulder biomechanics in situ, with some sample calculations included. Fourth, in vivo and in vitro techniques for verifying computational models will be briefly reviewed. Finally, a summary of emerging trends will indicate the clinical impact that computational modelling can be expected to have in progressing our understanding of shoulder complex movement and its fundamental biomechanics.

INTRODUCTION

Computer modelling and simulation of human body kinetics has advanced rapidly in recent years, as it is thought that this approach can provide more quantitative explanations of how the neuromuscular and musculoskeletal systems interact to produce movement. Modelling human pathological movement is also of interest, particularly to rehabilitation scientists in the research and development of musculoskeletal and neurological therapies (Zajac, 1993). Models of human movement have always been desirable due to the high cost of experiments with human cadavers and/or anthropomorphic dummies. Driven by ramping up of computer speed and performance, models of greater complexity are being used to study musculoskeletal function at levels much deeper than have been possible to date.

The study of biomechanics of the shoulder complex constitutes perhaps the most challenging and least successfully modelled regions of the musculoskeletal structure of the human body (Tumer & Engin, 1989). Lack of an appropriate biomechanical database and the inherent anatomical complexity of this region have led to much speculation over the role of musculoskeletal coordination in upper limb movement. This chapter seeks to show how the development of accurate mathematical models of the shoulder complex and simulations of shoulder movement kinematics can be used in elucidating the roles and interactions of specific muscles. Clarifying the complex nature of shoulder musculature function and control would directly enhance clinical predictions in shoulder stability treatment, joint replacement surgery, and rehabilitation.

Shoulder pain after exposure to high or sustained biomechanical strain in occupational situations has shown an increasing incidence in society (Hogfors, Karlsson, & Peterson, 1995). Common problems in assessing pain syndrome etiology in the human shoulder is that, due to the convoluted nature of the shoulder complex, it is often difficult to relate pain localised in certain structures to the actual loads applied to these structures in specific working conditions. For example, the case history of working with hands at or above shoulder level has been shown to constitute a significant occupational risk factor (Bjelle, Hagberg, & Michaelson, 1981). In order to identify occupational risk factors to prevent shoulder pains and the development of degenerative lesions, load distribution in the shoulder must be subject to comprehensive analysis. Such assessments are greatly enhanced with biomechanical models of the shoulder complex, particularly models embracing both the kinematics of the bones involved and the muscle forces between them.

To describe the mechanical behaviour of a musculoskeletal system such as the shoulder complex, and in developing a stable and manageable model of the system, many of its physiological characteristics must be considered. In particular, knowledge of musculoskeletal parameters is essential to understanding and modelling a muscle's force generating capability, for example, physiologic cross-sectional area (PCSA) relates directly to the maximum force a muscle can produce (Zajac, 1989). Anthropometric measurements and values of muscle-specific parameters are normally determined by cadaveric studies, through in vivo techniques such as MRI and fluoroscopic imaging, or computed using parameter estimation algorithms (Bao & Willems, 1999). This chapter reveals how studies of musculoskeletal parameters become essential in providing guidelines and limitations in computational simulations of shoulder movement.

A number of attempts have been made to develop computational models of the shoulder. Some have been elementary two-dimensional models, or have addressed only certain components such as the glenohumeral joint (Poppen & Walker, 1978). Others have been more ambitious, such as Garner and Pandy (2001) who produced a complete three-dimensional 13 degree-of-freedom musculoskeletal model of the entire upper limb. Typical biomechanical quantities computed by models may include articulating forces, moment arms, and torque, as a function of joint angle. Subsequent verification therefore becomes an important step in determining the accuracy of a model and substantiating its output. In vitro and in vivo experimentation is commonly used to obtain useful information to support or invalidate results of modelling human movement and biomechanics. This chapter addresses current techniques in computational and mathematical shoulder model verification, and their limitations.

A BACKGROUND IN ANATOMY AND BIOMECHANICS OF THE SHOULDER COMPLEX

In order to understand and model the function of the complex group of articulations that constitute the shoulder joint, it is necessary to review their anatomy and modes of action.

Articulations of the Shoulder Complex

The shoulder complex consists of a chain of bones connecting the humerus to the trunk, forming a group of four independent articulations that together constitute the shoulder girdle — the sternoclavicular joint, where the clavicle articulates with respect to the manubrium of sternum. The acromioclavicular joint, where the clavicle meets the acromion process of the scapula, the glenohumeral joint, where the humerus articulates with the glenoid cavity of the scapula, and lastly, the scapulothoracic joint, which describes the motion of the scapula about the thorax (Figure 1). While each of these is an independent entity, capable of independent motion, all contribute their relative motion simultaneously to the total upper limb movement (Inman, Saunders, & Abbott, 1944).

The sternoclavicular articulation is a double arthrodial joint formed between the sternal end of the clavicle with the upper and lateral part of the manubrium sterni and the cartilage of the first rib (Gray, 2000). The two sternoclavicular synovial cavities are separated by a fibrocartilaginous articular disk. The articular cartilage allows the joint to exhibit three degrees of freedom whose movement axes are in the sagittal and frontal planes, as well as rotation along the bone axis of the clavicle (Engin, 1980).

The acromioclavicular articulation is an arthrodial joint between the acromial end of the clavicle and the medial margin of the acromion of the scapula (Gray, 2000). The joint is comprised of a synovial articulating surface covered by fibrocartilage whose articulating surfaces are sloped to favour over-riding of the acromion by the clavicle. The movements of this articulation are of two kinds, a gliding motion of the articular end of the clavicle on the acromion, and rotation of the scapula forward and backward upon the clavicle (Inman et al., 1944; Inman, Saunders, & Abbott, 1996; Lucas, 1973).

Figure 1. The human shoulder girdle

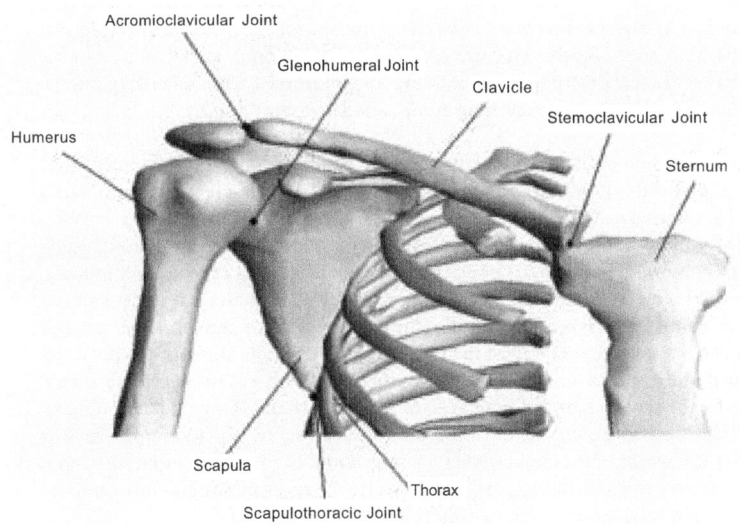

The scapulothoracic articulation is not a true joint as the two bones do not directly articulate. The joint instead is formed by the gliding of the scapula along the rib cage, and is anchored to the ribs by muscle rather than ligament. The lack of true articulation in the joint allows for wide range of motion of the shoulder including protraction-retraction, elevation-depression and rotation of the scapula. The range of glenohumeral to scapulothoracic motion is generally 2:1 (Inman et al., 1944).

The glenohumeral joint is an enarthrodial ball and socket articulation formed between the surfaces of the shallow glenoid fossa of the scapula and the hemispherical head of the humerus. The relatively free three degree-of-freedom movement of the glenohumeral joint is limited by the short muscles about the shoulder, as well as by the tubercles and the over hanging acromion process. The glenohumeral joint is character-ized by its relative lack of bony constraint, relying heavily on the congruent articulating surfaces and surrounding soft tissue envelope for static and dynamic stability (Sharkey, Marder, & Hanson, 1993). Effective articular function is achieved by a complex interac-tion between the capsule, ligaments, and labrum, as well as by dynamic restraints including the rotator cuff muscles (Apreleva et al., 1998).

Stability and Mobility of the Shoulder Complex

Shoulder stability and mobility are maintained by three prime topologically distinct muscle groups:

1. **The scapulohumeral group** (muscles passing from the scapula to the humerus) comprising the supraspinatus, infraspinatus, teres minor, subscapularis, deltoid and teres major.
2. **The axioscapular group** (muscles passing from the torso to the scapula) comprising the trapezius, rhomboids, serratus anterior, and levator scapulae.
3. **The axiohumeral group** (muscles passing from the torso to the humerus) comprising the pectoralis major, pectoralis minor, and latissimus dorsi.

Shoulder scapulohumeral musculature, which produces humeral movement with respect to the glenoid cavity, can be classified into the shorter muscles that act primarily to retain the humerus in its socket and rotate it therein, and, the longer muscles responsible for much of the free movement between the humerus and the glenoid cavity. Active glenohumeral movement can be achieved directly through the scapulohumeral musculature. Flexion of the humerus is brought about by contraction of the anterior portion of the deltoid, the clavicular portion of the pectoralis major, the choracobrachialis, and the biceps branchii. Extension of the humerus is carried out through the posterior fibers of the deltoid, the latissimus dorsi, the sternocostal fibers of the pectoralis major, and weakly by the teres major and long head of triceps branchii (Gray, 2000). Humeral abduction is carried out chiefly by the middle deltoid and supraspinatus, and adduction by the pectoralis major, teres major and latissimus dorsi. Medial or internal rotation is brought about primarily by the subscapularis, while the teres minor and infraspinatus carry out external rotation of the humerus (Hollinshead, 1976).

Active stabilisation of the glenohumeral joint during flexion and extension is brought about by the rotator cuff muscles, the subscapularis, infraspinatus, teres minor and supraspinatus. These muscles act to centre the humeral head in the congruent glenoid fossa through the midrange of this motion when the capsuloligamentous structures are lax (Halder, Kuhl, Zobitz, Larson, & An, 2001).

Movements of the humerus with respect to the scapula during elevation of the arm in flexion and abduction, are simultaneously accompanied by movement of the scapula itself. Scapula motion increases the force of arm movements, and by tilting of the glenoid cavity, can increase the range of movement of the free limb. Coordination of both elements is often termed scapulohumeral rhythm, and accordingly, specific muscles may act on either the scapula or humerus, or both. Of the muscles acting on the scapula, some act directly through their explicit attachments, while others act primarily through their humeral attachments. The muscles capable of scapula elevation typically belong to the axioscapular group, the trapezius, levator scapula and the two rhomboids (Gupta & Van der Helm, 2004). In depression, axiohumeral muscles, the pectoralis minor, latissimus dorsi, and pectoralis major, as well as the subclavius, may all participate. Upward rotation of the scapula is carried out by the combined actions of the trapezius and serratus anterior, while downward rotation is brought about through the action of the rhomboids and the levator scapulae (Hollinshead, 1976). Protraction of the scapula is brought about by the serratus anterior and by the pectoralis major and minor, while retraction involves the trapezius, rhomboids and latissimus dorsi.

Important Features of Shoulder Complex Models

Features of any model of human kinetics should depend entirely on the intended use of that model. For example, if the long-term goal is to understand shoulder articulating forces through combined muscle coordination, a useful model should accordingly include all relevant joints and muscles. Thus, models with joints actuated by joint torques instead of muscles forces are unlikely to be beneficial in muscle coordination studies as these models represent the overall effect of the muscles about each joint (Pandy, 2001).

Structures contributing to joint stiffness, including ligaments, menisci, cartilage and capsule, are usually excluded from multi-joint models used to study movement. This level of detail imposes inherent complications in model design and is usually not required for studies of muscle functionality. Ligament action is seldom modelled, but may be represented in the form of passive joint torques. Cartilage and menisci are scarcely ever included as these components serve to reduce articular contact stresses by increasing the contact areas between the bones (Shrive, O'Connor & Goodfellow, 1978).

Useful simulations of movement of the shoulder complex should include the following: (a) a model of the shoulder girdle or skeletal arrangement, (b) a model of the muscle paths, (c) a model of the muscle actuation, (d) a model of the excitation-contraction coupling, and (e) a model of the outcome of the motor activity.

TECHNIQUES IN MODELLING BIOMECHANICS OF THE SHOULDER COMPLEX

Countless models of varying complexity have been developed in an attempt to describe and explain the function of muscles, ligaments, and bones during human movement. The most popular approach in modelling upper limb biomechanics has been to isolate and examine specific aspects such as shoulder, elbow, and wrist joint function. No work to date has produced a complete musculoskeletal model of the entire upper limb including all the major joints, muscles, and degrees of freedom (DOF) of movement. Further, almost all research papers accept the following hypotheses: (a) bones and their soft tissue can be approximated as rigid bodies with fixed centres of gravity for all motion, and (b) joints can be approximated as ideal frictionless kinematic joints with fixed axes or centres of rotation. As a result of musculoskeletal structure simplification, computational models reported in the literature are often limiting in their scope and capacity to reflect in situ biomechanics. The following section presents a perspective on some popular techniques in modelling shoulder biomechanics.

Modelling the Shoulder Girdle

Mathematical and computational models allow workers to simplify the study of the kinematics and biomechanics of shoulder movement, chiefly by reducing the number of variables involved (de Leva, 1996). In modelling human movement, it is often helpful to model joints as simple hinges and ball-and-socket joints, where for each joint a geometrical centre is assumed to exist. These points, defined as "joint centres," are assumed to hold a fixed three-dimensional position relative to the segments forming the joint. This simplification also allows workers to easily reference segment positions and centres of mass. Reasonable estimates of joint centre locations can be estimated and obtained by measuring the positions of specific bony landmarks on a subject. This can be achieved using imaging techniques such as MRI and fluoroscopy. Unfortunately computational accuracy is invariably sacrificed as no human joint can perfectly meet these assumptions; one will always encounter a certain degree of joint laxity and subsequent translation. In reality, under physiological and external loading, each human articulating joint is actually capable of displaying a total of six DOF: three rotational DOF and three translational DOF.

Scapulothoracic movement simultaneously accompanies elevation of the upper limb in flexion and abduction, an arrangement which allows for enhanced shoulder muscle power and increased range of humeral movement. Motion of the scapula is constrained by the clavicle and by its muscle attachment to the thorax forming a "scapulothoracic-gliding plane." This mechanism forms a closed chain, resulting in three-dimensional movement of the glenoid. Due to the sliding nature of the scapula over the rib cage, the scapulothoracic joint is not considered a true articular joint, and so must be modelled distinctly. Workers have modelled the scapulothoracic joint in up to five DOF as a simplification to allow constrained rotations of the scapula to be handled about the acromioclavicular joint only (Maurel & Thalmann, 2000). In more advanced models, scapula movement is modelled as one or two four-DOF gliding points constrained to move about an ellipsoidal thorax (Garner & Pandy, 1999; van der Helm, 1994). This approach provides a very close approximation to scapulothoracic movement in situ.

The sternoclavicular, acromioclavicular and glenohumeral joints are commonly modelled in the form of three-DOF ball and socket joints (Bao & Willems, 1999; Chadwick, van Noort, & van der Helm, 2004; Engin & Tumer, 1989; Hogfors, Peterson, Sigholm, & Herberts, 1991; Maurel & Thalmann, 2000; Raikova, 1992; van der Helm, 1994; Van der Helm, Veeger, Pronk, Van der Woude, & Rozendal, 1992) to represent the required three-dimensional movement of the shoulder complex and its constituents. Models describing humeral motion in a single plane, or with respect to a fixed scapula (De Duca & Forrest, 1973; Poppen & Walker, 1978) are oversimplified and cannot be expected to add to the comprehension and understanding of overall shoulder function. The acromioclavicular and sternoclavicular joints are practically always modelled with centres of rotation fixed relative to the bones, since the movements about the instantaneous centres of rotation are small. The sternoclavicular and acromioclavicular joints are surface contact joints, so in modelling their corresponding centres of rotation, axes of rotation are normally taken at the centres of these respective surfaces. The sternoclavicular joint displays three-DOF motion whose movement axes are in the sagittal and frontal planes as well as along the bone axis of the clavicle. The acromioclavicular joint similarly exhibits three-DOF motion, two rotational axes perpendicular to the acromioclavicular-glenohumeral line and one along the bone axis of the clavicle. It is uncommon to model the sternoclavicular and acromioclavicular joints in any more than three DOF.

Most computational models of the glenohumeral joint do not incorporate humeral head translations or details of the glenohumeral articular surfaces and surrounding joint capsule. The glenohumeral joint is principally a sphere-in-a-sphere joint, with the difference between the radii of the glenoid cavity and the humeral head being around 5mm. Thus the glenohumeral joint is typically modelled as a three-DOF ball-and-socket joint. In a cadaveric study, Veeger (2000) showed that the kinematic rotation centre of the glenohumeral joint is essentially the geometric rotation centre described by the centre of a sphere fitted through the glenoid surface. It has been shown however that the humeral head translates relative to the glenoid during upper limb movement due to joint laxity inherent in the normal shoulder (Novotny, Nichols, & Beynnon, 1998). While the influence of glenohumeral translation on shoulder kinematics and biomechanics remains largely un-investigated, elucidation of shoulder structure function in the lax and pathological lax glenohumeral joint would have significant clinical implications, for example in the diagnosis and treatment of common anterior glenohumeral instability.

Modelling Muscle Paths of the Shoulder Complex

For models of the shoulder complex and other human musculoskeletal systems, the mechanical effect of a muscle is represented by force vectors about a muscle's line of action. This line of action stems from muscle attachment points, passing from the anatomical origin to insertion. The muscle tendons are normally assumed to insert at a single point on the bones, however when the mechanical effect of muscles with large attachment sites are modelled, muscle is usually divided into two or more bundles (Van der Helm & Veenbaas, 1991). Large muscles represented by single lines of action are not likely to yield accurate estimates of muscle moment arms and, therefore, joint torques.

Paths of muscles are often approximated using straight-line, centroid-line, or via-point techniques. The straight-line model represents a muscle as a line connecting the origin and insertion of muscle attachment sites. This method often proves unsatisfactory, as muscle lines of action can cross joint axes of rotation. The centroid line technique better represents a muscle path as a line passing through the locus of cross-sectional centroids of the muscle. This approach assumes that the muscle line of action at a certain cross-section of the muscle is representative of the forces transmitted by that muscle at that specific cross-section. Therefore, this method is often difficult to implement as it is often not feasible to obtain cross-sectional centroid locations for muscles, nor is it feasible to determine how a muscle's path may vary with joint position. This issue has been partially corrected with the via-point technique by inserting bone fixed attachment sites or via points about a joint. In this approach, a muscle line of action can be approximated by a straight or curved line connecting any number of via points. As the via points remain fixed relative to bones even as the joint is moving, a muscle wrapping effect can be simulated by making via points active or inactive, depending on joint configuration. This method has shown to be effective for muscles spanning one-DOF joints, but can lead to discontinuities in computed moment arms when joints exhibit more than one-DOF (Garner & Pandy, 2000).

Introduced by Garner and Pandy (2000), the obstacle-set method allows simulation of muscle freely sliding over anatomical constraints such as bones and other muscle, as the configuration of the joint changes. Anatomical structures may be represented by regular-shaped rigid bodies such as spheres and cylinders. Because the computed muscle path is not constrained directly by anatomical constraints, the obstacle-set method produces smooth and realistic moment arm and joint angle curves. This technique can be applied to more complex muscle paths which, for example, may span multiple multi-DOF joints such as the deltoid. The obstacle-set method is illustrated in Figure 2.

Figure 2. Obstacle set model of an arbitrary muscle

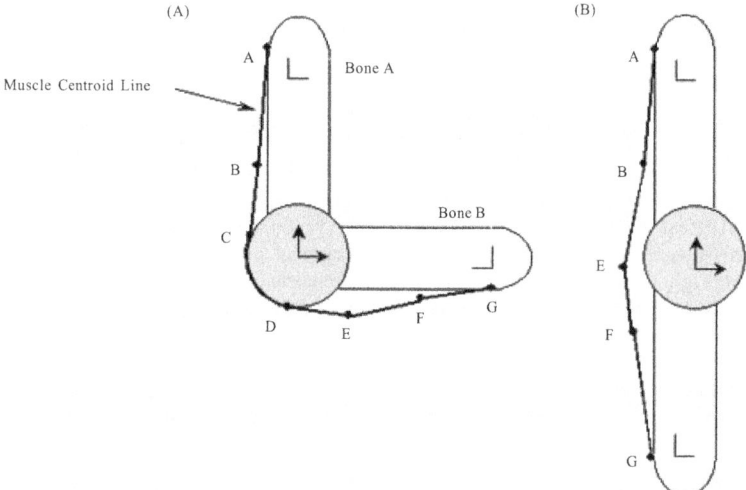

Bones A and B are connected by a muscle spanning their single connecting joint. Rotation of the joint causes the centroid path of the muscle to deviate. The model contains 7 via points. Points A, B and E, F, G are fixed via points, rigidly attached to bones A and B respectively. Points B and E are bounding-fixed via points, and points C and D are obstacle via points. The obstacle set is represented by points B, C, D, E, and the connecting muscle path. The obstacle shown is spherical, with diameter chosen to reflect the physiological shape of the constrained muscle path. At some joint configuration during pivoting between position (A) and (B), points C and D become inactive. Because the paths of all segments of the muscle are known, except those between the bounding-fixed via points B and E, calculating the positions of the obstacle via points C and D then fully describes the muscle path.

Modelling Shoulder Muscle Actuation

Hill-type models are often used to represent the mechanical properties of muscles in computational models of the shoulder complex. In a Hill-type model, each musculotendon actuator is described by a three-element muscle in series with an elastic tendon. This contractile element is capable of modelling a muscle's force-length-velocity property, as well as its active and passive stiffness. The Hill-type model is often used to inter-relate muscle force and velocity due to its computational efficiency and efficacy in simulating human movement. Studies have shown that models that fail to include a non-linear elastic element in series with the contractile actuator, nor a sufficient non-linear description of this contractile actuator, are of little use in biomechanics (Winters & Stark, 1985, 1988).

A muscle's force producing properties can be properly described by four parameters: peak isometric force, optimal muscle fibre length (the length at which active muscle force peaks), pennation angle (the angle at which muscle fibres insert on tendon when the fibers are at their optimal length), and the maximum shortening velocity of muscle. Parameters such as fibre length and pennation angle are normally obtained from detailed cadaver dissections. As it is not possible to determine peak muscle force or optimum muscle fibre length in vivo, optimisation procedures are commonly used to estimate these parameters (Garner & Pandy, 2003)

Tendon is often represented as an elastic element, although it has been shown that when a tendon is stretched from its initial length, force varies nonlinearly with length (Kubo, Kawakami, Kanehisa, & Fukunaga, 2002). The assumption of purely elastic behaviour can overestimate the amount of strain energy stored in tendon, but will not normally affect actuator performance. Values of tendon rest length and optimum contractile length are critical in determining actuator performance. If not determined accurately, these values will offset the magnitude of the maximum force developed by the actuator, and the joint angle at which this force occurs.

Modelling Excitation-Contraction Coupling in the Shoulder Complex

When a human skeletal muscle fibre is activated by a nerve impulse, an action potential spreads from the endplate region along the entire fibre. The impulse enters the transverse tubular network and induces the release of calcium from the sarcoplasmic reticulum. The released calcium binds to troponin and activates muscle force by initiating the cross-bridge cycle. This process of muscle stimulation and cross-bridge cycling is often referred to as excitation-contraction coupling. Because muscle cannot be activated or relaxed instantaneously, delays in excitation-contraction coupling become significant in human musculoskeletal coordination studies of movement. Adaptation of mathematical excitation-contraction coupling to computationally simulated shoulder movement is rare. Most upper-limb modelling studies comprehend simple shoulder simulations typically under static or quasi-static conditions. Modelling of excitation-contraction coupling is likely to be valuable in future research for elucidating neuromuscular control of shoulder complex function, particularly for dynamic coordination simulations, such as throwing.

The process of excitation-contraction coupling has proven difficult to study experimentally. As a result, complete descriptions of how a cell couples surface membrane and intracellular signal transduction proteins to achieve stable regulated intracellular calcium release are lacking. No single model is yet capable of describing all features of cardiac excitation-contraction coupling. However, current mathematical modelling has emphasised the importance of requirements for microscopic sub-cellular regions or "local control," so that one can separate local behaviour from the overall cell averaged "common-pool" behaviour during muscle excitation. Consequently the micro-architecture of the neuromuscular cleft has been shown to be a key factor in determining local calcium transients (Soeller & Cannell, 2004). In order to progress modelling of cardiac excitation-contraction coupling, an experimental database of time dependent intracellular calcium dynamics must be established. In particular, time dependence of processes such as ryanodine receptor function and sarcolemmal calcium influx to release sarcoplasmic reticulum calcium should be quantified.

Simple mathematical models frequently seek to represent physiological excitation-contraction coupling in an equation of muscle activation as a function of net neural drive. Other models may formulate muscle activation as a function of parameters such as motor neuron recruitment, stimulation frequency, muscle mass and fibre composition. Implicit of all such models of excitation-contraction coupling are the critical process rise and fall time-constant values. Values of these parameters vary widely in the literature and have a marked effect on movement coordination predictions (Pandy, 2001).

Modelling Motor Tasks and Determining Solutions to Redundant Problems in Shoulder Biomechanics

Like most human musculoskeletal joint systems, the shoulder complex is an indeterminate mechanical system. That is, the number of load-transmitting elements far exceeds the number of available force and moment equilibrium equations for the clavicle, scapula, and humerus. In principle, there are an infinite number of distinct combinations of muscle, ligament, and articular contact forces that can be executed to produce a given shoulder movement (Collins, 1995). The indeterminacy dilemma has led to analytical procedures being accepted as a standard means of resolving muscle forces and other parameters non-invasively.

Two approaches are generally taken to solve indeterminate problems in biomechanics - the reduction method, and the optimisation method. In the reduction method, the number of unknown forces acting at a joint is reduced to the number of equilibrium equations. This technique usually requires making gross assumptions and simplifications of the number of musculoskeletal structures about a joint, for example, by ignoring muscles, or grouping muscles of similar function. As a consequence, the scope of data output will be considerably limited, and error is invariably induced because the mechanical action of individual muscles is obscured (Dul, Johnson, Shiavi, & Townsend, 1984). One of the most effective ways to solve indeterminate mechanical problems is to formulate an objective function to solve using optimisation techniques. In biomechanics, the underlying principle is that a being must somehow distribute its muscle forces among muscle groups so as to create and control its movement optimally. The assumption is that during learned activities the human body selects a unique way of distributing its internal forces, and that this distribution can be effectively governed by certain physiological criteria. For example, it has been stated that synergistic muscles share their load in proportion to their size or capacity, and further, muscle capacity relates directly to a muscle's cross-sectional area (Zajac, 1989). Optimisation problems typically involve minimising an objective function that relates to the specific performance criteria. For instance, common optimisation techniques seek to minimise the sum of the squared or cubed muscle forces, muscle stresses, or muscle energy expenditure. Because solutions to optimisation problems are not unique, their output can vary considerably in the literature. Other common limitations of these techniques include the inability to account for ligament control of joint movement and antagonistic muscle activity.

Optimisation may involve inverse dynamics (static optimisation) or forward dynamics (dynamic optimisation) techniques. Inverse dynamics techniques use measurements applied to body segments such as external forces, position, velocity and acceleration, as inputs to calculate muscle forces. Static optimisation is an analysis-based approach, which solves a unique optimisation problem at each instant during motion. This technique requires segment kinematics to compute a static force analysis, so it cannot be used to formulate a complete model of the motor task. Because position and orientation of body segments during motor activity normally depends on the previous history of applied forces, and not just the current muscle state, future consequences of body movement remain largely unaccounted for during the static optimisation process. Progress in development of inverse dynamics models complex enough to study muscle coordination has been limited by the ability to collect precise biomechanical and kinesiological data (Zajac, 1993). Poor estimations of body segment velocity and acceleration leads to significant errors in calculated values of net joint torques and therefore muscle force. For this reason it has also been difficult to include muscle physiology in the formation of static optimisation problems because parameters such as muscle length and contraction velocity are wholly dependent on accurate estimates of segment position, velocity, and acceleration.

Forward dynamics techniques use muscle excitations or activations as inputs to calculate corresponding body motions, thus dynamic optimisation techniques can be used to study how muscle forces affect motion. Dynamic optimisation techniques are more powerful than static optimisation techniques: dynamic optimisation solves a single optimisation problem for an entire cycle of a motor task, and so a model of the goal of the task can be included in the framework of the problem. Despite the inherent benefits, dynamic optimisation has gained little attention until recently, as its computational processing is significantly more exhaustive than the equivalent inverse dynamics.

Static and dynamic optimisation solutions have been shown to lead to very similar results (Anderson & Pandy, 2001; Happee, 1994). But which method is most suitable for resolving muscle forces in the human shoulder? Static optimisation is considerably more popular due to the inherent benefits in computation time. Common applications of static models may, for example, seek to estimate instantaneous muscle moments about joints from experimental kinematic data. If, however, one seeks to study how changes in muscle parameters affect function and performance of a motor task, dynamic optimisation is more appropriate because entire motor tasks can be formulated when body motion measurements remain unknown.

SAMPLE CALCULATIONS OF A MUSCULOSKELETAL MODEL

As discussed, musculoskeletal models of the shoulder are a useful tool in determining muscle moment arms, forces, torques, and joint forces. Models of the upper extremity, such as the one developed by Garner and Pandy (1999, 2001), have made these calculations possible.

The musculoskeletal model developed by Garner and Pandy, referred to as the model for the remainder of the section, is a musculoskeletal model of the entire arm. It contains 13 degrees of freedom and seven bones, including the scapula, clavicle, humerus, ulna, radius, wrist, and hand. The model has a total of 26 muscles and 42 muscle bundles. Muscles with a broad origin or insertion, like the deltoid and trapezius, were split into separate bundles, in order to more accurately demonstrate their anatomy (Figure 3).

Figure 3. (A) The muscle paths for the three portions of the pectoralis major, (B) The muscle mesh for the pectoralis major

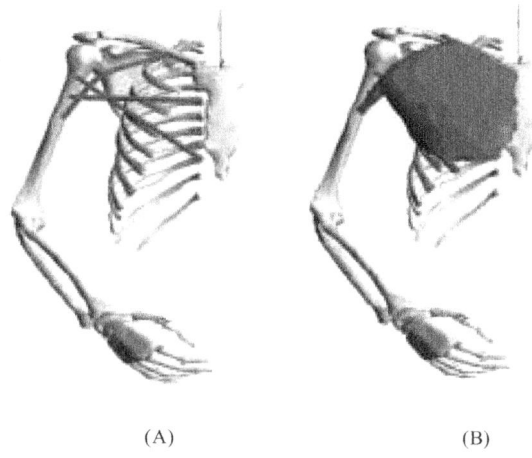

(A)   (B)

Figure 4. The graphical user interface for the arm musculoskeletal model developed by Garner and Pandy (1999, 2001)

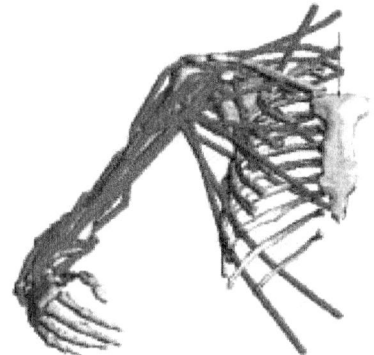

Sixteen of those muscles, and 31 of the muscle bundles, cross the glenohumeral joint. The muscle paths follow the cross-sectional centroid of the muscles and wrap around other muscles and bony protrusions according to the obstacle set method (Garner & Pandy, 2000). The muscle tendon actuators are three-element Hill-type muscles in series with a tendon. The muscle's peak isometric force was assumed to be proportional to the physiological cross-sectional area (PCSA) with the maximum muscle stress set at 330 kPa. The force was determined by the length of the muscle in relation to position on the normalized force-length curve (Garner & Pandy, 2003).

Because the model contains more elements than it does degrees of freedom, an optimisation routine was used to find the solution. The optimiser minimized the sum of the squares of all muscle stresses, in theory doing as the brain would by causing as little stress on the body as possible. The system also operated under additional limits including two holonomic constraints to ensure the scapula does not pierce the thorax. In order to validate the model, moment arms from cadaveric research were compared with those from the model during maximum strength tests (Garner, 1998). Figure 4 shows the model graphical user interface that allows the user to visualize arm position and active muscles during a particular motion.

The model was also used to develop solutions for maximum isometric calculations. In this instance, the muscle forces in the arm during maximal abduction from 0-90 degrees (Figure 5) show that the middle deltoid produced more force than any other muscle. Other significant contributions to the force production came from the muscles of the rotator cuff — the infraspinatus, subscapularis, and supraspinatus.

In addition to muscle force, the model can also determine how much torque the muscles produce. Although there were other muscles that produced high forces, the torque of the arm was produced mostly by the middle deltoid (Figure 6).

The large contribution of the middle deltoid to the total arm torque was due to its large moment arm. Moment arm calculations can also be made from the model by

Figure 5. Maximum isometric abduction forces from 0-90 degrees for the anterior deltoid (DeltC), middle deltoid (DeltA), clavicular portion of the pectoralis major (PmajC), supraspinatus (Supr), infraspinatus (Infr), and subscapularis (Subs)

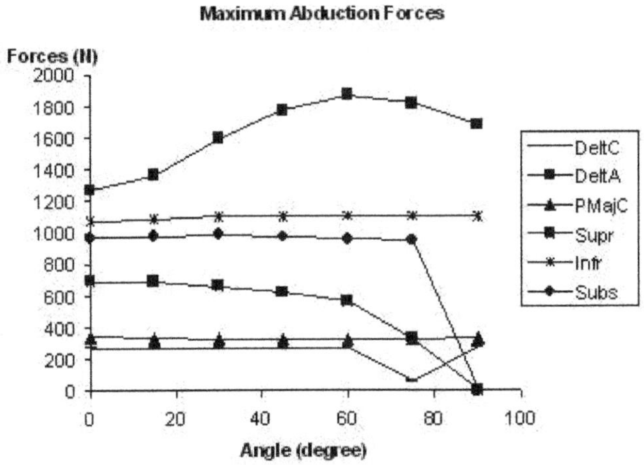

Maximum Abduction Forces

Figure 6. Maximum isometric abduction torques from 0-90 degrees for the anterior deltoid (DeltC), middle deltoid (DeltA), clavicular portion of the pectoralis major (PmajC), supraspinatus (Supr), infraspinatus (Infr), and subscapularis (Subs)

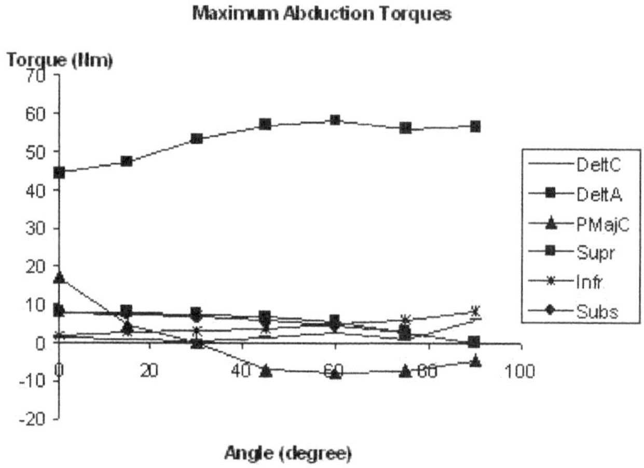

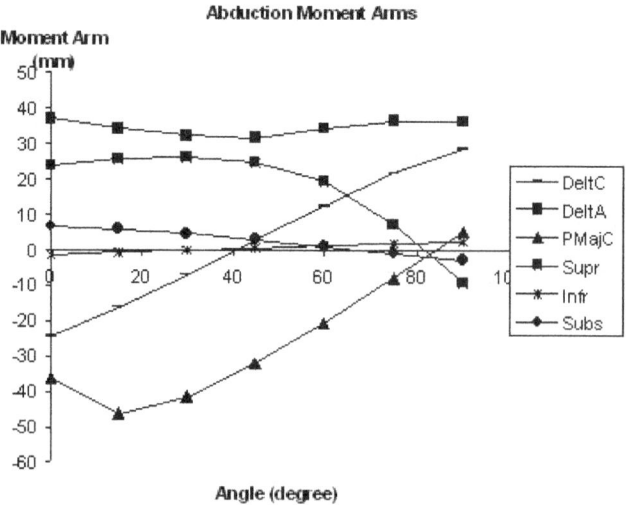

Figure 7. Abduction moment arms from 0-90 degrees for the anterior deltoid (DeltC), middle deltoid (DeltA), clavicular portion of the pectoralis major (PmajC), supraspinatus (Supr), infraspinatus (Infr), and subscapularis (Subs)

Figure 8. Forces calculated for the three portions of deltoid during flexion in the sagittal plane

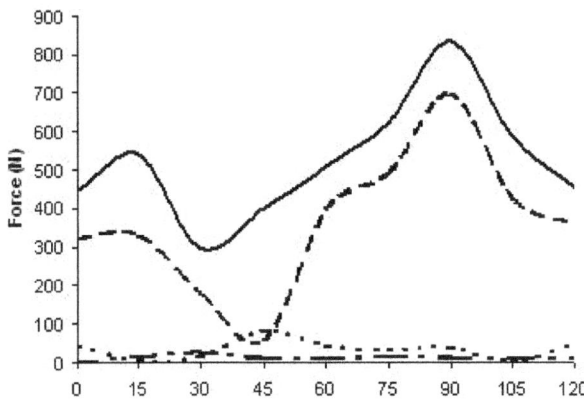

Also shown is the resultant force acting at the glenohumeral joint (solid line) (Goodwin et al., 2003).

mimicking a tendon excursion experiment as done in cadavers. The middle deltoid had a large abducting moment arm across the range of motion (Figure 7), causing the more significant amount of torque from the middle deltoid despite the large forces produced in several other muscles.

As well as giving muscle estimates for maximum isometric contractions, the model can estimate muscle force and torque in arm elevation, holding the arm up against only gravity. This time the model used the optimiser to find the pattern of muscle activations that minimizes muscle stress. In arm flexion, the middle deltoid remained the main contributor in force production (Figure 8) (Goodwin, Pandy, Yanagawa, & Shelburne, 2004). The optimiser favoured muscles with large physiological cross-sectional areas (PCSA) because these muscles are less stressed when producing great force. The middle deltoid had the largest PCSA of the muscles crossing the glenohumeral joint, so along with its large flexion moment arm, it became a favourable choice to produce torque. In addition to the muscle force, the force transmitted by the bones at the glenohumeral joint is shown. The joint force changed in relation to changes in the middle deltoid force, the most contributing muscle in elevation in the sagittal plane (Figure 8).

In addition to studying the normal function of the shoulder, the model can be modified to study the changes that occur when the shoulder is reconstructed, like with a reverse shoulder prosthesis (RSP). When a RSP is used, the rotator cuff muscles are removed and the joint centre of the shoulder moves medially (Figure 9).

Figure 9. The normal anatomy of the shoulder (left) with the anatomical joint center (1) and the anatomy of the reverse shoulder prosthesis (right) with the new, displaced joint center (2)

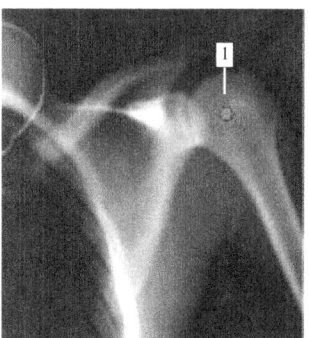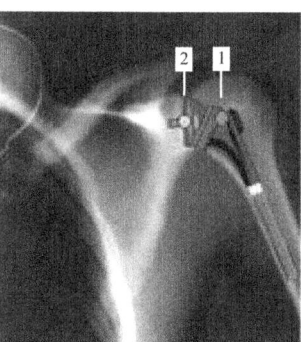

Figure 10. Middle deltoid abduction moment arms for the normal shoulder (anatomical) and during glenohumeral joint displacement in the medial and lateral direction

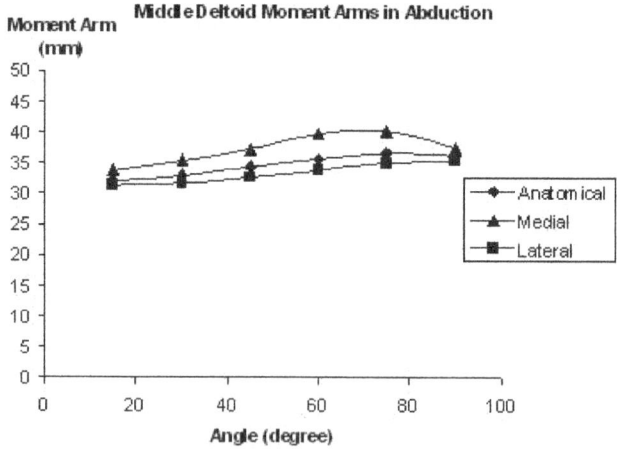

In a very simple study to address shoulder joint reconstruction, all the muscles were removed from the shoulder except for the middle deltoid. Because the middle deltoid is the prime mover of the arm and is responsible for almost all of the arm-raising torque, studying the function in this muscle during reconstruction is most telling. After leaving

Figure 11. Middle deltoid abduction forces for the normal shoulder (anatomical) and during glenohumeral joint displacement in the medial and lateral direction

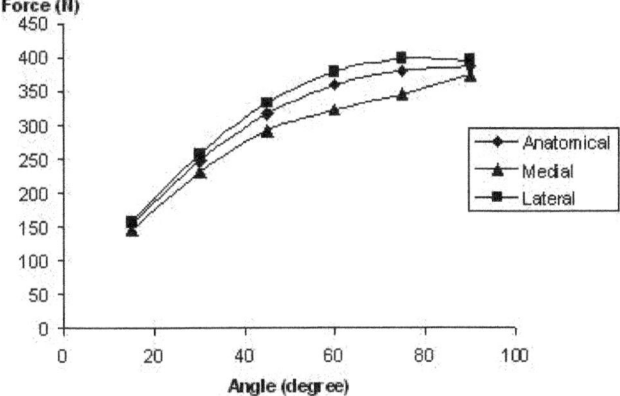

just this muscle in place, the reconstruction was simulated by displacing the centre of the glenohumeral joint. In one study, the joint was displaced medially 1.0 cm and laterally 1.0 cm. The medial displacement caused an increase in the middle deltoid moment arm (Figure 10). The increase in moment arm led to a decrease in force production compared to the normal functioning shoulder (the anatomical model) (Figure 11). Therefore, in order to produce the same amount of torque to raise the arm in the scapular plane, the muscle does less work when the joint is displaced medially.

We believe the results obtained from the model are accurate estimates of in vivo measurements that are otherwise invasive or impossible to obtain. The middle deltoid emerges as the most torque producing muscle in abduction. Similar studies can be done with other motions, like flexion (i.e., elevation of the humerus in the sagittal plane). In flexion, the middle deltoid is also responsible for most of the arm raising torque. In addition, alterations can be made to the model in more complex fashions to simulate reconstruction. These studies allow engineers to develop more biomechanically advantageous prostheses and surgeons to place them more efficiently.

TECHNIQUES IN VERIFICATION OF SHOULDER MUSCULOSKELETAL MODELS

The long term aim of computational and mathematical modelling of the shoulder complex is to explain how neuromuscular and musculoskeletal systems interact to produce shoulder movement. Models of shoulder movement therefore seek to predict quantities such as joint torques, muscle forces and moment arms, and articulating forces under a variety of different shoulder positions and load situations. But how can one be confident of selecting a set of musculoskeletal input parameters for such models? Further, how can one ascertain the degree to which a computational model portrays in situ shoulder biomechanics during simulated motor tasks? Because of the inherent impracticalities in measuring muscle forces directly, a number of in vitro and in vivo approaches have been taken in an attempt to provide a practical basis for musculoskeletal parameter estimation and model verification.

Muscle activity and coordination have very often been studied using electromyography either with surface electrodes or intramuscular fine-wire electrodes. Combining, for example, muscle PCSA and moment arm data obtained from the literature, EMG-based models have been produced to estimate shoulder muscle forces (Laursen, Jensen, Nemeth, & Sjogaard, 1998). Comparison of muscle force predictions with EMG recordings is made difficult by the unknown muscle force-length relationship and the muscle length dependency of EMG. Therefore, despite the large contribution of muscle EMG data to the literature, EMG amplitude can only be used to qualitatively validate muscle force predictions in shoulder complex models.

Models of the shoulder complex are commonly validated by comparing computed joint moments with external torque measurements determined in vivo. Dynamometers and custom designed instrumented testing rigs are used to measure shoulder joint torques during maximum voluntary isometric contractions of shoulder muscles e.g. for humeral flexion-extension. Joint torques measured over the full range of movement of each joint can also serve to provide model input parameters such as joint torque-angle data. However, if one assumes an idealised condition during maximum voluntary contraction of maximum agonist and no antagonist muscle activation, computed joint moments are likely to be overestimated (Kronberg, Nemeth, & Brostrom, 1990).

Researchers have also attempted to verify the roles of individual muscles of the shoulder in studies of intramuscular pressure. Although this approach gives a qualitative assessment of muscle effort, the relationship between intramuscular pressure and muscle force is unclear (Karlsson & Peterson, 1992). Other in vivo methods have involved determining torque-producing roles of individual muscles by blocking their nerves with local anaesthesia. This approach has proven difficult to implement for all shoulder components due to prominent anatomical constraints.

Validation of predicted joint forces at the shoulder has proven problematic due to the difficulties of directly measuring articular contact loads. In the hip (Davy et al., 1988; Graichen & Bergmann, 1991; Taylor et al., 1997) and knee (D'Lima, Townsend, Arms, Morris, & Colwell, 2004; Kaufman, Kovacevic, Irby, & Colwell, 1996), articulating forces have been measured in vivo using instrumented telemetric endoprostheses. No such data has been obtained at the glenohumeral joint as yet. Custom designed shoulder cadaver testing equipment however is now allowing researchers to simulate in vivo shoulder kinematics while directly measuring parameters such as joint reaction force and humeral translation in vitro. Debski et al. (1995) developed a dynamic shoulder cadaver testing apparatus to simulate muscle action and measure corresponding tendon excursion and joint motion. Actuators were used to apply forces to tendons of the shoulder joint to flex and extend the humerus. A universal force moment sensor (UFS) was later integrated to measure glenohumeral articulating force magnitude and direction (Apreleva et al., 1998; Apreleva, Parsons, Warner, Fu, & Woo, 2000; Debski et al., 1995; McMahon et al., 1995; Parsons, Apreleva, Fu, & Woo, 2002; Thompson et al., 1996). This apparatus, as with others (Wuelker, Wirth, Plitz, & Roetman, 1995) come close, but are still limited in their ability to replicate in situ shoulder biomechanics. Firstly, dynamic muscle load sharing, which occurs during upper limb movement, is usually not simulated during these experiments. Secondly, muscle lines of action of shoulder muscles with large attachment sites are often represented by just one force vector, and these may not accurately reproduce the moment arms of muscles crossing the glenohumeral joint. Finally, as with many other in vitro shoulder studies (McMahon, Chow, Sciaroni, Yang, & Lee, 2003; McMahon et al., 1995; Warner et al., 1998), scapulohumeral rhythm is not accounted for. In summary, joint reaction force data obtained from most current in vitro studies is likely to be of qualitative use only. Further development of shoulder cadaver testing equipment is progressing, and in the future will provide a more meaningful biomechanics database with which to verify computational models.

FUTURE TRENDS IN MODELLING THE SHOULDER COMPLEX

The future of shoulder modelling is one of continuous growth and complexity. The past decade has delivered an increasing knowledge base, which has been the foundation for significant improvements in the accuracy of mathematical and computational models. As computer processor speed and performance improves, greater model complexity will be facilitated, providing deeper insights into neural control of musculoskeletal movement. The result of the capacity to understand and model the complex biomechanics interactions that produce musculoskeletal movement will have direct clinical benefits. For example, in the accurate prediction of outcomes following shoulder joint replacement surgery and rehabilitation, or in the reduction of shoulder pain. Currently clinical applications, such as the modelling of shoulder reconstructions (Ahir, Walker, Squire-Taylor, Blunn, & Bayley, 2004; Wood, Meek, & Jacobsen, 1989a, 1989b) and joint replacements, are already receiving interest from model developers. However, before shoulder models progress to complex clinical applications, the current shoulder models, which are mostly quasi-static, must begin dynamically simulating shoulder motion, like the walking models of Anderson and Pandy (2001), Zajac et al., (2003), and others.

Accurate representation of human anatomy is the foundation for a useful computational model of movement. Model complexity has progressed only with the aid of sophisticated techniques in representing muscle and bone properties (Chao, 2003). For example, digital cadavers have been made available in the public domain. Full digital renderings of male and female subjects can allow workers to develop entire three-dimensional images of the human musculoskeletal system, to be used in computational models (Garner & Pandy, 1999; B. A. Garner & Pandy, 2000). Similarly, computed tomography (CT) scans and magnetic resonance imaging (MRI) allow for improved imaging of the musculoskeletal system compared to classical modelling techniques such as straight-line body segment representations.

The capacity for a computational shoulder model to represent in situ biomechanics is strongly dependent on the accuracy of its parameters, or the effectiveness of its output as substantiated by verification techniques. For this reason, in vivo and in vitro experimental techniques are likely to be developed with more rigour. In particular, the design of instrumented prostheses and dynamic shoulder cadaver testing rigs, as these devices comprise the only known means to directly measure joint reaction forces during shoulder complex motion. Looking further into the future, more sophisticated methods for determining muscle forces, such as muscle tracking systems, may also be developed and implemented in modelling studies.

It is anticipated that the principles and methods used in elucidating the neuromuscular and musculoskeletal control of shoulder movement could be applied to all other functionally significant musculoskeletal regions of the human body.

CONCLUSION

The shoulder remains one of the least successfully modelled regions of the human body due to the complexity of its constituents and the sophisticated nature of its neural control. This chapter describes current techniques in modelling movement and biomechanics of the shoulder complex. The strengths and weaknesses of these techniques were assessed by analysing the biomechanical principles and assumptions behind these methods. This chapter has shown that for a model of the shoulder to be of any clinical relevance, it must embrace a detailed model of the shoulder girdle, a representation of muscle paths, a model of the muscle actuation, an assessment of muscle excitation-contraction coupling, and an optimisation procedure, in order to solve the indeterminate problem with respect to the motor task. In vivo and in vitro verification techniques were shown to be a key factor in determining the validity of these computational models. Assessment of future trends concluded that as computer power increases, models of greater complexity will be developed, providing deeper insights into the principles of neuromuscular and musculoskeletal control of movement. Further, clarifying the complex nature of shoulder musculature function and control will have immediate clinical implications including improvements in predictions of shoulder stability treatment, pain reduction, joint replacement surgery, and rehabilitation.

REFERENCES

Ahir, S. P., Walker, P. S., Squire-Taylor, C. J., Blunn, G. W., & Bayley, J. I. (2004). Analysis of glenoid fixation for a reversed anatomy fixed-fulcrum shoulder replacement. *Journal of Biomechanics, 37*(11), 1699-1708.

Anderson, F. C., & Pandy, M. G. (2001). Static and dynamic optimization solutions for gait are practically equivalent. *Journal of Biomechanics, 34*(2), 153-161.

Apreleva, M., Hasselman, C. T., Debski, R. E., Fu, F. H., Woo, S. L., & Warner, J. J. (1998). A dynamic analysis of glenohumeral motion after simulated capsulolabral injury. A cadaver model. American *Journal of Bone Joint Surgery, 80*(4), 474-480.

Apreleva, M., Parsons, I. M. t., Warner, J. J., Fu, F. H., & Woo, S. L. (2000). Experimental investigation of reaction forces at the glenohumeral joint during active abduction. *Journal of Shoulder and Elbow Surgery, 9*(5), 409-417.

Bao, H., & Willems, P. Y. (1999). On the kinematic modelling and the parameter estimation of the human shoulder. *Journal of Biomechanics, 32*(9), 943-950.

Bjelle, A., Hagberg, M., & Michaelson, G. (1981). Occupational and individual factors in acute shoulder-neck disorders among industrial workers. *British Journal of Independent Medicine, 38*(4), 356-363.

Chadwick, E. K., van Noort, A., & van der Helm, F. C. (2004). Biomechanical analysis of scapular neck malunion — a simulation study. *Clinical Biomechanics (Bristol, Avon), 19*(9), 906-912.

Chao, E. Y. (2003). Graphic-based musculoskeletal model for biomechanical analyses and animation. *Medical Engineering and Physics, 25*(3), 201-212.

Collins, J. J. (1995). The redundant nature of locomotor optimization laws. *Journal of Biomechanics, 28*(3), 251-267.

Davy et al. (1988). Telemetric force measurements across the hip after total arthroplasty. American *Journal of Bone Joint Surgery, 70*(1), 45-50.

De Duca, C. J., & Forrest, W. J. (1973). Force analysis of individual muscles acting simultaneously on the shoulder joint during isometric abduction. *Journal of Biomechanics, 6*(4), 385-393.

de Leva, P. (1996). Joint center longitudinal positions computed from a selected subset of Chandler's data. *Journal of Biomechanics, 29*(9), 1231-1233.

Debski, R. E., McMahon, P. J., Thompson, W. O., Woo, S. L., Warner, J. J., & Fu, F. H. (1995). A new dynamic testing apparatus to study glenohumeral joint motion. *Journal of Biomechanics, 28*(7), 869-874.

D'Lima, D. D., Townsend, C. P., Arms, S. W., Morris, B. A., & Colwell, C. W. (2005). An implantable telemetry device to measure intra-articular tibial forces. *Clinical Orthopaedics & Related Research, 440,* 45-49.

Dul, J., Johnson, G. E., Shiavi, R., & Townsend, M. A. (1984). Muscular synergism—II. A minimum-fatigue criterion for load sharing between synergistic muscles. *Journal of Biomechanics, 17*(9), 675-684.

Engin, A. E. (1980). On the biomechanics of the shoulder complex. *Journal of Biomechanics, 13,* 575-590.

Engin, A. E., & Tumer, S. T. (1989). Three-dimensional kinematic modelling of the human shoulder complex — Part I: Physical model and determination of joint sinus cones. *Journal of Biomechanical Engineering, 111*(2), 107-112.

Garner, B. A. (1998). *A musculoskeletal model of the upper limb based on the medical image dataset of the visible human male.* University of Texas, Austin.

Garner, B. A., & Pandy, M. G. (1999). A kinematic model of the upper limb based on the visible human project (VHP) image dataset. *Computer Methods in Biomechanical and Biomedical Engineering, 2*(2), 107-124.

Garner, B. A., & Pandy, M. G. (2000). The obstacle-set method for representing muscle paths in musculoskeletal models. *Computer Methods in Biomechanics and Biomechanical Engineering, 3,* 1-30.

Garner, B. A., & Pandy, M. G. (2003). Estimation of musculotendon properties in the human upper limb. *Annals of Biomedical Engineering, 31*(2), 207-220.

Goodwin, C. J., & Pandy, M. G. (2004). Computation of muscle and joint reaction forces at the shoulder during forward flexion. *Recent Res. Devel. Biomechanics, 2.*

Graichen, F., & Bergmann, G. (1991). Four-channel telemetry system for in vivo measurement of hip joint forces. *Journal of Biomedical Engineering, 13*(5), 370-374.

Gray, H. (2000). *Anatomy of the human body* (20th ed.). New York: bartleby.com.

Gupta, S., & Van der Helm, F. C. T. (2004). Load transfer across the scapula during humeral abduction. *Journal of Biomechanics, 37,* 1001-1009.

Halder, A. M., Kuhl, S. G., Zobitz, M. E., Larson, D., & An, K. N. (2001). Effects of the glenoid labrum and glenohumeral abduction on stability of the shoulder joint through concavity-compression: An in vitro study. *American Journal of Bone Joint Surgery, 83-A*(7), 1062-1069.

Happee, R. (1994). Inverse dynamic optimization including muscular dynamics, a new simulation method applied to goal directed movements. *Journal of Biomechanics, 27*(7), 953-960.

Hogfors, C., Karlsson, D., & Peterson, B. (1995). Structure and internal consistency of a shoulder model. *Journal of Biomechanics, 28*(7), 767-777.

Hogfors, C., Peterson, B., Sigholm, G., & Herberts, P. (1991). Biomechanical model of the human shoulder joint — II. The shoulder rhythm. *Journal of Biomechanics, 24*(8), 699-709.

Hollinshead, W. H. (1976). *Functional anatomy of the limbs and back* (4th ed.). Philadelphia: W.B. Saunders.

Inman, V. T., Saunders, J. B., & Abbott, L. C. (1944). Observations on the function of the shoulder joint. *Journal of Bone and Joint Surgery (American), 58*(A), 1-30.

Inman, V. T., Saunders, J. B., & Abbott, L. C. (1996). Observations of the function of the shoulder joint. 1944. *Clinical Orthopedics* (330), 3-12.

Karlsson, D., & Peterson, B. (1992). Towards a model for force predictions in the human shoulder. *Journal of Biomechanics, 25*(2), 189-199.

Kaufman, K. R., Kovacevic, N., Irby, S. E., & Colwell, C. W. (1996). Instrumented implant for measuring tibiofemoral forces. *Journal of Biomechanics, 29*(5), 667-671.

Kronberg, M., Nemeth, G., & Brostrom, L. A. (1990). Muscle activity and coordination in the normal shoulder. An electromyographic study. *Clinical Orthopedics* (257), 76-85.

Kubo, K., Kawakami, Y., Kanehisa, H., & Fukunaga, T. (2002). Measurement of viscoelastic properties of tendon structures in vivo. *Scandanavian Journal of the Medicine and Science of Sports, 12*(1), 3-8.

Laursen, B., Jensen, B. R., Nemeth, G., & Sjogaard, G. (1998). A model predicting individual shoulder muscle forces based on relationship between electromyographic and 3D external forces in static position. *Journal of Biomechanics, 31*(8), 731-739.

Lucas, D. B. (1973). Biomechanics of the shoulder joint. *Archives of Surgery, 107*(3), 425-432.

Maurel, W., & Thalmann, D. (2000). Human shoulder modeling including scapulothroracic constraint and joint sinuc cones. *Computers and Graphics, 24*, 203-218.

McMahon, P. J., Chow, S., Sciaroni, L., Yang, B. Y., & Lee, T. Q. (2003). A novel cadaveric model for anterior-inferior shoulder dislocation using forcible apprehension positioning. *Journal of Rehabilitation Res Development, 40*(4), 349-359.

McMahon, P. J., Debski, R. E., Thompson, W. O., Warner, J. J., Fu, F. H., & Woo, S. L. (1995). Shoulder muscle forces and tendon excursions during glenohumeral abduction in the scapular plane. *Journal of Shoulder and Elbow Surgery, 4*(3), 199-208.

Novotny, J. E., Nichols, C. E., & Beynnon, B. D. (1998). Normal kinematics of the unconstrained glenohumeral joint under coupled moment loads. *Journal of Shoulder and Elbow Surgery, 7*(6), 629-639.

Pandy, M. G. (2001). Computer Modeling and Simulation of Human Movement. *Annual Review of Biomedical Engineering, 3*, 245-273.

Parsons, I. M., Apreleva, M., Fu, F. H., & Woo, S. L. (2002). The effect of rotator cuff tears on reaction forces at the glenohumeral joint. *Journal of Orthopedic Res, 20*(3), 439-446.

Poppen, N. K., & Walker, P. S. (1978). Forces at the glenohumeral joint in abduction. *Clinical Orthopedics* (135), 165-170.

Raikova, R. (1992). A general approach for modelling and mathematical investigation of the human upper limb. *Journal of Biomechanics, 25*(8), 857-867.

Sharkey, N. A., Marder, R. A., & Hanson, P. B. (1993). *The role of the rotator cuff in elevation of the arm.* Paper presented at the 39th Annual Meeting, Orthopaedic Research Society, San Francisco.

Shrive, N. G., O'Connor, J. J., & Goodfellow, J. W. (1978). Load-bearing in the knee joint. *Clinical Orthopedics,* (131), 279-287.

Soeller, C., & Cannell, M. B. (2004). Analysing cardiac excitation-contraction coupling with mathematical models of local control. *Prog Biophys Mol Biol, 85*(2-3), 141-162.

Taylor, S. J., Perry, J. S., Meswania, J. M., Donaldson, N., Walker, P. S., & Cannon, S. R. (1997). Telemetry of forces from proximal femoral replacements and relevance to fixation. *Journal of Biomechanics, 30*(3), 225-234.

Thompson et al. (1996). A biomechanical analysis of rotator cuff deficiency in a cadaveric model. *American Journal of Sports Medicine, 24*(3), 286-292.

Tumer, S. T., & Engin, A. E. (1989). Three-dimensional kinematic modelling of the human shoulder complex—Part II: Mathematical modelling and solution via optimization. *Journal of Biomechanical Engineering, 111*(2), 113-121.

van der Helm, F. C. (1994). Analysis of the kinematic and dynamic behavior of the shoulder mechanism. *Journal of Biomechanics, 27*(5), 527-550.

Van der Helm, F. C., Veeger, H. E., Pronk, G. M., Van der Woude, L. H., & Rozendal, R. H. (1992). Geometry parameters for musculoskeletal modelling of the shoulder system. *Journal of Biomechanics, 25*(2), 129-144.

Van der Helm, F. C., & Veenbaas, R. (1991). Modelling the mechanical effect of muscles with large attachment sites: application to the shoulder mechanism. *Journal of Biomechanics, 24*(12), 1151-1163.

Warner, J. J., Bowen, M. K., Deng, X. H., Hannafin, J. A., Arnoczky, S. P., & Warren, R. F. (1998). Articular contact patterns of the normal glenohumeral joint. *Journal of Shoulder and Elbow Surgery, 7*(4), 381-388.

Winters, J. M., & Stark, L. (1985). Analysis of fundamental human movement patterns through the use of in-depth antagonistic muscle models. *IEEE Trans Biomed Eng, 32*(10), 826-839.

Winters, J. M., & Stark, L. (1988). Estimated mechanical properties of synergistic muscles involved in movements of a variety of human joints. *Journal of Biomechanics, 21*(12), 1027-1041.

Wood, J. E., Meek, S. G., & Jacobsen, S. C. (1989a). Quantitation of human shoulder anatomy for prosthetic arm control — I. Surface modelling. *Journal of Biomechanics, 22*(3), 273-292.

Wood, J. E., Meek, S. G., & Jacobsen, S. C. (1989b). Quantitation of human shoulder anatomy for prosthetic arm control — II. Anatomy matrices. *Journal of Biomechanics, 22*(4), 309-325.

Wuelker, N., Wirth, C. J., Plitz, W., & Roetman, B. (1995). A dynamic shoulder model: reliability testing and muscle force study. *Journal of Biomechanics, 28*(5), 489-499.

Zajac, F. E. (1989). Muscle and tendon: properties, models, scaling, and application to biomechanics and motor control. *Criicalt Review of Biomedical Engineering, 17*(4), 359-411.

Zajac, F. E. (1993). Muscle coordination of movement: a perspective. *Journal of Biomechanics, 26 Suppl 1*, 109-124.

APPENDIX: GLOSSARY

A

Acromioclavicular. The joint in the shoulder complex that connects the clavicle to the acromion of the scapula. It is often represented by a three degree of freedom joint in mathematical and computational models.

Alignment. This term defines the process of computing the initial values of roll, pitch and yaw for the initial attitude matrix.

Alzheimer's Disease. A neurodegenerative disorder characterized by the progressive and irreversible loss of neurons located in specific regions of the brain. The effects of this disease include impairment of memory, cognition and behavior.

Ambulatory Monitor. A portable system that can be used to continuously record multiple physiological and biomechanical variables during functional activities, e.g., walking.

Ambulatory Monitoring. Technique using body fixed sensors to monitor human biological and biomechanical signals.

Anesthesia. Characterized by unconsciousness, muscle relaxation and loss of sensation over the whole body. Results from the administration of a general anesthetic agent.

Artificial Neural Network (ANN). A technique used in artificial intelligence and machine learning applications to classify a pattern. Using weights and connections derived from pervious training examples.

Auditory Brainstem Response. A laboratory test that records an objective potential in response to acoustic stimuli.

Automatic Control. Autonomous control by machines without human involvement.

Axiohumeral. The muscles that pass from the torso to the humerus. For example, the pectoralis major, pectoralis minor and latissimus dorsi.

Axioscapular. The muscles that pass from the torso to the scapula. For example, the trapezius, rhomboids, serratus anterior and levator scapulae.

B

Back-Propagation Network. This class of neural networks is named after the method how these networks are trained. The backpropagation training algorithm calculates the error in each step of the training based on the anticipated output and the current output then adjusts the weights of the various layers backwards from the output layer all the way back to the input layer.

Back-Propagation Through Time (BPTT). The BPTT algorithm is an extension of the standard backpropagation algorithm. It may be derived by *unfolding* the temporal operation of a recurrent neural network into a layered feedforward network, the topology of which grows by one layer at every time step.

Beat Detection. Term commonly used in computerised processing of the electrocardio-gram reflecting the process of detecting each cardiac cycle in each recording lead and in addition location of the temporal reference points for each interwave component.

Bi-Group Neural Network. A standard multi-layered perceptron with one node in the output layer. Capable of discriminating between two classes.

Biomechanics. Study of the effects of internal and external forces on the human body in movement and rest.

Biomimetic Control. Control systems that mimic biological control mechanisms.

Body Surface Potential Mapping. An approach to capture the electrical activity of the heart as reflected on the entire surface of the torso.

Bradykenisia. Slowness of movement.

Brain-Computer Interface. Operates on the principle that non-invasively recorded cortical EEG signals contain information about an impending task and that such information can be identified by an ANN.

C

Calibration. Process of optimizing the parameter values to match forward dynamics joint moments to those of inverse dynamics.

Center of Mass. Whole body point of mass representing the weighted summation of multiple body segment masses.

Center of Pressure. Central point of contact of the functional base of support with the ground. A weight summation of all force contact points beneath the feet.

Central Nervous System (CNS). Parts of the nervous system in brain and spinal cord.

Cerebral Blood Flow Oscillation. Oscillation of the cerebral blood flow is a common feature in several physiological or pathophysiological states of the brain.

Chromosome. The bit string which represents the individual solution to an optimization or control problem. The chromosomes can encode the values of tuning variables in multidimensional optimization problems.

Clinical Gait Analysis. A process that uses objective methods to describe a patient's walking with a view to understand the underlying pathomechanisms and to inform the clinical decision making process which aims to improve the patient's gait through various forms of interventions.

Cluster. Compound of neurons that build a closed network area and contain similar information.

Concept Learning. Also known as novelty detection or one-class classification. This is a tool to describe one class of objects based on their common features and distinguish from all the other objects.

Cooperative Control. Two or more control systems operating on the same plant to achieve a common control objective.

Coxarthrosis. Osteoarthritis in hip joint.

Cross-Correlation. Provides an estimate of the similarity of waveforms.

Cross Validation. Used as a tool for optimizing network parameters and architectures where remaining residual generalisation errors can be reduced by invoking ensembles of similar networks.

Crossover. Creating new chromosomes by combining parts of parent chromosomes.

Crouch Gait. Walking with bent knees during the stance phase of gait. The knee never extends beyond 30 degrees of flexion. One of the many causes is a weak/long Achilles tendon.

D

Decision Line. A numeric value used to discriminate a specific data point into one category or the other.

Decision Tree. Classification algorithm using values of features which can be represented as a tree structure and which is easily understood by an expert.

Discrete Fourier Transform (DFT). Signal transform that transforms the discrete time signal to the frequency domain. The discrete-time Fourier transform (DTFT) of a discrete time signal is a representation of the signal in terms of complex exponential signal $e^{-j\omega n}$, where ω is a real frequency variable. The values of DFT are evenly spaced samples of the DTFT.

Dual Task Paradigm. Adding an attention demanding task during walking.

DyCoN (dynamically controlled network). Self-organizing map with individually learning neurons. This enables the network to dynamically adapt to changing distributions over the input space.

Dynamic Adaptation. Physiological or learning process where a time-series of external stimuli effects a change of the affected object's state or behaviour.

Dyskinesia. Involuntary movement which occurs in Parkinson disease patient due to levodopa therapy.

E

Electrocardiogram (ECG). A recording of the heart's electrical currents obtained with the electrocardiograph. This is an instrument designed for recording the electrical currents that traverse the heart and initiate its contraction.

Electroencephalography (EEG). Technique used to detect electrical activity of the brain produced by functioning neurons.

Electromyography (EMG). Electrical potentials released during voluntary or involuntary contractions of muscles. The magnitude of an EMG signal represents the activation level of muscles.

Epilepsy. A disorder caused by excessively powerful activity of certain neurons in the brain. Can result in fainting, involuntary violent shaking or brief unconsciousness.

Evolutionary Computation. Computational methods that simulate the mechanism of natural evolution.

F

Fast Fourier Transform (FFT). A technique used to derive frequency and phase information from a time waveform.

Feature Extraction. Information compression method used to reduce the size (i.e., dimensions) of the signal under investigation. It is a central problem of signal processing to select proper information compression method in order to extract the important features of the signal.

Fitness Function. A function that quantifies the relative fitness, quality or optimality of an individual chromosome. In most cases, the goal is to find a chromosome with maximum fitness.

Forward Dynamics. Modeling approach that estimates joint torque from neural command.

Freezing. A phenomenon appearing during gait in Parkinson disease patient where the subject is unable to move the foot forward.

Functional Electrical Stimulation (FES). Application of electrical current to paralyzed muscles of the spinal cord injured patients to restore the lost movements.

G

Gait. Walking or running pattern. Gait analysis is routinely used to diagnose musculo-skeletal problems in the lower limb and also for evaluation of treatment outcomes.

Gait Cycle. The period of time from an event (normally initial contact) of one foot to the next occurance of the same event with the same foot.

Gait Inter-Cycle Variability. Gait cycle time variation estimated from the standard deviation of gait cycle time during walking which shows walking stability.

Gait Velocity. The average anterior velocity of the whole body during a complete gait cycle.

Gaussian Mixture Model (GMM). A statistical model in which the overall probability distribution is synthesized from a weighted sum of individual Gaussian distributions. With GMM, arbitrarily complex distributions can be approximated with a parametrically controlled amount of precision.

Genetic Algorithms. The most popular type of evolutionary computation that encodes the solution of a problem by a binary string. A population of strings (chromosomes) are then evolved by application of genetic operators such as mutation, crossover, and reproduction to improve their fitness in solving the problem.

Genetic Programming. Unlike genetic algorithms that evolve the data or parameters of a program or circuit, genetic programming evolves the program structure or circuit architecture themselves.

Glenohumeral. The joint of the shoulder complex that is the articulation of the humeral head on the glenoid fossa of the scapula. The glenohumeral joint or 'shoulder joint' is represented almost exclusively as a ball-and-socket, three degree of freedom joint.

GPS. Global position sensing.

H

Harris Hip Scores. Clinical score used for outcome evaluation in orthopaedics.

Hearing Threshold. Lowest level at which a person can perceive a sound.

Heel-Strike. The initial contact of the heel during walking.

Hidden Markov Model (HMM). This is an extension of a Markov model. It represents a stochastic process generated by an underlying Markov model and a set of observation distributions associated with its hidden states.

Hill Model. Muscle-tendon unit model based on physiological relationships.

Huntington's Disease. An autosomal dominant neurodegenerative illness characterized by disorders of movement, cognition, behaviour and functional capacity.

Hybrid Model. A software approach that attempts to combine data from complimentary sources, in order to provide better classification.

I

Inverse Dynamic. Determination of joint forces and moments from the known kinematics of the human body.

Inverse Dynamics. Modeling approach that estimates joint torque from position and external reaction forces.

Isointegral Map. Representation technique for the display of body surface potential distributions.

Isopotential Map. Representation technique for the display of body surface potential distributions.

J

Jewett Waveform. A waveform providing a characteristic pattern of peaks and troughs, indicating that the subject perceived an acoustics stimulus.

K

Kinematics. Descriptive study of motion, body segment displacements and their relation to time. The typical variables measured in gait analysis are linear and angular displacements, velocities and accelerations.

Kinetics. Study of forces that produce movement. The typical variables measured in gait analysis are joint moment of force and joint power.

Kohonen Neural Network. (Self-organising map, SOM). An unsupervised learning neural network which clusters data on the basis of similarities by projecting high-dimensional data onto a two dimensional surface. Consists of an input vector and an output matrix (typically 2D) of nodes which are highly inter-connected. Training of the net involves the iterative presentation of input data through the input vector which changes the weights characterising the connections between nodes.

L

Learning. Different learning paradigms are used for network training. Two main classes can be discriminated, the supervised and the unsupervised. Hybrid systems which are based on both strategies have also been developed.

M

Machine Interface. Parts of the man-machine system (hardware or software) that are concerned with interacting with a human user such as knobs, levers, keyboards, displays and graphical user interfaces.

Man-Machine System. A system in which the functions of a human operator and a machine are integrated.

Manual Control. Control of machines by hands.

Maximum Isometric Contraction. The act of holding the arm in such a position that the length of the muscle is not changing but the muscle is fully activated.

Maximum Likelihood (ML). The method of maximum likelihood is a general method of estimating parameters from observed data by values that maximize the likelihood function.

Membership Function. A curve that defines how each point in the input space is mapped to a membership value.

MEMS (micro-electro-mechanical systems). The integration of mechanical elements, sensors, actuators and electronics on a common silicon substrate through microfabrication technology.

Moment Arm. The perpendicular distance between a muscle's line of action and a joint axis of rotation.

Movement Disability. Physical impairments that interfere with or prevent normal movement.

Movement Rehabilitation. The process of helping patients with movement disability to improve their functional movements by therapeutic measures and training.

Multilayer Perceptron (MLP). Further development of a perceptron in which inputs are propagated through the net in forward direction only. Usually consists of an input layer, one or more hidden layers and an output layer.

Muscle Force. Amount of stress on the muscle divided by the physiological cross sectional area of the muscle.

Muscle Moment Arm. Shortest distance between the line of action of the muscle-tendon unit and the joint center.

Musculoskeletal Model. A mathematical model that includes bones as rigid bodies and muscles as actuators that elicit joint torques.

Mutation. Random changes in the chromosome.

Myoelectric Teleoperation System. An architecture to remotely operate automated and robotic systems using EMG signals via communication network.

N

Neural Gas. A neuronal clustering algorithm that solves several points of weakness in classical self-organizing Kohonen maps. The neural gas has no predefined topology which determines the neighbourhood relation between the neurons. No determination of the net's dimensions is needed a priori.

Non-Contact Impedance Control Method. A robot motion control method without using contact force. This method controls the mechanical impedance properties between the end-effector and environments.

Normalisation. Proportional scaling of a range of variables into another range.

Novelty Detection. See concept learning.

O

One-Class Classification. See concept learning.

One-Class Support Vector Classifier. A concept learning model. Its goal is to estimate the support of the data from one class.

Optimization. Process of finding a solution to a problem where the number of load-transmitting elements far exceeds the number of available force and moment equilibrium equations that can be derived. Static optimisation uses inverse dynamics to solve the problem while dynamic optimisation uses forward dynamics techniques.

Outcome Evaluation. Evaluation of a medical treatment.

Outlier. In concept learning, all the data that are different from the target are regarded as outliers. See to target.

P

Pennation Angle. The angle between the line of action of the muscle fibers and that of the tendons.

Petri Net. A directed graph interconnecting places and transitions on which multiple tokens are permitted to travel. A Petri net is widely used for describing an event-driven task model in the robot motion planning.

Prediction. Using optimized parameter values to estimate muscle forces and joint moments.

Pre-Processing. Preparing an object to be processed to adapt it to the processing tool.

Probabilistic Neural Network (PNN). A type of neural network using kernel-based approximation to form an estimate of the probability density functions of classes in a classification problem.

Process. A sequence of time-depending activities that changes the state of an object.

Proprioception. The sense of limb movement and position by receptors in muscles, tendons and joints.

Prosthetic Feedback. The sense of limb movement by artificial sensors.

Q

Qualitative Information. Complex information consisting of features like numbers, graphics, patterns, pictures, etc., where not the single measurable quantity but the correlation of features is of interest.

R

Recording Site. Anatomical position of electrocardiogram acquisition electrode.

Reproduction. Production of new generation of chromosomes from existing members where chromosomes with higher fitness have higher chances for selection.

Residual Voluntary Control. Parts of the voluntary control system not affected by injury or disease.

Reverse Shoulder Prosthesis. A glenohumeral joint replacement arthroplasty that reverses the ball-and-socket anatomy of the glenohumeral joint.

RoboCup. An international research initiative attempting to foster methods of artificial intelligence and robotics by providing standard problems, mainly inspired by soccer matches.

Rotator Cuff. Muscles that cross from the scapula to the humerus and act primarily to provide glenohumeral joint stability by concavity compression. The muscles include the supraspinatus, infraspinatus, teres minor and subscapularis.

Rule Extraction. Although artificial neural networks may recognise patterns in many different areas of interest, they do not give a direct insight in to how they use the information from input for classification. Therefore some methods have been developed to extract human comprehensible explanations from the trained networks.

S

Scapulohumeral. Group of muscles that pass from the scapula to the humerus. For example, rotator cuff muscles, deltoid and teres major.

Scapulothoracic. The joint of the shoulder complex that defines the motion of the scapula relative to the thorax. This joint is often represented in mathematical and computational models as a constrained gliding joint.

Self-Organizing Map (SOM). A special kind of unsupervised neural network which can be used to map high-dimensional continuous input signals to lower-dimensional discrete output signals in an adaptive way. The most famous one is the Kohonen feature map.

Shoulder Complex. Four articulations that make up the shoulder girdle. Includes the sternoclavicular, acromioclavicular, glenohumeral and scapulothoracic joints.

Simulated Annealing. Global optimization method based on random perturbations.

Sleep Apnea Syndrome. Occurs if patient breathing stops for a certain period of time, if the magnitude of the respiration movements decreases for at least 10 seconds to less than 5% of the physiological values. In medical literature the mild version of apnea is called *hypnea*, where movements decrease below half of the normal values.

Spatial Gait Parameters. Gait parameters related to the different distance and angle and their variations during walking

Spinal Cord Injury (SCI). Injury to the spinal cord resulting in paralysis and loss of sensation below the level of injury.

Step Width. The lateral distance between feet during a single step.

Sternoclavilcular. The joint in the shoulder connecting the sternum to the clavicle, often represented in mathematical and computational models as a three degree of freedom joint.

Strapdown. The condition when inertial sensors are directly strapped (without gimbals) to a moving body, to sense its linear and angular motion.

Strapdown Integration. The numerical procedure implemented on digital computers and microcomputers in order to estimate the attitude matrix and the position (in the navigation frame) of a moving body to which inertial sensors are strapped to.

Stride Length. The anterior distance traveled during a gait cycle, as defined by successive heel strikes of the same foot.

Stride Time. Time necessary to complete a single stride.

Support Vector Machine. A learning model based on statistical learning theory, applicable to both classification and regression.

T

Target. The training data of concept learning models which come from the same class. It is assumed that no information is used about the other class. See to outlier.

Temporal-Distance Parameters. Numeric descriptions of gait, describing time and distance values as they relate to the gait cycle.

Temporal Gait Parameters. Gait parameters related to the different time phase of walking.

Terminal Attractor (TA). A kind of steady state in a dynamical system. The idea of a TA is based on a violation of the Lipschitz condition at a fixed point. Since the Lipschitz condition is violated, the fixed point becomes a singular solution which envelopes the family of regular solution.

Testing. One of the application phases of an ANN. Weights and biases of the net are not changed any more. In this phase the ability for generalisation is evaluated.

Tilt Compensation. This term defines the computational procedure needed to perform the estimation of heading by means of a compass, when the case of which is inclined relative to the horizontal plane.

Toe-Off. The terminal contact of the toe during walking.

Torque. The product of the muscle force and the moment arm about which the muscle is acting.

Training. One of the application phases of an ANN. During the training phase the ANNs adjust their weights and biases according to the algorithm, which was chosen for adaptation.

Trajectory. Geographical representation of a process, where each state is represented by a geometric location in an abstract space.

U

UPDRS. Used for outcome evaluation in PD.

V

VAS. Visual analog scale which is scaled usually from 0 to 10 used for clinical evaluation.

Vector Matching. This term defines the computational procedure which estimate an attitude solution by matching no less than two non-zero, non-colinear vectors that are known in one coordinate frame, i.e., the navigation frame, and measured in another one, i.e., the body frame.

W

Wavelet Transform. Signal transformation that transforms the time signal to the frequency domain. The wavelet transform is based on a decomposition of a signal using an orthogonal family of basis functions derived from the so-called mother wavelet function.

About the Authors

Rezaul Begg received the BSc and MSc Eng degrees in electrical and electronic engineering from Bangladesh University of Engineering and Technology (BUET), Dhaka, Bangladesh, and the PhD degree in biomedical engineering from the University of Aberdeen, UK. Currently, he is an associate professor within the Biomechanics Unit of the Centre for Ageing, Rehabilitation, Exercise and Sport at Victoria University, Melbourne, Australia. Previously, he worked with Deakin University and BUET. He researches in biomedical engineering, biomechanics and machine learning areas, and has published over 100 research papers in these areas. He is a regular reviewer for several international journals, and was on the TPC for a number of major international conferences. He received several awards, including the BUET Gold medal and the Chancellor prize for academic excellence.

Marimuthu Palaniswami received his BE (Honors) from the University of Madras, ME from the Indian Institute of science, India, MEngSc from the University of Melbourne and PhD from the University of Newcastle, Australia before rejoining the University of Melbourne. He has been serving the University of Melbourne for over 16 years. He has published more than 180 refereed papers. He was given a Foreign Specialist Award by the Ministry of Education, Japan in recognition of his contributions to the field of machine learning. He served as associate editor for journals/transactions including IEEE Transactions on Neural Networks and Computational Intelligence for Finance. His research interests include SVMs, sensors and sensor networks, machine learning, neural network, pattern recognition, signal processing and control. He is the convener for Australian Research Network on Sensor Network. He is the associate editor for *International Journal of Computational Intelligence and Applications* and serves on the editorial board of *ANZ Journal on Intelligent Information Processing Systems*. He is also the subject editor for *International Journal on Distributed Sensor Networks*.

* * * * *

David C. Ackland is a graduate of mechanical engineering and science (physiology) at The University of Melbourne, Australia, and is currently a PhD student in bioengineering at the Department of Mechanical and Manufacturing Engineering (MAME), University of Melbourne. Under Professor Marcus Pandy, Mr. Ackland is researching biomechanics of the shoulder complex with an emphasis on glenohumeral biomechanics and shoulder arthroplasty. By combining novel cadaveric experimental techniques with sophisticated computational modelling, Mr. Ackland's PhD seeks to improve understanding of shoulder function in the normal upper-limb, and after implantation of common prostheses. During his studies, Mr. Ackland has designed biomedical equipment for the Department of Optometry and MAME. Mr Ackland currently tutors, demonstrates and lectures on a casual basis at the University of Melbourne.

Kamiar Aminian received the electrical engineering degree in 1982 and the PhD degree in biomedical engineering in 1989 from Ecole Polytechnique Fédérale de Lausanne (EPFL), Switzerland. He worked as a research associate in the Metrology Laboratory, EPFL and as an assistant professor at Sharif University of Technology, Tehran, Iran. In January 2002, he joined the School of Engineering Sciences and Techniques (STI) of EPFL where he is the head of the Laboratory of Movement Analysis and Measurement. He teaches in the areas of sensors, electrical systems and medical instrumentation. His research interests include transducers, movement analysis, ambulatory systems and biomedical signal processing.

Gabor J. Barton graduated from the University of Pecs Medical School, Hungary in 1993. Following a three years research programme at Liverpool John Moores University, he took up a position at Alder Hey Children's Hospital in Liverpool where he had been running the Clinical Gait Analysis service for five years. At present he is a senior lecturer in Biomechanics at Liverpool John Moores University, UK. His research activity is related to the application of neural networks to support decision making in clinical gait analysis. He is also associated with establishing the Movement Function Research Laboratory which allows a combination of balance, posture, gait, biomechanics, motor control and virtual reality. He is a member of ESMAC and CMAS.

Daniel N. Bassett is a research assistant in the Center of Biomedical Research Engineering at the University of Delaware, USA, where he is currently a graduate student in the Department of Mechanical Engineering. He received his Bachelor of Science in engineering with a mechanical concentration from LeTourneau University. His research interests include gait analysis and musculoskeletal modeling with applications to impaired individuals, particularly children.

Russell Best is a senior lecturer in Biomechanics at Victoria University, Australia, where he is the co-coordinator of the Biomechanics Unit. Following an Honours degree in sport science in 1985, he graduated in 1996 with a PhD from the Department of Aeronautical & Mechanical Engineering at Salford University, splitting his research between Salford and Manchester Metropolitan University. His research interests in gait analysis revolve around long-term gait function (e.g., analysing up 3000 consecutive strides) and includes research into gait variability, long and short-term gait control (chaos applications) and modelling of an individual's probability of tripping and slipping while walking. He also maintains a strong interest in sports biomechanics, especially javelin throwing, Olympic shooting and golf biomechanics.

Thomas S. Buchanan is professor and chair of the Department of Mechanical Engineering at the University of Delaware, USA. His academic training was at the University of California at San Diego, Northwestern University, and MIT. Dr. Buchanan is a fellow of the ASME and is editor-in-chief of the *Journal of Applied Biomechanics*. His research interests are in neural control of human movement, biomedical imaging, and biomechanical modeling applied to the musculoskeletal system.

Li-Shan Chou graduated with a BS degree in mechanical engineering from the Tatung Institute of Technology, Taiwan, in 1987 and received his PhD degree in mechanical engineering from the University of Illinois at Chicago in 1995. He was subsequently trained as a post-doctoral research fellow in the area of biomechanics at both the University of Chicago and Mayo Clinic before his arrival at the University of Oregon, USA, in 2000. Dr. Chou's teaching focus is in the areas of biomechanical analysis of human movement and orthopedic biomechanics. In his research he applies engineering and mechanical approaches to enhance the understanding of mechanisms underlying the increased incidences of falls in the elderly or patients with movement/balance disorders.

Peter Dauscher is research associate at the Institute of Computer Science at the Johannes Gutenberg-University Mainz, Germany. He received his diploma degree in physics in 1999 and his PhD degree in 2003. In his diploma thesis he investigated discrete traffic models and their attractor structure. His PhD thesis deals with theoretical aspects of evolutionary algorithms. His current research interests focus on intelligent data analysis, especially parameter estimation of complex dynamical systems by means of evolutionary computation methods.

Rahman Davoodi received BS in mechanical engineering and MSc in biomechanical engineering both from Sharif University of Technology, Tehran, Iran and PhD in biomedical engineering from University of Alberta, Edmonton, Canada. He is currently a research assistant professor in the Biomedical Engineering Department of the University of Southern California, Los Angeles, USA. His interests focus on the use of functional electrical stimulation (FES) to restore activities of normal daily living to the paralyzed, such as standing, reaching, grasping, and exercise. He has developed several software tools for musculoskeletal modeling and control and is currently leading a large team of researchers and developers to develop the next generation software for clinical fitting of FES systems. His research involves musculoskeletal modeling, FES control of movement, motor control and learning, cooperative control of man-machine systems in rehabilitation, virtual reality training for movement rehabilitation, reinforcement learning, and evolutionary computation.

Arthur M. Farley has an educational background in mathematics and computer science, receiving his PhD in computer science from Carnegie-Mellon University in 1974. Since then he has been on the faculty of the University of Oregon, USA, where he has pursued research in graph theory applied to communication network design and in artificial intelligence applied to reasoning and learning about complex systems.

Joseph D. Gardinier is a research assistant in the Center of Biomedical Research Engineering at the University of Delaware, USA, where he also received a Masters of Science in biomechanics and movement science. Prior to this he received a Bachelor of Science in engineering with a mechanical concentration at LeTourneau University. His research interests include gait analysis and musculoskeletal modeling of the lower extremity.

Cheryl J. Goodwin received her Bachelor of Science in biomedical engineering at Louisiana Tech University. She is currently a graduate student at The University of Texas at Austin, USA, where she is studying biomedical engineering focusing on computational biomechanics. Her supervisor is Professor Marcus Pandy. She is currently working on a computational musculoskeletal model of the upper extremity developed from the Visible Human Male. The focus of her work is on validation and clinical applications of the model, including studying shoulder reconstruction using reverse shoulder prostheses.

Michael E. Hahn's academic background is in biological science and human movement science. He received his PhD from the University of Oregon in 2003. He has taught at Montana State University, USA, since then, advising pre-physical therapy students and mentoring graduate students in biomechanical research. His research interests have been focused in the areas of upper extremity joint function, and balance control during locomotion in older adults.

Gerald E. Loeb, is director of the Project on Implantable Nanoengineered Sensory Cilia for Therapeutic Monitoring. This is one of several projects in the Medical Device Development Facility, which he directs for the Alfred Mann Institute at University of Southern California, USA. Dr. Loeb pioneered many methodologies for biomedical research and clinical use, many of which are widely available commercially. These include multichannel cochlear implants, injectable RFID transponders, nerve cuff electrodes and Parylene-coated microelectrodes for neurophysiology. He has consulted widely for the medical device industry, including serving as chief scientist for Advanced Bionics Corp. (1994-1999). He is a fellow of the American Institute of Medical and Biological Engineers, holder of over 40 issued U.S. patents and author of over 200 scientific papers. Most of Dr. Loeb's current research is directed toward neural prosthetics to reanimate paralyzed muscles and limbs using a new technology that he and his collaborators developed called BIONs®. This work is supported by an NIH Bioengineering Research Partnership for which he is PI and is one of the test beds in the NSF Engineering Research Center on Biomimetic MicroElectronic Systems for which Dr. Loeb is deputy director. These clinical applications build on his long-standing basic research into the properties and natural activities of muscles, motoneurons, proprioceptors and spinal reflexes. Dr. Loeb is the founding director of a professional master's program in medical device & diagnostic engineering at USC, which draws on the broad activities of the Mann Institute in evaluating and developing novel medical products, including intellectual property law, biomaterials and design, preclinical and clinical testing, regulatory compliance and entrepreneurship.

Kurt T. Manal is an assistant professor of mechanical engineering at the University of Delaware, where he also received his PhD in biomechanics and movement science following previous academic work in exercise science at McGill University, the University of Montreal and the University of Massachusetts. His research interests include gait analysis, musculoskeletal modeling, and rehabilitation engineering.

Marcus G. Pandy is appointed as chair of mechanical and biomedical engineering in the Department of Mechanical and Manufacturing Engineering at the University of Melbourne, Australia. Professor Pandy received a PhD in mechanical engineering from Ohio State University in Columbus (1987). He then completed a two-year post-doctoral fellowship in the Department of Mechanical Engineering at Stanford University. In 1990, he was appointed as an assistant professor at the University of Texas at Austin. He was promoted to associate professor in 1995 and to full professor in 2002. In 2002, he was appointed as the Joe J. King professor in the Department of Biomedical Engineering at the University of Texas at Austin. He is a fellow of the American Institute of Medical and Biological Engineering and a fellow of the American Society of Mechanical Engineers. Professor Pandy's research interests are in biomechanics and control of human movement. Much of his research is aimed at using computer models of the human body to study muscle, ligament, and joint function in the normal, injured, and diseased states. He has published nearly 200 scientific papers on this topic. He is a key participant on a number of currently active research grants, including a 5-year VESKI fellowship to support research on the development of advanced patient-specific computer models of the human musculoskeletal system and a 5-year multi-institutional research grant funded by the NHMRC to support a new Centre for Clinical Research Excellence in Gait Analysis and Rehabilitation.

Aftab E. Patla graduated with a BTech (Honors) in electrical engineering from Indian Institute of Technology, Kharagpur, MSc Eng in electrical engineering from University of New Brunswick, Canada, and PhD in Kinesiology from Simon Fraser University, Canada. Since 1982 he has been a faculty member in the Department of Kinesiology at the University of Waterloo, Canada, where he currently holds a rank of full professor. His research interests are in understanding the sensory contributions to the control of human gait over complex terrains, control of posture and balance and the effects of pathologies and aging on mobility and balance. He has published over 100 referred journal papers in these areas, and is a co-author on a text on signal processing and modeling for movement science and edited a book on adaptability of human gait. He has been a past president of the Canadian Society of Biomechanics and the International Society for Posture and Gait Research, executive editor for Journal of Motor Behavior and Associate Editor of Gait & Posture, and sits on the editorial board for Experimental Brain Research. He is the recipient of Pease Family Scholar Lecturer Award from the University of Iowa in 2005, William Evans Fellow, University of Otago, New Zealand in 2001, and

The Herbert & Jean Barron Visiting Scholar Award, The Center for Locomotion Studies, Pennsylvania State University, in 1990.

Jürgen Perl started his studies in mathematics in 1963, received his PhD in 1971. After a professorship for applied computer science he became a full professor for computer science at the University of Mainz in 1984. He started a biennial international symposium-series on computer science in sport in 1997. Currently, he is president of the International Association of Computer Science in Sport (IACSS) that was established in 2003. His main working area is modelling and simulation with a focus on sport science and medicine. He is author respectively co-author of over 100 scientific articles and textbooks.

Stephen D. Prentice is an associate professor in kinesiology at the University of Waterloo, Canada. He graduated in 1990 with a Hon BSc in kinesiology and in 1995 a PhD in kinesiology (biomechanics), both from University of Waterloo, Canada. He then completed a 3-year postdoctoral fellowship in neurophysiology at the University of Montreal before in 1997. His research interests lie in the control and biomechanics of gait and posture with specific interests in modelling the neural and mechanical elements involved in the coordination of movement and posture during various locomotor activities. Currently, he is the president of the Canadian Society for Biomechanics and holds memberships in the International Society of Biomechanics, International Society for Postural and Gait Research and the Society for Neuroscience.

Angelo M. Sabatini received the DrEng degree in electrical engineering from the University of Pisa, Italy in 1986, and the PhD degree in biomedical robotics from Scuola Superiore Sant'Anna, Pisa in 1992. From 1987-1988 he was with Centro E. Piaggio, Faculty of Engineering, University of Pisa. During the summer of 1988 he was a visiting scientist at the Artificial Organ Lab, Brown University, Providence, RI, USA. From 1991-2001 he was an assistant professor of biomedical engineering at Scuola Superiore Sant'Anna, USA, where he has been an associate professor of biomedical engineering since 2001. His main research interests are design and validation of intelligent assistive devices and wearable sensor systems, biomedical signal processing and quantification of human performance.

Alistair Shilton received his combined BSc / BEng degree from the University of Melbourne, Australia, in 2000, specializing in physics, applied mathematics and electronic engineering. He is currently pursuing the PhD degree in electronic engineering at the University of Melbourne. His research interests include machine learning, specializing in support vector machines; signal processing, communications and differential topology.

Bharat Sundaram has a bachelor's degree in electrical engineering from the Indian Institute of Technology, Kanpur. He has four conference publications, one of which won the Best Paper Award at ICISIP 2004. He is now pursuing a research degree at the University of Melbourne, Australia, in the Department of Electrical and Electronic Engineering under A. Prof. M. Palaniswami. His current research centers on support vector machine based localization in sensor networks.

Index

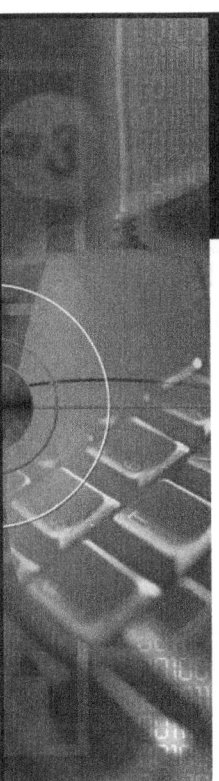
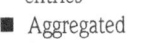

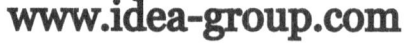

CPSIA information can be obtained at www.ICGtesting.com
Printed in the USA
BVOW10*1231011213

337354BV00005B/16/P